This is the errata for the first edition of:
The Cellular Radio Handbook by Neil Boucher
published June, 1990, ISBN #0-930633-17-2.

The corrections that should be noted are:

1. Page 126: The photo is upside down.

2. Page 164: Figure 12.1 shows a colinear antenna with four dipoles. The dipoles should be of equal length.

3. Page 281: Figure 19.1, the magnetic base antenna gains should be: in column 1 row 2, **-2 to 0 db** instead of -6 to 0db and column 1 row 3 should be **-4 to -2 db** instead of -10 to -4 db.

4. Page 243: The photo should be rotated 90 degrees clockwise.

5. Page 437: The photo caption should read: Notice that it consists of only three 600mm equipment racks together with two tape drive racks.

QUANTUM PUBLISHING INC.
MENDOCINO, CA TEL. 707-937-4488

THE
CELLULAR
RADIO
HANDBOOK

THE
CELLULAR
RADIO
HANDBOOK

A Reference for
Cellular System Operation

Neil J. Boucher

QUANTUM PUBLISHING

Published by Quantum Publishing, Inc.
P.O. Box 310
Mendocino, California 95460

Produced by Online Press Inc.
14320 N.E. 21st Street, Suite 18
Bellevue, Washington 98007

First American Edition – June 1990

Photos by the author unless otherwise indicated

ISBN 0-930633-17-2
Library of Congress Catalog Card Number 90-61485

Printed in the United States of America

Contents

Chapter 10 Base-Station Control and Signaling   137

Call to Mobile Station • Mobile-Originated Call • Call Supervision • Setting Up a Call Between Two Cars • Handoffs • AMPS Signaling Format • Signal Strength Parameters

Chapter 11 Cellular Repeaters   153

Cell-Extender Repeaters • Enhanced Cell Extenders • Traffic Capacity of the Simple Repeaters • Cell-Replacement Repeaters • Traffic Capacity of Cell-Replacement Repeaters • Tunnels

Chapter 12 Antennas   163

Omnidirectional Antennas • Sector Antennas • Polarization • Antenna Materials • Mounting • Drainage • Intermodulation • Measuring VSWR

Chapter 13 Microwave Links   173

Fresnel Zone • Fading Depth • Losses in Antenna Coupling • Calculation of Outage Time • System Gains • Gain Measurements • Feeder Losses • Interference • Margins • System Capacities • Advantages of Digital Systems • Modulation Systems • Rack Space • Microwave Links in Cellular Systems • US and Japan • Rest of the World • Drop and Insert • Factors in Choosing Microwave • Survey • Trunking • Standby • Split Routes

Chapter 14 Base Station Maintenance   193

Maintaining Quality of Coverage • Air-Conditioning • Transceivers • Using Statistics • Customer Complaints • Line Up Levels • Test Mobile • Site Audio Test Loops • Interaction with the Switch • Site-Log Books • Call-Out Procedures • Equipment • Quality and Calibration of Test Equipment • Test Sets • Quantifying Coverage Problems • Co-Channel and Adjacent-Channel Interference • Third-Party Interference • Spare Parts

Chapter 20 Towers and Masts   285

Monopoles • Guyed Masts • Towers • Soil Tests • Other Users
• Lightning Protection • Antenna Platforms • Tower Design
• How Structures Fail • Tower, Mast, and Monopole Maintenance
• Inspection • Stiffness • Repair • Tower Inspection Checklist

Chapter 21 Installations   309

Training • The Operator's Responsibility • Acceptance Testing
• Commissioning • Moving away from Turnkey Installation
• Acceptance-Test Sheets • Power Rectifiers • Batteries and
Distributions • External Plant • Internal Plant • Items to Be
Checked

Chapter 22 Equipment Shelters   327

Basic Considerations • Building Description • Internal Finishes
• External Finishes • External Supply • Electrical Power Outlets
• External Emergency Plant • Essential Power • Air-Conditioning
• Typical Switch Room • Base-Station Housing • Grounds and
Paths

Chapter 23 Budgets   343

Equipment Requirements • Typical Work-Hour Requirements
• Costs • Digital Radio Systems (DRS Single Hops) • Billing
System (Including Computer and Software) • Costs of a "Typical"
Cellular System • An Exercise

Chapter 24 Billing Systems   353

Resellers • Billing Houses • Do-It-Yourself Billing • Billing
Service • Management Information Systems (MIS)

Chapter 30 Privacy   429

Privacy Methods (Scrambling, Encryption)

Chapter 31 Rural Applications of Cellular Radio   435

Environmental Limitations • Small Switches and Repeaters in Rural
Areas • Cellular Mobiles as "Fixed" Telephones • Cellular Pay
Phones • Rigidly Mounted Versus Mobile Rural Units • Networks
Without External Terminal Equipment Antennas (Type I)
• Networks with External Terminal Equipment Antenna (Type II)
• Calling Rates and Customer Density • Hybrid Systems

Chapter 32 Preparing Invitations to Tender   445

Technical Preparation • Sample Tender Offer

Chapter 33 Digital Cellular   467

Digital Caveat Emptor • No Duplex Filter • Battery Talk Time
• GSM Pan-European Cellular • US Digital Cellular Options
• US Digital Compared to GSM • Japanese Digital • The Future
of Analog Cellular in a Digital World • Future Trends • Handheld
Telephones • Digital Advantages • Digital Cellular Operations
• Personal Communications Network • Second-Generation Digital
and Beyond • The Potential of GHz Cellular • GHz Base-Station
Antennas • Mobile Units • Satellite Systems • Universal Mobile
Radio

Chapter 34 Other Mobile Products   495

Public Mobile Radio • Paging • Voice Messaging/Voice Mail
• Packet Radio

Chapter 35 Glossary of Terms   507

Index   515

Chapter 1

WHAT IS CELLULAR RADIO?

Cellular radio is the fastest growing area of telecommunications. In almost every installation throughout the world, the operators' initial forecasts of customer demand have needed constant upward revision. Market surveys have been spectacularly unsuccessful because the public is unfamiliar with the product and has difficulty visualizing the need for yet another communications device. The mobile telephone, however, largely sells itself. Most customers report that they first learned about mobile telephones through word of mouth from friends and acquaintances.

The dramatic growth in the industry has lead to a worldwide shortage of skilled personnel, and with the industry doubling in size every two years, this problem has no solution in sight.

The introduction of digital cellular in late 1991 will bring an attendant increase in hardware and system complexity. The urgency with which digital cellular is being brought into service has been caused by the lack of spectrum and the inability of the analog systems to keep up with the demand foreseen over the next few years. Few in the industry doubt the need for digital cellular. Still, as with any technology that is introduced too hastily, digital cellular is bound to experience early problems that only time will rectify.

Today, more than 20 percent of all new telephone installations in countries with mature analog cellular systems are mobile. Forecasts for the next decade indicate that most new installations will be mobile.

To date, the major manufacturers have borne most of the responsibility for establishing new installations and training staff. In these endeavors, they have generally done a commendable job. The strain on their resources, however, is beginning to show, and it is obvious that, in the future, operators will need to become more self-sufficient.

There are numerous mobile systems in existence today that provide telephone access. Cellular radio, as a concept, was originally conceived as a means of providing high-density mobile communications without consuming large amounts of spectrum. The earliest cellular-like proposals date back to the American Telephone and Telegraph (AT&T) proposals of the 1940s for high-density mobile systems. In 1968, AT&T submitted a proposal for a cellular system to the FCC.

The original concept involved containing a group of frequencies within a "cell," reusing the frequencies in the same vicinity, but separating them in space to allow reuse without serious interference. The hardware needed to implement such a system did not become available until the late 1970s and by then the "cellular" concept (that is, frequency reuse in cells) was accepted as a sound frequency planning tool.

Prior to the first cellular telephone systems, a number of automatic mobile telephone systems existed, usually having only one transmit-and-receive site. Such systems sometimes had 2000 subscribers and were characterized by high sites, with relatively high transmit power and deviation. These systems usually operated in the VHF or low UHF bands, and from single cells, could achieve a radius of operation of 12 miles or more. They were not cellular systems in the modern sense even though they were often fully automatic.

Cellular radio is different things to different people. To many investors it is seen as a potential gold mine. High growth rates, decreasing costs, and value to the community at large have caused the value of many companies that own cellular franchises to rise substantially in the last few years. But even with highly regulated tariffs, few cellular companies have been able to achieve a positive cash flow. Achieving profitability, however, is relatively easy.

Recently, particularly in the US and Australia, expansion to rural areas has begun on a large scale. However, there is considerable evidence that the demand in those areas is significantly lower on a per capita basis than in cities. In the US, where an ever expanding array of investors is opening up new cellular markets, there is an underlying suspicion that the bubble may be about to burst. Companies with little experience in design and operation may find rural markets unprofitable.

To many, cellular radio is still a rich man's toy and possibly just the latest status symbol. To the majority of users, however, most of whom are not rich, it is an indispensable asset to their business. Market studies worldwide have shown that the typical cellular user is a person who operates a small business and uses cellular radio to conduct business from a car or from locations other than the office.

Regardless of what is paid for a cellular service (and start-up prices vary from about $200 to $5,000), the cellular phone is a business asset.

To the cellular operator, cellular radio is a whole new world where the technologies of radio, switching, transmission, and computing merge into a single system. But the operator must be proficient in all of these areas or pay dearly for expertise when it is required. There is also a high cost for mistakes and bad decisions caused by a lack of experience. New operators often find the spectrum of skills required to be a successful player quite daunting.

Once the technical skills have been addressed, the new operator must then face the areas of marketing, finance and accounting, advertising, and public relations. The successful operator will be the one who also masters these basic business skills.

Cellular radio is no place for amateurs. A minimum of two years full-time experience is needed to produce an effective cellular engineer or marketing professional. It takes a lot longer to become an expert. Getting a wide spectrum of experience is difficult. Large operators can rotate their staff to widen employee exposure. Small operators often lack the necessary available positions for employees to move within the company.

Mobile installation enterprises are also likely to find cellular radio a challenge. Most mobile installers have another (dominant) business, with cellular installation typically being about 35 percent of the business turnover. The diversity of mobile models on the market, falling prices, and intense competition mean that maintaining market share requires considerable business skill.

EARLY CELLULAR SYSTEMS

Early cellular radio systems were designed for frequency reuse and had low capacities, which were thought adequate for future demand. The earliest cellular system, NAMTS, the Tokyo metropolitan system that started service in 1979, came with a basic capacity of 4000 subscribers and was expandable to 8000 subscribers. In 1981, the NMT450 system was introduced with a basic capacity of 10,000 subscribers and was expandable to about 20,000 subscribers.

Each of these systems operated in the 400-MHz band and used 180 frequencies. It was possible to have about 4000 subscribers without frequency reuse.

In 1967, the Japanese Telecommunications Laboratory (NTT) developed a 400-MHz cellular frequency plan for cellular radio, but it was never implemented commercially.

Later systems—AMPS (Chicago, 1983), TACS (1985), and NMT900 (1987)—were designed to operate in the 800- and 900-MHz bands and use from 666 to 1000 frequencies. It was assumed that they would usually operate in a multiple frequency-reuse environment (that is, interference environment). Therefore, these systems came with an inherent protection against interference.

MOBILE AND TRUNKED RADIO

Because of the high initial costs of implementing a full cellular network, these early systems targeted the top of the market (car telephones). These networks were the beginning of modern cellular mobile-phone systems. They could reuse frequencies and could "hand-off" a mobile call from one cell to a more appropriate one as the call moved out of range of the original cell. Until 1987, these two features were sufficient to define a cellular telephone system.

However, in recent times, the emergence of powerful trunked radio systems has made this distinction inadequate. A trunked radio system is one that allows dynamic assignment of radio access channels to a group of mobiles.

A simple mobile radio system consists of a mobile and a repeater (see Figure 1.1 on the following page). Car A transmitting on frequency F_1 is received by the repeater and rebroadcasts to all other mobiles on frequency F_2.

A trunked radio system works in much the same way, except that the mobiles, instead of having only two pairs of frequencies, have a group of frequencies and some logic control to ensure the right frequency is used. Logic in the base and in the mobile enables a free channel to be selected and switched to automatically (see Figure 1.2 on the following page).

Consider a typical mobile-originated call:

1. The mobile calls (by a data transfer) on a common channel (a signaling-only channel) and requests the use of a free channel.
2. The controller either assigns a free channel, indicating that the channel is assigned to the mobile via the common channel, or the controller places the request in a queue until a channel becomes available.
3. The mobile switches to the assigned channel and conversation can take place, as with a simple mobile repeater.

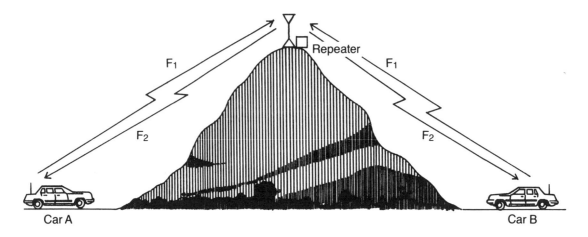

Figure 1.1   Duplex mobile repeater

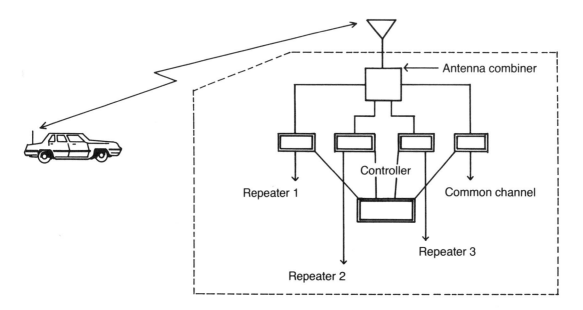

Figure 1.2   A typical trunked system

As trunked mobile systems evolved, they used more sophisticated hardware and software so that frequency reuse and call handoff are now sometimes a feature of a trunked radio system. Furthermore, automatic telephone access and calls strictly to individual mobiles are also common.

The main distinction between a cellular system and a sophisticated trunked radio system is that the latter is usually 2-channel simplex while a cellular system is full duplex. Another distinction is that cellular systems were designed to operate in an interference environment where frequency reuse was seen as a limiting factor. Trunk mobile radio wasn't designed with this feature.

Trunked radio systems permit group calling; that is, the simultaneous calling of all mobiles in a particular group, which can be quite large. Cellular systems don't permit group calling. This is not a technical limitation but one imposed by the FCC and other regulatory commissions to limit competition between cellular and land mobile systems. Therefore, cellular systems are designed to handle conventional telephone traffic and trunked systems are designed primarily to handle dispatched mobile traffic.

CELLULAR SYSTEMS

A cellular radio system is structured differently than a land mobile system. A basic cellular system is illustrated in Figure 1.3.

The heart of a cellular system is the cellular switch. It is called a full-availability switch (it can connect any inlet to any outlet). The cellular switch connects base stations to the PSTN (Public Switched Telephone Network) and base stations to each other as required. The cellular switch makes these connections using trunk routes to the PSTN.

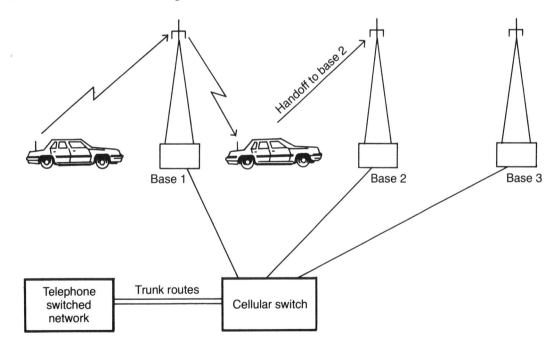

Figure 1.3   A basic cellular system

A feature of cellular architecture is the continuous monitoring of the call progress and the ability to reconfigure the system quickly so that switching occurs without disturbing the user.

Handoff times vary from about 0.5 to 1 second in older systems, to about 100 milliseconds (which is barely noticeable) in new systems. Therefore, the mobile shown in Figure 1.3, traveling from base 1 to base 2, will eventually drive out of range of base 1. The system senses the decreased field strength and instructs the surrounding bases to look for the mobile and report back on field strength. If one base reports a higher field strength, the system can instruct the mobile to change channels and "handoff" to the best base (in this case, base 2).

Early systems used SPC (Stored Program Control) switches, which were designed primarily as land-line switches. Generally, it was found that the processor load had been underestimated and that a processor upgrade was necessary to handle the extra demands of the cellular system. Modern cellular switches have processors that are well suited to the system demands and the number of ports supported.

The cellular switch can either be a purpose-designed switch or a modified telephone switch. The switch is usually processor-controlled. The processor handles many functions other than switching, including customer validation, call monitoring, system diagnostics, and interconnection with other cellular switches and base stations. Base stations have some local intelligence but they are essentially controlled repeaters.

Most major systems today use analog voice transmission in FM mode, although control channels (data channels) use Manchester or FFSK (Fast Frequency Shift Keying) digital signaling. Future systems (1992) will be digital.

No strict definition of cellular radio exists, but all cellular systems have the following features:

- Frequency reuse
- Ability to handoff a mobile from cell to cell according to signal field strength and/or noise requirements
- Multicell and multibase configurations
- Access to a fixed telephone network with mobiles receiving calls on an individual basis only (group calling is only available using a switched network)
- Ability to work in a controlled interference environment

The familiar hexagon pattern that has popularly become associated with cellular radio is shown in Figure 1.4. To understand cellular radio, it is important to understand what it isn't. Therefore, it is worthwhile exploring the hexagon theme a little further.

It would be possible to create this hexagon concept if:

- All sites were to have the same antenna systems (that is, all omni or all sectored)
- The terrain were perfectly flat (no forests or large buildings)
- All antenna heights were identical

This type of terrain is called a desert! A "desert" cellular system that satisfies these conditions is illustrated in the cell pattern in Figure 1.5. With an analog cellular system, this is as close as you can get to an hexagonal pattern. Digital systems permit rigid definitions of boundaries by triangulating the position of the mobile.

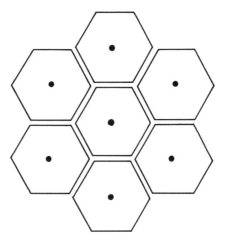

Figure 1.4   The hexagon frequency pattern of cellular radio

Because of the extreme symmetry of this type of terrain, it has been possible for mathematically inclined engineers to write volumes on how to obtain high-density cellular systems by regular cell division and range contraction. Various cell plans have been devised that optimize frequency reuse (minimize the ratio of carrier to interference) for these theoretical systems. It is assumed that this approach can be translated directly into a real-world environment so that the most efficient concept on paper will also be the most efficient in practice.

The hexagon approach can be used as a starting point, but the real environment should determine the actual system configuration. The 4-cell model requires fairly close compliance to the cell patterns. More tolerance is built into the 7- and 12-cell plans.

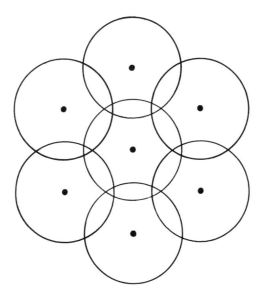

Figure 1.5   Cellular patterns on a flat earth

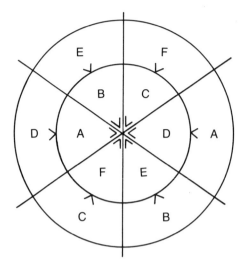

Figure 1.6   The Stockholm-Ring cell pattern

Many other configurations are possible, the most successful competitive model is the "Stock-holm Ring," illustrated in Figure 1.6. However, it is only applicable to low deviation NMT systems.

This model allows all channels to be used at a common high-density center (usually a city center). It was used in the Stockholm NMT450 system. This model has since been used successfully elsewhere.

Chapter 2

WORLD SYSTEM STANDARDS

There are a number of different cellular systems used around the world today. The lack of a uniform standard has hampered the development of cellular systems and of roaming (that is, using the mobile phone in other than the home network) in particular. GSM, the pan-European digital system, is at least a regional attempt to solve the uniformity problem. Roaming between Europe and the US is not possible because the two continents reserved different blocks of frequency for mobile telephone services. Completely world-compatible cellular systems will probably have to be realized in the 1-GHz+ band, but even agreement on a suitable band has not yet been reached.

The total number of users in June 1989 was 4.6 million. Table 2.1 shows the market shares of the different world systems. The ETACS system is the same as TACS except that it operates over a wider bandwidth and is slightly different in frequency. It also has additional channel capacity.

SYSTEM	NAME/COUNTRY	SHARE
AMPS	Advanced Mobile Phone System	57%
NMT450	Nordic Mobile Telephone (450 MHz)	14%
TACS	Total Access Cellular System	15%
NMT900	Nordic Mobile Telephone (900 MHz)	6%
NTT	Japanese NTT 800 MHz & 450 MHz systems	6%
C System	German digital system	2%
RMTS	Italian system	1%
ACS	Comvik system	<1%

Table 2.1   Market shares of world systems

TECHNICAL COMPARISONS

The earliest systems, released around 1980, operated on 450 MHz. These systems had a relatively small number of channels—usually 180 to 225—and were not designed for extensive frequency reuse. The maximum capacity of such systems is thus between 20,000 and 30,000 users in most city areas.

These 450-MHz systems were not designed to use a handheld, but at least one manufacturer produces one and transportables are readily available. Generally, installations do not include any 400-MHz systems, except where spectrum is not available for 800/900-MHz systems.

AMPS was the first 800-MHz system produced; it was released in late 1983. TACS is essentially an "enhanced" AMPS system. AMPS holds the major share of the market (especially outside of Europe) leaving TACS and NMT, plus at least five minor systems, to share less than one half of the market. The NMT900 system is currently making a significant impact but only in Scandinavia.

Although costs vary considerably, AMPS mobile systems are generally about 10 percent cheaper than the other systems, due mainly to volume of production and competition. The switch and base-station costs (except in rural areas) are approximately the same for all major systems. Many countries have recently announced new systems, resulting in the further dominance of AMPS.

Table 2.2 compares the AMPS, TACS/ETACS, NMT900, and NMT450 systems

NET$_Z$ B & C WEST GERMANY

Germany uses a non-cellular, 160-MHz car telephone system (B-Net$_Z$) (without handoff) alongside its digital cellular C-Net$_Z$, 450-MHz system. The C-Net$_Z$ has 237 channels and a capacity of about 150,000 customers. In September 1989, there were 120,000 customers, and considerable problems have been associated with that many subscribers.

The C system is digital, and it uses propagation delay times to control handoffs so that they occur within predefined boundaries. Similarly, a mobile system locks onto the control channel of its local cell by measuring propagation delays, as shown in Figure 2.1. In a similar way, base stations can determine the distance to any vehicle, making it possible to tailor the handoff so that it occurs at any given distance from the base station.

JAPAN

Japan started an 800-MHz cellular system in 1979 in the Tokyo metropolitan area. The system had 600 channels and used 25-KHz channel spacing.

In order to cater to the ever-increasing demand for capacity, a second system using 6.25-KHz channel spacing with 2400 channels was later introduced. This new 6.25-KHz system includes these additional features:

• Diversity reception in mobile and base stations
• Dynamic channel assignment

SPECIFICATION	AMPS	TACS/ETACS	NMT900	NMT450
TX Band	800 MHz	900 MHz	900 MHz	450-470MHz
Channel separation	30 KHz	25 KHz	25/12.5KHz	25/20 KHz
Duplex separation	45 MHz	45 MHz	45 MHz	10 MHz
Channels	832	920*	1000(1999)	180/225
Modulation type	FM	FM	FM	FM
Peak deviation	±12 KHz	±9.5KHz	±4.7KHz	±4.7KHz
Compander	2:1 Syllabic	2.1 Syllabic	2.1 Syllabic	No
Possible cell plans	4, 7, 12	4, 7, 12	7, 9, 12	7
Control channel modulation	FSK	FSK	FFSK	FFSK
Control channel deviation	±8 KHz	±6.4 KHz	±3.5 KHz	±3.5 KHz
Control channel code	Manchester	Manchester	NRZ	NRZ
Ctrl. chn. capacity (subs)	77,000	62,000	13,000	13,000
Transmission rate	10 Kbit/s	8 Kbit/s	1.2 Kbit/s	1.2 Kbit/s
Competitive operators allowed**	Yes	Yes	No	No
Interexchange handoff	Yes	Yes	Under developement	
Diversity	Yes	Yes	Yes	No
Subscribers in service 1988	2.614 M	0.650 M	0.270 M	0.600 M
Voice privacy available	Yes	Yes	No	No
Roaming between different service area	Yes	Yes	Yes	Limited due to different channel spacing and frequencies of operation
* Excludes GSM reserve channels. ** Due to control channels being exclusively available to each operator.				

Table 2.2   Technical comparisons of four cellular systems

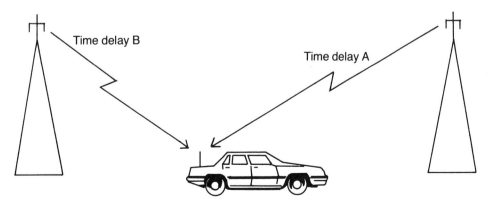

Figure 2.1   By measuring the arrival time of the time synchronization bits, the mobile unit can determine the physically nearest cell rather than the one with the highest field strength.

- Compatibility with the earlier wide-band system (through dual-mode mobile systems)
- Digital speech encryption

The system originally used a 17-cell omni cell plan, but this was later reconfigured to $N = 14$. The base stations have a 5-watt transmitter power and are installed 16 to a rack.

By late 1988 there were 520 mobile bases and 4200 radio channels. Over half of all subscribers are in Tokyo.

NMT450

This system became available in 1981 and featured fully specified system interfaces. It specified, to the level of the switch-to-base station and switch-to-switch, a standard signalling protocol (as well as the base station-to-mobile interface). This standardization, missing from other systems, enables equipment from different manufacturers to be interconnected. Thus, a base station from one supplier could be directly connected to a switch from another supplier.

But soon after its introduction, NMT450 systems were installed in various countries on different frequencies and with different channel spacings. This made them incompatible from a subscriber's viewpoint. Different frequencies were chosen to suit local spectrum availability. The channel spacing, initially 25 KHz, was altered to 20 KHz in some systems to obtain more channels (an increase from 180 to 225) in the same spectral bandwidth.

NMT900

The NMT900 system is essentially an NMT450 system moved in frequency to 900 MHz. Although many more channels are available (1000 or 2000 using interleaving), the paging channel baud rate is not increased so that these channels can be used effectively.

Some improvements were made, including the addition of receiver antenna diversity and a voice compander. The low deviation of 4.7 KHz was retained, however, resulting in significantly poorer performance in rural and in-building environments, as well as reduced interference immunity when compared to the other 800- and 900-MHz systems.

ITALY

Italy started its wide-area mobile network in 1974 and introduced a cellular system known as RMTS in 1985. This network had 70,000 subscribers at the end of 1989 and operates in the 450-MHz band. The RMTS has countrywide coverage provided by 350 base stations. The system is reaching saturation in the Milan and Rome areas. To allow for future demand, a 430-base-station TACS network is to be placed into service in 1990.

ACS (ADVANCED CELLULAR SYSTEM)

The Comvik ACS system was placed in service in 1983 in Sweden and later introduced into Hong Kong. This system operates on the 400-and 800-MHz bands. It features distributed intelligence and relies on the mobile unit to control more of the switching functions than is the case in other cellular systems. For this reason the Comvik ACS system has small and relatively cheap base stations.

The handoff is initiated by the mobile unit, which begins to scan for a free channel once the field strength drops below 3 microvolts. When indicated, the mobile will begin a scan of free channels at other sites, and if a channel is available, it will request a handoff. The system is limited to 16 unidirectional channels per site.

When first turned on, a mobile will scan for a free Mobile Terminating (MT) channel and stay on this channel until a handoff is indicated. Calls are initiated by changing to a Mobile Originating (MO) channel.

PRE-CELLULAR SYSTEMS

IMTS

This mobile telephone system is used extensively in the US and has been used elsewhere. The system does not feature frequency reuse and is based on the concept of a large single cell. There is no handoff between cells, but because the cells are so large the user does not see this as a serious problem.

The number of channels allocated for IMTS across the US is variable, so that the customer capacity of the system also varies. As no frequency reuse is employed in a given geographical region, the spectrum efficiency of this system is poor.

In Chicago only 23 channels are available, and it is assumed that this number is sufficient to serve 1150 customers, or 50 customers per channel. It can be seen that the customer density is much higher than that used in cellular systems where 20 customers per channel would be more usual. This means that a lower grade of service (or higher call fail rate) is to be expected on the IMTS.

The attraction of the IMTS, at least until recently, was the wide area over which roaming was possible. With numerous inter-operator roaming agreements in-place today between cellular operators, this advantage has diminished.

RADIOCOM 2000

This French system is in many ways similar to the IMTS system. It operates in the 200- and 450-MHz bands. It was launched in 1985 and uses 256 channels. The UHF version is used countrywide, while the VHF version is used in Paris, Lyon, and Marseille.

In late 1989, an extension of this system into the 900-MHz band was announced. This extension was in response to the growing demand for service on the existing networks (capacity 170,000 at the end of 1989), which are unlikely to be able to hold the demand until such time as GSM is introduced.

WHERE AMPS CANNOT BE USED

Europe uses the AMPS band for UHF TV. (In fact, TACS evolved from placing AMPS on a new band using the European mobile-phone spectrum.) Thus, AMPS is not used in Europe.

Other countries may have the AMPS band allocated to fixed services. In general, the investment (measured in dollars per MHz of spectrum) is very low for fixed services when compared to mobile services; it is usually economical to buy out the fixed link operators. Where the AMPS band has been allocated to other mobile services, the investment per MHz will probably be quite high and recovery may not be practical.

CALL CHANNEL CAPACITY

The paging capacity of the systems differs substantially. Table 2.3 shows the paging capacities where mobile call messages are assumed to be single words occupying 50 percent of the control channel availability.

The difference between the TACS and AMPS paging rate (of 8 and 10 KBPS) is due to the difference in channel bandwidth (25 and 30 KHz, respectively).

NMT900 uses idle voice channels for paging; it does not use dedicated paging channels. This is advantageous in rural areas with small systems where it is not necessary to allocate a dedicated paging channel; in systems with high traffic, this results in reduced system capacity. The lack of dedicated control channels precludes competitive operators using the same system in any one area, because operator-specific control channels are not available.

SYSTEM	PAGES/HOUR
AMPS	77760
TACS	62000
NMT900/450	13000

Table 2.3   Paging capacities

FREQUENCY BANDS

The major systems differ in two basic ways: range and spectrum efficiency. The first of the "new" (800- to 900-MHz) systems designed was AMPS; it was released in late 1983. TACS followed in 1985 and NMT900 in early 1986. The "newer" systems (TACS and NMT900) emphasized conservation of frequency and the availability of many channels. Various techniques were used but, most significantly, the frequency deviation of the FM modulation was progressively reduced. For operation above the noise threshold, this decrease in deviation was accompanied by a decrease in range, as shown in Table 2.4.

Table 2.4 compares the range from co-sited bases of the four major systems. Although operation below the threshold negates the advantages of higher deviation, it is generally expected that cellular mobile systems will operate in high-quality S/N (signal-to-noise) conditions. Notice that in the later TACS and NMT900 systems, the range progressively decreases.

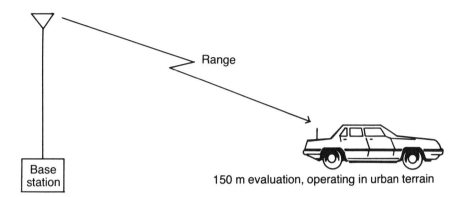

150 m evaluation, operating in urban terrain

SYSTEM/TYPE	RANGE (km)	RANGE AS % OF AMPS	COVERAGE (km^2)	DEVIATION (KHz)
AMPS	10	100	312	4.7/12
NMT450/TACS	8	85	200	9.5
NMT900	6	60	113	4.7

Table 2.4   Effect of deviation on coverage. Range determined by Okumura method for urban regions for high voice quality (30 dB S/N).

An additional but relatively minor propagation difference is caused by the smaller aperture of a 900-MHz antenna relative to an 800-MHz antenna of the same gain. This gives a net overall gain of about 0.6 dB to the 800-MHz system, which results in a small increase in range.

NMT450 does not have high deviation but, because it operates at about one half the frequency of the other systems, it has a usable path loss 6 dB greater than those systems. Its range is consequently similar to AMPS and TACS.

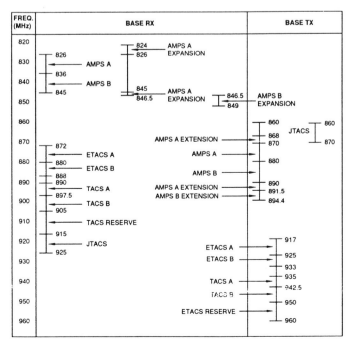

Table 2.5   Frequency bands used by cellular radio systems

SPECIFICATION	AMPS	TACS	NMT450	NMT900
Channels	832	1320	180	1000/2000
TX frequency	824-849	872-905	453-457.5 (& other bands)	872-905
Channel separation	30 KHz	25 KHz	25/30KHz	25KHz/12.5
Frequency stability	±2.5 ppm	± 2.5 ppm	±5 ppm	±5 ppm
TX power	3 watts	2.8 watts	15 watts	3 watts
Voice deviation	12 KHz	9.5 KHz	4.7 KHz	4.7 KHz
Data deviation	8 KHz	6.4 KHz	3.5 KHz	3.5 KHz
Receiver sensitivity*	12dBc @ -116dBm	20dBp @ -113dBm	20dBp @ -113dBm	20dBp @ -113dBm
Adjacent channel selectivity	60 dB	55 dB	70 dB	70 dB
Spurious rejection	60 dB	65/55 dB	70 dB	70 dB
Intermodulation	65 dB	65/55 dB	67 dB	67 dB
Signalling method	Manchester	Manchester	FFSK	FFSK
Speed	10 KBPS	8 KBPS	1.2 KBPS	1.2 KBPS
* Note: Because of differing measurement techniques these figures cannot be directly compared. Sensitivity is about equal for all systems.				

Table 2.6   Subscriber units specifications compared

Table 2.5 shows the spectrum used by the various 800- and 900-MHz cellular systems. Notice in particular the frequency overlays between the ETACS and AMPS/AMPS extended band, which involves 10.4 MHz in the region 894.4 to 872 MHz.

Table 2.6 gives the specifications of the various system subscribers units. These specifications are constrained only by the state-of-the-art of production equipment.

Tables 2.7 through 2.12 show the customer bases of various systems worldwide. The total number of units in service is 4,597,625. You can see the dominance of the US. You can also see that 20 countries represent the bulk of the cellular network, and that a growing interest worldwide is strongly evident.

THE IMPACT OF DIGITAL

The specification of the GSM (European digital) frequency to be the same as TACS and NMT900 has limited future expansion of those systems. Countries not currently operating in those bands, and that have those bands reserved or used by a small number of subscribers, are unlikely to use TACS and NMT900 if a digital system is contemplated in the near future.

For example, France recently decided to install a version of the technologically obsolescent NMT450 system rather than use the 900-MHz band it has reserved for GSM. This system, along with all analog 900-MHz systems installed between now and the introduction of GSM, is considered an "interim system." The urgent need for more capacity has caused France recently to reconsider the application (additionally) of a 900-MHz digital system. The current plan is to introduce a new Radiocom 2000 system onto the 900-MHz band. Compatibility with future digital cellular systems should be the major consideration when using current systems.

COUNTRY	CUSTOMERS	ANNUAL GROWTH (new units)
American Samoa	200	
Argentina	(starts 1989)	
Australia	90,000	60,000
Bahamas	?	
Bermuda	?	
Bolivia	?	
Brazil	(starts 1989)	
Brunei	1,000	
Canada	242,438	104,000
Cayman Is.	?	
Chile	(starts 1990)	
Columbia	(starts 1989)	
Costa Rica	150	
Dominician Republic	400	
Gabon	?	
Guatemala	(starts 1990)	
Hong Kong	30,000	12,000
Indonesia	(at tender)	
Israel	7,000	
Jamaica	?	
Kenya	?	
Korea	30,000	
Malaysia	(at tender)	
Mexico	1,000	
Neth Antilles	(starts 1989)	
New Zealand	9,000	
Pakistan	(starts 1990)	
Panama	?	
Philippines	600	* 13,000
Puerto Rico	?	
Singapore	9,000	12,000
St. Maarten	(starts 1989)	
Taiwan	* 30,000	
Thailand	10,000	4,200
US	2,193,197	800,000
Venezuela	?	
Zaire	?	
Total	2,621,275	980,200
* Capacity order		

Table 2.7   AMPS-system customer base as of June 1989

COUNTRY	CUSTOMERS	ANNUAL GROWTH (new units)
Austria	(being installed)	
Algeria	?	
Bahrain	2,000	
Barbados	?	
China	10,000	
Greece	(starts 1990)	
Hong Kong	30,000	18,000
India	(not in service)	
Indonesia	1,000	nil
Ireland	7,500	
Italy	30,500	
Kenya	(starts 1990)	
Kuwait	17,000	4,000
Macao	700	
Malaysia		* 5,000
Malta	(starts 1990)	
Mauritius	200	
Spain	(being installed)	
Sri Lanka	250	
UAE	(starts 1989)	
United Kingdom	586,000	250,000
Total	681,150	277,000
* Capacity order		

Table 2.8   TACS-system customer base as of June 1989

COUNTRY	CUSTOMERS	ANNUAL GROWTH (new units)
Algeria	(starts 1990)	
Cyprus	1,500	500
Finland	47,000	20,000
Moscow	(starts 1990)	
Netherlands	(being installed)	
Scandinavia	173,000	86,000
Switzerland	35,000	37,000
Turkey	14,000	5,000
Total	270,500	148,500

Table 2.9   NMT900-system customer base as of June 1989

COUNTRY	CUSTOMERS	ANNUAL GROWTH (new units)
Austria	38,000	11,600
Belgium	24,000	11,400
Denmark	53,000	0
Finland	94,000	18,700
Iceland	8,000	1,300
Indonesia	10,000	2,600
Luxembourg	400	200
Malaysia	30,000	10,000
Netherlands	28,000	3,000
Norway	12,000	9,900
Saudi Arabia	12,000	12,000
Sweden	200,000	36,500
Thailand	20,000	1,000
Total	**637,400**	**127,000**

Table 2.10   NMT450-system customer base as of June 1989

COUNTRY	CUSTOMERS	ANNUAL GROWTH (new units)
Africa (South)	2,300	
Germany (West)	112,000	
Portugal	?	
Total	**114,300**	

Table 2.11   NETZC-system customer base as of June 1989

COUNTRY	CUSTOMERS	ANNUAL GROWTH (new units)
Australia	15,000	nil
Hong Kong	4,000	nil
Japan	250,000	?
Singapore	4,000	nil
Saudi Arabia	?	
Total	**273,000**	

Table 2.12   JTACS/NAMTS-system customer base as of June 1989

The major installations of TACS are in the UK and Hong Kong. Both the UK and Hong Kong are experiencing problems with capacity and the existing demand is sufficient to ensure that they will switch to GSM as soon as the system is available. Other users of the band (for example, NMT900) may not have such an urgent need to switch, but it is unwise to expand into the GSM band once digital equipment becomes available. Thus production of TACS and NMT900 units will likely fall dramatically soon after the introduction of GSM (predicted in 1992).

Where the pressure of spectrum is not high and the considerations are purely financial, it is unlikely that GSM (or other digital systems) will be competitive before 1995. The use of analog systems will probably be extended, provided that extension of the analog systems does not interfere with the long-term prospects of the introduction of digital systems.

The digital route being taken by the US is far less a threat to the future of AMPS than GSM is to the 900-MHz systems. The US will use TDMA (Time Division Multiple Access) techniques to overlay 3 x 10-KHz digital channels on the existing 30-KHz AMPS channels. Digital systems will share the same switch and a significant amount of base-station hardware with the analog system.

Dual-mode (analog and digital) mobile units are expected to be available and have been specified so that a gradual transition from analog-to-digital can be made.

Chapter 3

BASIC RADIO

R adio was first postulated in 1873 by Maxwell, demonstrated in 1888 by Hertz, and used for practical communications in 1895 by Marconi. Radio is an electromagnetic phenomenon and radiates as photons. It belongs to the family of radiation that includes X-rays, light, and infrared (heat) waves. The different categories of radiation differ in frequency, as shown in the Figure 3.1. They also differ in energy and ability to propagate those different media.

BASIC ELEMENTS

All practical radio systems can be reduced to the basic scheme shown in Figure 3.2 on the following page.

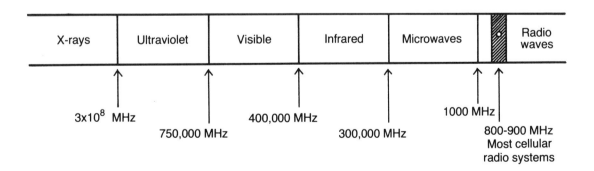

Figure 3.1   Electromagnetic spectrum

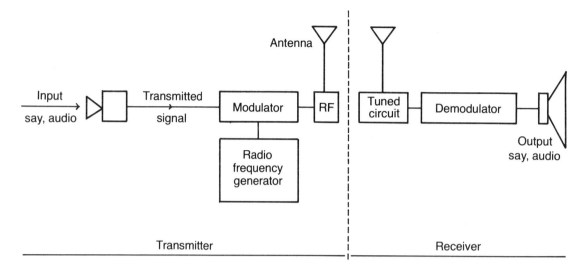

Figure 3.2   Analog transmitter with an audio input

TRANSMITTER

The transmitter consists of two basic parts: a modulator and a carrier. Figure 3.2 shows an analog transmitter with an audio input that is converted into the form to be transmitted (in this case, via a microphone to the modulator). A radio frequency generator generates the radio energy that will carry the signal. This generally consists of an oscillator (which produces the initial signal) and a number of amplifier stages (which amplify the level to that required at the antenna). A modulator mixes the signal to be transmitted with the radio frequency signal (called the carrier) in such a way that the signal can be decoded at a distant receiver.

RECEIVER

The receiver gets a signal from its antenna, which also receives a number of unwanted signals. The tuned circuit tunes out all but the wanted signal, which is then demodulated (decoded) by the demodulator.

MODULATOR

The modulation system used in most cellular radio systems is known as FM (Frequency Modulation). In this type of modulation the frequency of the carrier is varied proportionally to the signal to be transmitted. A typical FM modulator is shown in Figure 3.3 on the following page.

The audio input varies the bias on the varactor (a solid state variable capacitor, illustrated in Figure 3.3), which in turn changes the frequency of the tuned circuit. The maximum amount that the frequency can deviate from its central carrier frequency is called the peak deviation.

The S/N performance of FM systems is very high, provided the noise level is reasonably low. FM systems with wide deviation have better S/N performance than those with narrow deviation.

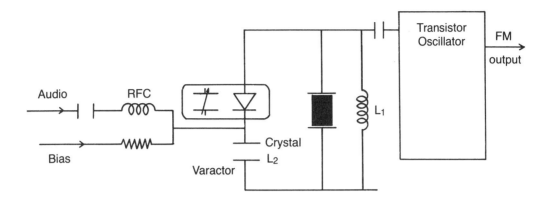

Figure 3.3   Typical FM modulator

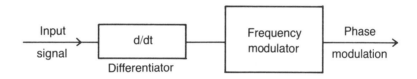

Figure 3.4   Phase modulation from frequency modulation

Some systems use phase modulation, particularly for data transmission. Phase modulation is closely related to frequency modulation and can be derived from it by passing the signal through a simple differential circuit before frequency modulation (see Figure 3.4).

Frequency Shift Keying (FSK), at relatively low speeds, (for example, 8 Kbits Manchester code for TACS and AMPS on the control channel) is often used for data, because it has better S/N performance than FM at low signal levels. This enhances signaling in areas of poor reception.

As the signal level increases, the quality of FM (in noise performance terms) rises fairly rapidly, whereas the quality of FSK does not. Because cellular systems are designed to operate in relatively high signal environments (low noise), FM is chosen for the voice path.

DYNAMIC CHANNEL ALLOCATION

Because cellular systems use from 180 to 2000 channels, it is necessary for the mobile to automatically switch to the correct channel. This is done by sending an instruction (data) that indicates the channel number required. The mobile system then switches to the channel indicated by using synthesized tuning, a system where the frequency of the oscillator is numerically compared to the required frequency and adjusted by a "phase-locked loop" until the two frequencies match.

NOISE AND SIGNAL-TO-NOISE PERFORMANCE

All radio systems are ultimately limited in range by noise. When the intrusion of noise is such that an acceptable signal can no longer be obtained, then the system is said to be noise-limited.

Medium wave (broadcast band) and shortwave broadcasts operate in a very noisy environment; background noise limits the performance in the broadcast/shortwave bands. The VHF (Very High Frequency) and UHF (Ultra High Frequency) bands where cellular radio operates are relatively much quieter and most of the noise is generated in the radio frequency preamplifier of the receiver itself. Regardless of how well designed the receiver is, there is a theoretical noise power level which, at a given temperature, cannot be improved upon. This is because of the thermal noise, generated by the movement of atomic particles (in the receiver and most particularly in the first radio frequency amplifier). This noise is proportional to the operating temperature. Hence, the antenna and RF amplifier stages will generate thermal noise continuously. For this reason high-quality receivers, such as radio telescopes, operate their input stage RF amplifiers at liquid-nitrogen temperatures.

In order to perceive a relatively noise-free signal, the incoming signal must exceed the noise level by a respectable margin, known as the signal-to-noise ratio. For cellular radio systems, this level is usually regarded as 12 dB for marginal reception and 30 dB for good quality conversations.

Signal-to-noise ratio is usually expressed as:

$$S/N = \frac{Signal\ level}{Noise\ level} \ (usually\ expressed\ in\ dB)$$

Because modern receivers closely approach the theoretical noise limits for their operating temperatures, it can easily be deduced what minimum received signal level is required to achieve satisfactory signal-to-noise ratio.

Thus when we speak of a 39 dBμV/m boundary level for an AMPS system, this is equivalent to specifying the point at which the signal-to-noise ratio is regarded as satisfactory.

A "satisfactory" level of signal-to-noise for cellular systems is usually regarded as one where the noise is just noticeable. For PMR (Public Mobile Radio), a usable but noisy level is often regarded as satisfactory. In FM systems, the noise usually occurs as sharp clicks (sometimes known as "picket fence" noise because it is similar to the sound produced by dragging a stick along a picket fence).

dBs

Humans perceive power logarithmically. For example, doubling the energy level of a sound pulse produces only a 3 dB increase in the perceived level—and that increase is only just noticeable. The term dB was introduced to define relative power levels logarithmically.

The term dB is used often in radio systems and can be a major source of confusion to the uninitiated because of the large number of different units of dB's. Essentially, the dB level is the log of a power ratio: dBm, the most common form of dB, is the power of the system measured compared to 1 milliwatt. Mathematically, this can be expressed as:

$$Power\ dBm = 10\ log\left[\frac{Power\ (in\ watts)}{(0.001)}\right]$$

$$Power\ dBm = 10\ log\left[\frac{Power\ (in\ milliwatts)}{1}\right]$$

$$Thus\ 1\ watt = 10\ log\frac{1}{0.001}\ dBm = 30\ dBm$$

dBμV/m is a unit of field strength which compares the measured level with 1 μV/m (1 microvolt per meter).

Mathematically, this is:

$$dB\mu V/m = 20\ log\left[\frac{field\ strength\ in\ microvolts\ per\ meter}{1}\right]$$

Note: 20 is the multiplying factor here because the terms being used are voltage, not power. Voltage squared gives the power ratio.

PROPAGATION

Radio propagates at the speed of light (299,800 km/sec, or approximately 300,000 km/sec). Medium- and high-frequency waves can propagate very long distances by reflecting off the ionosphere, as shown in Figure 3.5.

At higher frequencies (above about 50 MHz), the troposphere/ionosphere absorbs the waves

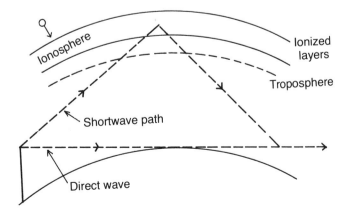

Figure 3.5   Waves reflecting off the ionosphere

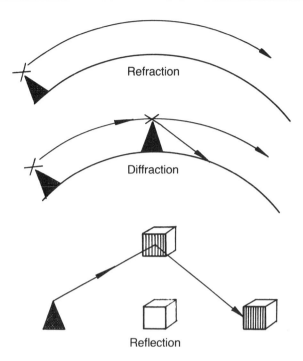

Figure 3.6   Modes of propagation at cellular radio frequencies. Super-refraction, or ducting, is a sporadic phenomenon responsible for propagation over very large distances under certain atmospheric conditions.

instead of reflecting them so that the predominant mode is the direct wave. This is the mode that applies to cellular radio systems.

The direct wave is not limited to line of sight; in fact, a good deal of refraction (bending of the path of propagation) occurs. This enables the transmissions to extend well beyond line of sight. Diffraction (bending around obstacles) also occurs, allowing the path of the wave to extend around obstacles. The ability to refract and diffract decreases with increasing frequency but is still most significant at cellular radio frequencies (800–900 MHz).

A third property, reflection, is also significant at cellular radio frequencies. Coverage in high-density city areas is significantly enhanced by the ability of the radio system to reflect into most areas that are inaccessible via a direct path.

These three modes of propagation are summarized in Figure 3.6.

ANTENNAS

Antennas used in cellular radio are usually gain antennas, meaning that they have gain, compared to the simplest form of antenna, the dipole. (A dipole is shown in Figure 3.7. The simplest vehicle-mounted antenna with the same gain as a dipole antenna is the quarter-wave antenna, illustrated in Figure 3.8.) A gain antenna is usually easily recognized by its loading coil, as shown in Figure 3.9. The loading coil is usually visible as a "bump" in the antenna.

Mobile cellular antennas are usually between 3 and 4.5 dB gain. Base station omnidirectional antennas, which stack many radiating elements in series, are often 9 dB in gain; unidirectional base station antennas can have gains as high as 17 dB. Very high gain antennas are only practical

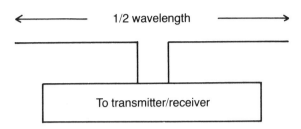

Figure 3.7   Dipole antenna

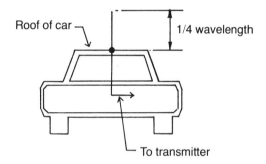

Figure 3.8   Quarter-wave antenna

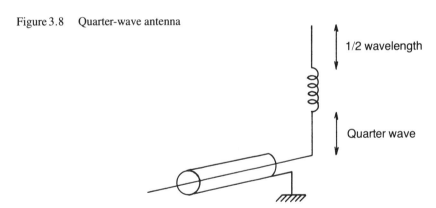

Figure 3.9   3 dB gain antenna for roof mounting

in fixed locations, because they are very large and must be exactly vertical to operate satisfactorily. Sometimes, however, they are deliberately tilted down to limit the base-station range.

MOBILE TRANSMIT POWER AND HEALTH

The radiation from a radio system is non-ionizing (distinguishing it from radioactive decay products); the main effect on the human body is a rise in temperature.

In 400-MHz systems, most mobile units have transmitter outputs of 10 to 15 watts; in 800–900 MHz systems, the power is usually limited to 3 watts for vehicle-mounted units and 0.6 watts for handheld units. It is generally believed that these powers and the subsequent radio frequency radiation levels do not pose any health hazard. However, it is not recommended to hold or touch the antenna of a mobile in use. Some authorities have expressed doubts about the wisdom of long-term use of handheld units.

Contrary to popular belief, it is not the high frequencies used by cellular radio (800–900 MHz) that are inherently most harmful to humans. At the lower frequencies—around 100 MHz—the body can become resonant and, therefore, very absorbent. Hence these lower frequencies are potentially more hazardous. Early experiments using small laboratory animals pointed to relatively more harmful effects at higher frequencies. However, the small size of the animals, which gave the animals a high resonant frequency, probably accounted for these results.

Chapter 4

CELL SITE SELECTION AND SYSTEM DESIGN

Cell site selection is the process of selecting good base-station sites. The process is half art, half science. To the customer, the most vital feature of a cellular system is its coverage. Therefore, it is important for the system to deliver what the customer wants—good coverage within a logically defined service area. The service provider is usually interested in both extensive frequency reuse and good coverage (except in small towns). The selection of the best cell sites is essential. Often this part of cellular-system design is poorly considered and results in poor system performance.

DESIGN OBJECTIVES

The design objective should be to cover the service area without any serious discontinuities as economically as possible; and, where applicable, to allow for future frequency reuse. In most technical journals, coverage is defined as 90 percent of the area covered for 90 percent of the time. Were this "objective" to be achieved literally, the coverage would indeed be poor (that is, 81 percent of the service area would provide adequate field strength for successful calls; 19 percent of the target area would be substandard). In fact, the 90 percent/90 percent coverage standard means something quite different. The 90 percent/90 percent definition means 90 percent of the area should be covered at any time or, alternatively, that 90 percent of the regions achieve a defined standard at any point in space and time.

The FCC has specified a field strength of 39 dBμV/m average (for AMPS) as the boundary of a cell. Although this figure is realistic, it is a compromise. In practice, the boundary field strength for acceptable service is a function of the terrain. The real objective is to obtain a S/N ratio comparable to a land-line telephone, which is usually accepted as 30 dB (some European

AREA	RECOMMENDED FIELD STRENGTH
Urban CBD	60 dBμV/m average
Suburban	39 dBμV/m average
Rural	34 dBμV/m average

Table 4.1   Boundary of cell sites for AMPS systems

authorities use 20 dB, which is below the quality of a very poor telephone line). Land-line systems, however, generally achieve 40 dB or more.

As a guide, Table 4.1 lists more realistic boundaries. No firm rule for Urban CBD can be formulated, but it is generally accepted that a 20–30 dB margin over mobile levels is required for handheld use in multistory buildings. Good handheld coverage is defined as a signal level yielding a voice quality that is usable without discomfort in buildings from the ground floor up, excluding elevators and their immediate vicinity (2 meters).

ASSUMPTIONS AND LIMITATIONS

All new systems need good handheld coverage, a fact that provides a good starting point. Individual cellular operators will probably have their own preferences for buildings to house cellular bases, such as telephone switching centers, buildings designated as radio telephone sites, or maybe just a few particular buildings to which access can easily be arranged. The cellular-system designer should accommodate these preferences, but not at the expense of sound design.

A cell plan (that is, N = 4, 7, or 12) should be chosen on the basis of the long-term customer density, noting that high density favors the 4-cell system. Initially, however, the 4-cell system is both more costly and less tolerant of site selections that do not approximate the theoretical cell plan.

DEFINED COVERAGE

Before a design is implemented, the number of bases to be used should be defined based on the techniques outlined in Chapter 23, "Budgets."

DEFINING BOUNDARIES

Radio waves propagate according to natural laws that have little to do with city boundaries and customer service areas. Therefore, there is little value in defining a precise area to be covered until the "natural" boundaries defined by the propagation characteristics are known. Of course, it is still practical to define certain areas as essential for coverage, leaving the "fine-tuning" of the actual boundaries as flexible as possible.

When boundaries must be decided, the decision should be made in consultation with engineering and marketing staff once alternative boundaries (those possible given the resource constraints) are known. At coverage boundaries, high sites become increasingly attractive in gaining an economic population of customers in a given service area. These sites are usually in low-density suburban or rural areas.

Usually, the physical boundaries of a city are poorly defined (in terms of customer density). At the "edges," a number of options are available. For example, one base to the north of a city may provide a potential 1000 km^2 coverage along a sparsely populated highway, while the same base to the south may cover only 300 km^2 but includes two important towns. For this reason it is important that the system designer have good communication with the marketing staff.

While the designer should be free to select sites that provide continuous coverage in high-density regions, the engineering staff must recognize that there are no clear-cut "best" designs and that many alternate solutions, with different coverage, could be equally viable. Indeed, detailed marketing studies can be undertaken to "resolve" the problem. Although these studies may yield an answer, they generally prove to be of little value. (For more information, see Chapter 25, "Marketing.")

Note: It is important to provide your marketing staff with a clearly marked map showing the expected coverage of the alternatives. Without this information, they have no foundation on which to draw their conclusions.

SUITABLE SITES

All selected sites must have reasonable access for installation and maintenance, and suitable accommodation (usually a minimum of 18 m^2) with a ceiling of 0.6 meters above the equipment rack. Equipment racks are usually 2.2, 2.7, or 2.9 meters high, depending on the manufacturer. It is important that equipment access is available and that provision is made to accommodate heavy crates of equipment. Access for maintenance must also be provided. If the site is located in a rural area, power and reasonable road access should be available. All of these improvements can be expensive if they have to be provided after initial installation.

Links to the base stations (either radio or physical) must be provided, and power is always essential. Consequently, the designer should undertake a site visit to ensure suitability before pursuing the design too far. If microwave links are used, the path back to the switch must be considered. Remember that base-site antennas can be large and 12 (or more) sector antennas with dimensions of 0.5 m x 3 m (approximately) may need to be fitted. Although it is usually impossible to guarantee the life of a site, some sites are more vulnerable than others and it is best to avoid those where over-building may occur.

The site needs access to commercial power (about 400 watts/channel, including air-conditioning) and usually some provision for an emergency power plant must be made. Hospitals are often good sites because they are usually relatively low buildings with good clearance around them (parking lots and gardens). They also have emergency power. On some tall buildings where emergency power is not available, it may be feasible to run a cable down the building to an external socket that can be powered by a portable generator. This saves cost, space, and possible objections to a permanent installation.

Equipment weight is a potential problem for installations on buildings. An "average" installation (about 40 channels) can weigh 6 tons. You must also add the weight of the shelters, towers, or support structures.

Because a cell site is a substantial shelter and, usually, a prominent antenna-support structure, the zoning of the site must be appropriate. Although it is often possible to get city planning approval to build base sites in residential areas, the process is uncertain and slow. Areas zoned for industrial or commercial use are usually easier to acquire for cellular purposes.

The availability of a site will be determined by local building codes, neighborhood environmental attitudes (particularly if a substantial antenna structure is involved), as well as any limitations imposed by the site owner.

If rental property is being considered, leases should be at least five years, preferably with an option to extend the lease. If possible, it is advisable to house on-site equipment in transportable buildings, which will minimize costs in the event of lease termination. Also, an installation should be done with the expectation that at some future date you will need to move the equipment.

Where an antenna support structure must be built, especially in a residential area, the following questions will be posed by the local residents:

- Will it interfere with the TV reception in the area?
- Is the tower safe?
- Will the microwave make me sterile?
- How big is the tower?

Of course, there are usually no problems, but the questions will be asked, and the design engineer should be ready with the answers.

GETTING A STARTING POINT FOR THE DESIGN

Inasmuch as the choice of suitable sites is an interactive one, the sooner a desirable central site can be identified, the better. A starting point for the design must be established. Usually, that point will be the primary site selected for the CBD coverage. The primary CBD site is a good starting point because it can be selected to cover the CBD regardless of other coverage requirements. Also, CBD sites are difficult to acquire and once one has been found, it is a good idea to make it fit into the final pattern.

In small towns, frequency reuse may not be necessary, so high "broadcast" sites can be chosen. Therefore, height can be used to select a good first site. This type of design is very different from high-density designs and is a good deal easier. Where handheld use is contemplated, it is often necessary to find a prominent central urban site to provide handheld coverage in CBD buildings. This is particularly true if all other sites are more than 4 km from the CBD.

In big cities the opposite is true; a selected site(s) should offer good handheld penetration while at the same time containing the coverage. It may be possible to use surrounding buildings as radio-path shields. If frequency reuse is a consideration, the ability to provide adequate frequency reuse should be a major consideration (see Figures 4.1 and 4.2). Frequency reuse would eliminate the tallest buildings (even considerable antenna downtilt will not help much) as CBD sites. A good choice might be to use a smaller building with a clear area of about 500 meters from the next obstructing building.

Site availability is a major limitation. Many ideal sites have uncooperative owners. The designer often has little choice, particularly with inner-city sites.

Range limited by downtilt and surrounding buildings

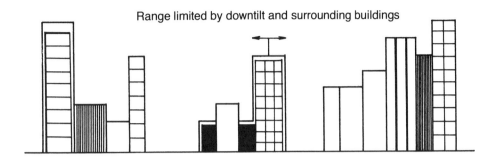

Figure 4.1   Preferred frequency reuse. Range is limited by downtilt and surrounding buildings.

Range limited by downtilt only

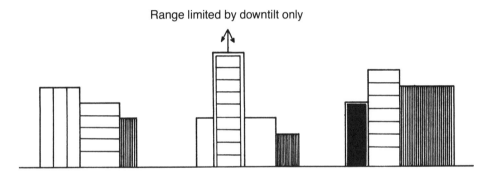

Figure 4.2   Poor frequency reuse. Range is limited by downtilt only.

For some rooftop installations, weight may also be a problem. As already mentioned, a 40-channel cell site weighs about 6 tons for the equipment and batteries. This does not include the weight of a shelter or tower structure.

SPECIAL CONSIDERATIONS

Significant reflections occur from large buildings. As a result, some of the front-to-back ratio immunity of sectored antennas will be lost. For this reason it is essential to survey sectored sites facing tall buildings, paying particular attention to the area outside the nominal cell area. The front-to-back ratio of an antenna that is nominally 22 dB can be reduced to 6–15 dB because of reflections (see Figure 4.3). This reduction can result in interference problems from behind the cell.

Figure 4.3 shows one way of exploiting natural or man-made boundaries to improve frequency reuse. Assuming that the buildings form long rows following the coastline, as shown in Figure 4.3, then the buildings can be effectively serviced from a seashore site. At the same time, the buildings shield the site to permit effective frequency reuse behind them. Experience shows that ship-to-land cellular communications are usually made via the highest base station rather than the one nearest the sea. This fact should be considered when deciding if reflected signals are likely

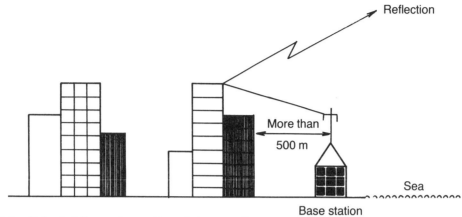

Figure 4.3   Using buildings and natural boundaries to confine coverage

to be a source of interference. Use by sea vessels is generally limited, and reflections are not likely to cause problems with sea coverage. Where significant sea traffic is anticipated, a high base should be dimensioned accordingly. Line-of-sight propagation losses (rather than mobile environment losses) should be assumed. Because of reduced multipath, a lower field strength of 32 dBμV/m will suffice. This lower figure results in significantly greater coverage. For the path loss calculations, the boat antenna can be considered to be at 3 meters.

These same reflections can, however, be a serious source of interference in high-density areas when the reflected signal reappears behind the cell site. Downtilt minimizes this problem.

Special problems occur when buildings are not built on flat ground and excavation was necessary. Figure 4.4 shows such a building. The building in Figure 4.4 is, in effect, underground from some directions. For example, the building is clearly underground with respect to base A. The first two levels probably will not have coverage from that base, but they will achieve coverage from base B. Many buildings are constructed this way, particularly in hilly cities, and they can make good handheld coverage extremely difficult.

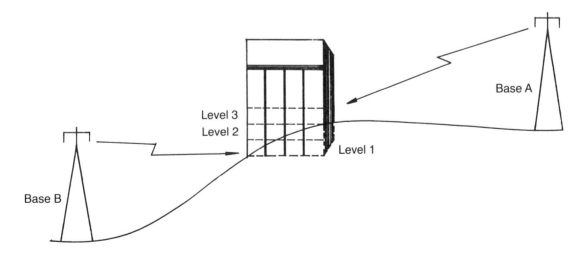

Figure 4.4   Land topography can cause dead spots. This building is partially underground from some directions. Note that the building may not be underground with respect to base B, but the first two levels of the building are underground with respect to base A.

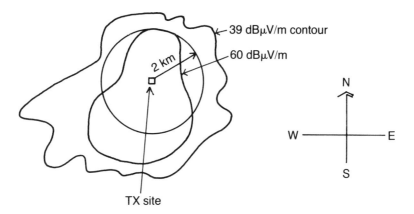

Figure 4.5   The surveyed field strength of the first selected site

Using these guidelines, a central site can be selected according to availability, cost, access, and the site's building potential. Having selected the pivotal site, the next step is to survey the site to determine its actual coverage. (For more information. see Chapter 5, "Radio Survey.")

In Figure 4.5, the poor city penetration in the northeasterly and northwesterly directions implies significant obstructions in these directions. Additional city sites may be necessary to remedy this situation. In large cities, if the 2-km boundary is greater than the 60-dBμV/m boundary, then you should use the 60-dBμV/m contour. Otherwise, 2 km or less may be used, depending on the expected traffic density. Sites should be selected from a combination of visual site inspections and map studies to provide continuous coverage.

MAP STUDIES

The first approximation of coverage is made from a map study or computer prediction. This process is only the first step in a design and should not be confused with the design itself. Cellular design is highly iterative.

The following approximation methods are recommended:

- CCIR. "Recommendations and Reports of the CCIR," 1982, Volume V, *Propagation in Non-Ionized Media*, Report 567-2
- Yoshihisa Okumura *et al. Review of the Electrical Communication Laboratory*, Volume 16, Nos. 9–10 September–October, 1968, NTT, Japan
- The program included in this chapter for range- and path-loss calculation
- Computer techniques

The first three methods are essentially based on similar empirical techniques and yield almost equivalent (but not identical) results. These methods use a series of curves that were derived from field studies of ranges obtained from transmitters on various frequencies and from various elevations. These methods have become classic studies that are widely used. Their main attraction is simplicity and proven utility.

Computerized methods using digitized maps can produce a higher degree of accuracy but are very costly. The accuracy of good computer techniques is still only ±6 dB (compared to ±10 dB for manual methods).

Computer prediction methods normally assume a two-dimensional path between the transmitter and receiver. In real-life propagation, however, contributions to the far-field pattern are made by reflected, scattered, and refracted paths that are in other planes. Wave scattering produces a spatial spectrum that is a complex holographic image of the surface causing the scattering. It is generally not practical to take these effects into account. Therefore, there is a limit to the degree to which calculations can reflect reality.

The detail needed to accurately determine the path profiles is also very large. If a city of 2000 km^2 is characterized by 100m x 100m areas, then there are 2000 x 10 x 10, or 200,000 such areas. A minimum representation would contain two pieces of information about each area, namely height above mean sea level and type of terrain (that is, urban, rural, open, or water), and thus would contain a total of 400,000 pieces of information. It is also desirable to store information about surface clutter (whether man-made, like buildings, or natural, like trees) and its height and distribution. This information can be the most difficult to obtain and to update. Moreover, for computer systems to be useful, they must not be limited to one city.

Thus, a large digital database and data storage in the high-megabyte range is necessary. The acquisition of this data is costly and is, ultimately, the limiting factor. Unless a database is available from other sources (for example, mapping authorities), the cost of producing one may be too high to be undertaken by a cellular operator. However, most computer-prediction databases lack surface-clutter information and are of limited use.

HANDHELDS

The measurement of field strength inside buildings reveals high standard deviations (meaning a large variability in the readings) and is not recommended for practical survey because it is difficult to do and does not yield useful results. However, on-site tests with handhelds, after the installation of bases, can be informative.

Street-level field strengths can be used to characterize levels inside buildings as an "average" building loss. Losses vary from 10 to 35 dB at ground level and are a function of floor level (decreasing by about 0.5 to 2 dB per floor). The loss measured from window or door increases as the building is entered, at a rate of about 0.6 dB per meter. The actual loss depends on the type of building, its size, and the amount of steel reinforcement used in its construction. Thus, in earthquake-, hurricane- and tornado-prone areas, losses are likely to be greater.

Some cities were planned with wide streets, most of which run at right angles to each other, with many open areas such as parks and squares. Other cities just grew with roads following cart tracks. Planned cities offer less impediment to cellular propagation, and spaces between buildings act as large and fairly efficient waveguides. In these cities, handheld penetration at considerable distance (4 km or more) can be achieved, particularly if the terrain is relatively flat. Propagation along the direction of a major road can be 6 dB better (at the same distance) than at right angles to that direction. In unplanned cities, a significant reduction in range can be expected.

For these reasons, it is not possible to have a universally applicable rule. It is advisable to survey each new city or region in order to get a clear understanding of the possible range before drawing conclusions from results obtained in other, different regions. Handheld coverage from 1 to 6 km has been reported for cities of similar size with different construction and features. A street-level field strength of 60 dBμV/m will generally be sufficient to ensure adequate handheld coverage inside normal office buildings at ground level.

Handhelds have introduced a completely new aspect into cellular design. With their low power outputs (0.6 watts), handhelds represent the real challenge of cellular RF design and also represent a new environment. Design rules need to account for the fact that handhelds are not usually used in high, multipath environments. Because handhelds are also the "weakest link" in the cellular path, a system designed for handhelds is generally adequate for vehicle-mounted units.

Because a handheld normally doesn't move at an appreciable speed, the low levels of multipath to a handheld make a big difference to two important parameters in cellular radio. First, the C/I (Carrier to Interference) ratio, usually quoted at 18 dB for mobiles in a multipath environment, can be reduced by 8 dB, allowing for cells to be placed significantly closer than they would be otherwise. Second, it also means that the higher deviation systems realize their full S/N enhancement in the handheld environment, which will result in a marked difference in handheld performance between systems that use high deviation (such as AMPS) and those that use low deviation (such as NMT).

BASE-STATION SENSITIVITY AND IMPROVED HANDHELD PERFORMANCE

The base-station sensitivity can be increased from -116 dBm to -122 dBm (a 6-dB improvement) to improve system balance for handhelds. This is sometimes done at omnidirectional bases. While effective, this method has been observed to increase intermodulation problems in receivers.

A PROGRAM FOR CALCULATING RANGE
OR PATH LOSS

A useful BASIC program is included in this chapter to enable the quick calculation of range and path loss. Based on CCIR Report 567-2, the program was written to enable determination of coverage from most cellular sites. The range of validity of the algorithm is limited to:

- Frequency: 450–1000 MHz
- Base station effective height: 30–200 m
- Vehicle antenna height: 1–10 m
- Range: 1–20 km
- Buildings/land: 3–50 percent

These limits should not cause any problems in the majority of cellular applications. Note that use beyond these limits (particularly for range) is not advised, as the results will not be accurate.

PROGRAM FOR RANGE AND PATH-LOSS CALCULATION

```
10 R = 5
20 RD = R
30 CLS
40 LOCATE 3
50 PRINT " THIS PROGRAM DETERMINES THE RANGE OR LOSS"
60 PRINT " OF A MOBILE RADIO SYSTEM GIVEN THE RESTRICTIONS"
70 LOCATE 7: PRINT " 1. THE RANGE 1-20 KM."
80 PRINT "            2. THE BASE HEIGHT IS 30-200 M."
90 PRINT "            3. THE VEHICLE HEIGHT IS 1-10 M."
100 PRINT "            4. THE FREQUENCY IS 450-1000 MHz."
110 PRINT "            5. THE % BUILDINGS TO LAND = 3-50%"
120 LOCATE 16,20: INPUT "RANGE(R) or LOSS(L) CALCULATION ?"; T$
130 REM frequency
140 F = 850
150 REM base station height
160 H = 30
170 REM % of buildings to land
180 BL = 15
190 REM vehicle height
200 V = 1.5
210 REM default for path loss
220 L = 148: D = 148
230 REM this software does not have logs to base ten
240 DEF FN BLOG(D) = LOG(D)/LOG(10)
250 REM fn blog is log base ten
260 CLS
270 LOCATE 3,50: PRINT " TO MAKE CHANGES ENTER THIS"
280 LOCATE 4,50: PRINT " NUMBER AND PRESS ENTER"
290 LOCATE 3,12: PRINT "THE DEFAULT VALUES ARE"
300 LOCATE 5: PRINT "_____          "
310 FOR N = 4 TO 20
320 LOCATE N,40: PRINT "|": NEXT
330 LOCATE 7,10: PRINT "frequency in MHz = "; F
340 LOCATE 9,10: PRINT "base stn height = ?"; H
350 LOCATE 11,10: PRINT " % buildings to land"; BL
360 LOCATE 13,10: PRINT "vehicle ht  = "; V
370 LOCATE 15,10: PRINT "terrain factor"; T
380 LOCATE 7,45: PRINT "change frequency type.......1"
390 LOCATE 9,45: PRINT "change base stn height type 2"
400 LOCATE 11,45: PRINT "change % buildings type....3"
410 LOCATE 13,45: PRINT "change vehicle ht type.....4"
420 LOCATE 15,45: PRINT "change terrain factor type.5"
430 LOCATE 17,45: PRINT "NO CHANGES TYPE............6"
440 LOCATE 19,45: PRINT "CHANGE ALL TYPE............7"
450 IF OPS = 7 GOTO 610
460 LOCATE 23,45: INPUT "change ?"; OPS
```

```
470 IF OPS = 0 THEN LET OPS = 6
480 LOCATE 18
490 ON OPS GOTO 500,520,540,560,580,610,500
500 INPUT "frequency?"; F
510 IF OPS <7 GOTO 260
520 INPUT "base stn height?"; H
530 IF OPS <7 GOTO 260
540 INPUT "% buildings to land"; BL
550 IF OPS <7 GOTO 260
560 INPUT "vehicle ht = "; V
570 IF OPS <7 GOTO 260
580 INPUT "terrain factor level suburban = 0dB"; T
590 IF OPS <7 GOTO 260
600 GOTO 260
610 S = 30 - 25 * (FN BLOG(BL))
620 A = (1.1 * (FN BLOG(F)) - .7) * V - 1.56 * (FN BLOG(F)) + .8
630 IF T$ = "r" OR T$ = "R" GOTO 690
640 LOCATE 18,2
650 PRINT "default range = "; R
660 LOCATE 19,10: INPUT "range"; R
670 IF R = 0 THEN LET R = RD
680 GOSUB 810
690 IF T$ = "L" OR T$ = "1" THEN GOTO 760
700 LOCATE 18,10: PRINT "PATH LOSS DEFAULT"; L
710 INPUT "path loss table needed y/n ?"; P$
720 IF P$ = "y" THEN GOSUB 880
730 INPUT "allowable loss = "; L
740 IF L = 0 THEN LET L = D
750 GOSUB 840
760 LOCATE 23,45: INPUT "more calculations y/n"; CAL$
770 IF CAL$ = "n" GOTO 800
780 OPS = 0
790 GOTO 260
800 END
810 L = 69.55 + 26.16 * (FN BLOG(F)) - 13.82 * (FN BLOG(H)) - A + (44.9 - 6.55
    *  (FN BLOG(H))) * (FN BLOG(R)) + S + T
820 PRINT "transmission loss = "; L
830 RETURN
840 REM calculates range
850 RA = 10^((L - S - T - 69.55 - 26.16 * (FN BLOG(F)) + 13.82 * (FN BLOG(H))
    + A)/(44.9 - 6.55 * (FN BLOG(H))))
860 LOCATE 21,45: PRINT "Range is"; RA
870 RETURN
880 PRINT "for 50 watt ERP"
890 PRINT "H/H to omni loss = 141 dB"
900 PRINT "car mounted loss = 148 dB"
910 PRINT "H/H to directional antenna loss = 138/148"
920 RETURN
```

MANUAL PROPAGATION PREDICTION

A number of empirical studies have produced algorithms that can be used to determine the far-field strength as a function of ERP (Effective Radiated Power) and range. Generally these results were produced after an extensive series of propagation tests in one or, at most, a few countries.

Studies done in different countries (with different terrains) revealed substantial variations in path attenuation over similar terrain. One of the main sources of discrepancy is in the description of topography. A hill in Venice may be compared to a molehill in San Francisco!

Thus, if a standard series of published curves (for example, as in Figure 4.6) are used to determine range, it is often wise to first "calibrate" the curves by measurement. In this process, a field-strength survey is undertaken to enable a detailed comparison between reality and the model. If a correction of more than 3 dB is necessary, additional measurements should be taken in various terrains to improve the accuracy.

ESTIMATING BASE-STATION RANGE

Let's assume you want to estimate the range for a typical cellular site; let's also assume a 20-watt TX transmitter power. The cellular system is designed for handheld coverage, and the system range is required in a suburban environment using an AMPS system. You can use Figure 4.6 to make the estimate. Proceed as follows:

1. Correct for actual ERP.

   The graph is drawn for 1000-watt ERP and so a suitable correction factor needs to be applied for the 50-watt ERP actually being used. The ERP of a 20-watt TX is about 50 watts (depending on feeder loss, antenna gain, etc).

   $$Therefore\ Factor = +10 \log \frac{1000}{50}$$

   $$= +13\ dB$$

   Thus, the 39-dBμV/m boundary (for AMPS) will be located on this curve at 39 + 13 = 52 dBμV/m.

2. Draw the graphs for the field strength attained at 50 percent of locations and times; that is, the field strength that is exceeded in 50 percent of all readings. Field strength is a log-normally distributed variable and is illustrated in Figure 4.7.

   If the standard deviation "σ" of the field strength is not known, it can be characterized as urban (σ = 8 dB to 12 dB), suburban (σ = 6 dB), rural (σ = 3 dB), or water paths (σ = 1.5 dB). Some propagation graphs show the field strength for 90 percent of locations and time. This can similarly be corrected to the mean using the relationship mean = (90 percent locations and time reading) + 1.28 x σ.

   Thus, to find the distance at which the mean is 39 dB using such a graph, it will be necessary

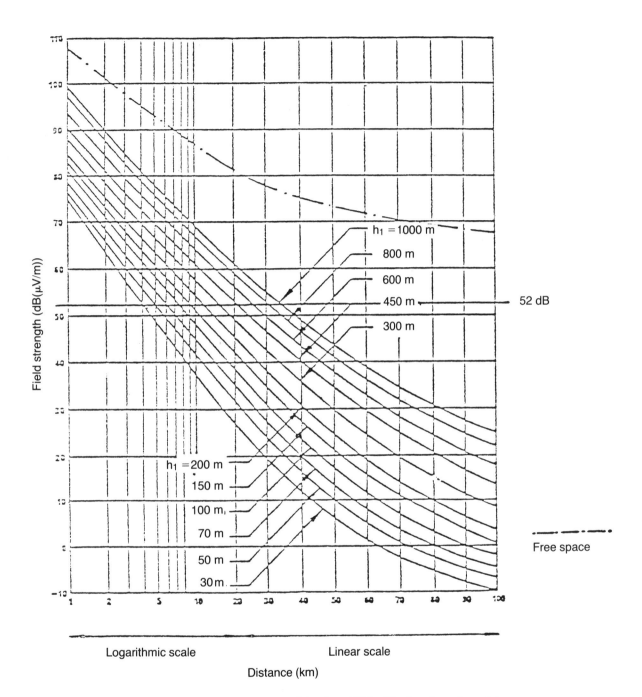

Figure 4.6   Field strength (dBμV/m) for 1 kw ERP in an urban area

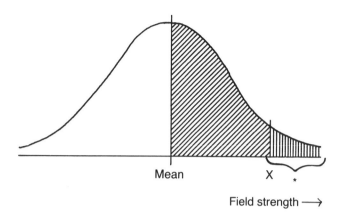

Field strength ⟶

Figure 4.7   Field strength is log-normally distributed. The shaded area represents the probability that field strength will exceed X. The value thus obtained is really the median value of the log of the field strength and will differ from the actual mean field strength. In practice, this discrepancy is generally ignored.

to look for the value of the field strength on the graph corresponding to 39 - 1.28 x σ. If σ = 6 dB, then 39 dBμV/m average = 31.3 dBμV/m on the 90 percent/90 percent graph.

This system is designed for AMPS handhelds, so for TACS add 2 dB more; for NMT900 add 8 dB (NMT450 systems use 450 MHz curves and 41 dBμV/m).

3. Draw a line through 52 dBμV/m on the graph to obtain the 39 dBμV/m contour. You can now obtain the range in an urban environment by reading off the range against this line. A correction must then be made for the actual environment (unless it is urban).

The graph is drawn for urban environments defined as 15 percent of the land occupied by buildings. In very dense CBDs, this percentage can be much higher, and in rural areas it can be zero. However, the formula is only valid to 2-percent occupancy. A correction factor, S = 30 - 25 log α, can be used, where α equals percent of building to total land area. This percentage can be applied to terrains different from suburban terrains.

If a correction factor is applied, the new level in Figure 4.7 for 39 dBμV/m (50 percent, 50 percent) is (39 dBμV/m - S). Notice that at α = 15 percent (that is, in an urban environment), S = 0 so that S may take positive or negative values.

Finer distinctions between terrain types are offered by the Okumura paper but because this whole technique is, at best, approximate, additional corrections are normally not needed except for very unusual terrain. Such terrain includes very flat land (reclaimed swamps, subtract 27 dB or S = 27), water or part water paths (see Okumura), and large hills (treat as absolute boundaries).

4. Find the curve corresponding to the base-station height (h_1) with respect to the surrounding terrain.

5. Read off the range corresponding to the type of terrain (Table 4.1); for example, Urban terrain, base-station height h_1 = 30 m, range = 4 km.

6. Plot this coverage on the map, making adjustments for local terrain; for example, hills form natural boundaries.

7. Plot the 39-dBμV/m (50 percent, 50 percent) contour (which is 52 dBμV/m on the graph) or other, as applicable.

Select sites that look likely to provide good continuous coverage with respect to the central site and then survey them. Note that this process is not used to select sites. It only determines which sites should be surveyed. From the survey results, it will become evident which sites are useful and which are not. The process is then repeated.

VARIABLE TERRAIN

Where the terrain is variable within one cell coverage area, the plot of the coverage should be done in sectors. Consider the prediction of the coverage for the area in Figure 4.8. There are three discrete types of terrain seen by the cell site. The propagation over each type will be quite different. If the CBD is substantial and the cell site is more than 4 km away, it should be regarded as an absolute barrier.

SEA PATHS

Sea paths offer the best radio paths. As boats move slowly (compared to wavelengths/second), they suffer very little from multipath noise. Boats should be equipped with antennas mounted higher than 1.5 meters (the level normally assumed for vehicles) unless, of course, handhelds are used. Even then, they would be used at heights greater than 1.5 meters. These sea paths can provide any given grade of service (measured as S/N) at a lower field strength than land paths.

A field strength of 32 dBμV/m can be regarded as adequate over sea, although a usable service

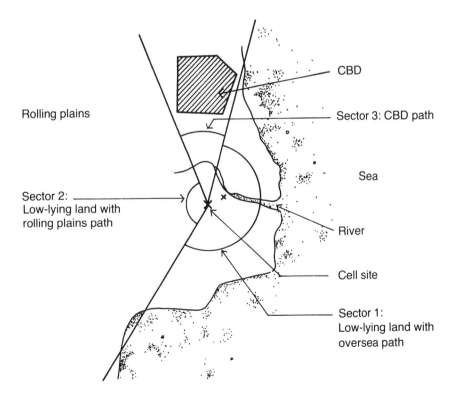

Figure 4.8   Dealing with variable terrain

is available at 25 dBμV/m and sometimes even lower (depending on antenna height). When at sea, the most prominent (highest) land site will be the one with the highest field strength rather than the one closest to the boat. Ranges of 50 km (over unobstructed paths) are typical and up to 100 km are not uncommon under favorable circumstances.

MULTI-CELL SYSTEMS

Having chosen and surveyed the central site, the next task is to plot its actual coverage. Using the expected subscriber density, determine the required cell radius for the adjacent cells. This procedure will provide a good indication of base-station height (lower bases for smaller cells).

Next, select six sites roughly equidistant from the central site and, by map studies or computer studies, plot the expected coverage of the new sites. The map study should be regarded as the first stage in site selection. That is, the map should be used to identify sites that are worth surveying and to eliminate those that are not. Because a survey takes about two weeks per site, this preliminary screening can save time and money.

Figure 4.9 shows the results of such a map study (dotted coverage contours). Some sites have

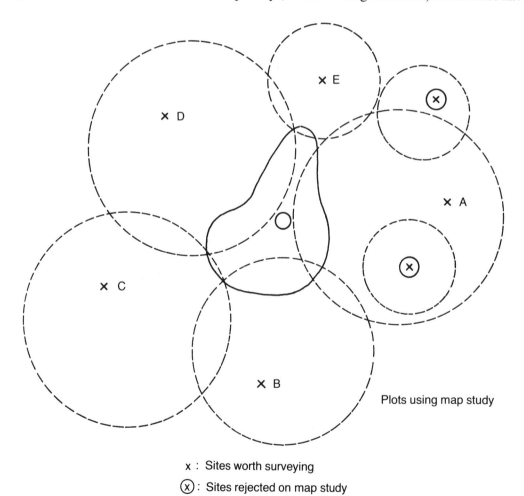

x : Sites worth surveying

(x) : Sites rejected on map study

Figure 4.9   Map studies of proposed sites

insufficient coverage to be worth consideration. The sites that hold promise, namely sites A, B, C, D, and E, are then surveyed and the results are plotted.

SURVEY PLOTS

Survey results are only as good as the equipment and techniques used. Therefore, large discrepancies should ring alarm bells that perhaps something is wrong with your equipment. Assuming that satisfactory explanations for any discrepancies can be found, the selection process is continued by selecting more sites to provide the remaining coverage, conducting more map (computer) studies and re-surveying until adequate coverage has been found.

Figure 4.10 shows that sites B, D, and E look promising but sites A and C are inadequate. However, before totally dismissing sites A and B, the large discrepancy between the predicted and actual coverage should be examined and the reason for the discrepancy identified.

Because continuous coverage is important, the design proceeds much like putting together a jigsaw puzzle—starting from a fixed point and working outwards. Placing six equal-coverage cells around a central cell gives a good approximation of the regular hexagon pattern. This method more closely approaches the theoretical "equal cell size" pattern than does the approach used by

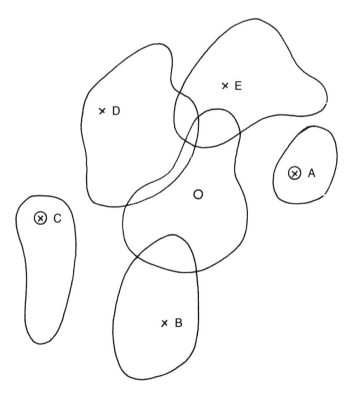

x : Suggested site with potential

ⓧ : Site not suitable

Figure 4.10   Survey results plotted

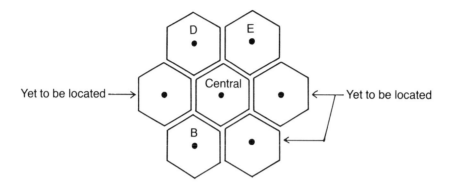

Figure 4.11   Hexagon pattern

some designers, who attempt to locate sites according to their physical location on the cellular grid.

Were sites to be precisely located (for example, on a flat surface) according to a cellular grid, they would only produce an approximate optimum pattern (equal hexagons). In the real world, however, obstructions such as hills, buildings, rivers, and foliage near sites that are located *exactly* according to their theoretical hexagonal pattern positions may produce patterns that do not resemble a regular configuration.

Site selection involves compromise, particularly with site availability. Fortunately, the designer can tailor coverage by using height, power, and antenna patterns to achieve efficient spectrum reuse.

Once the seven cells (or most of the seven cells) have been selected, the cell pattern and its orientation is established. That is, the hexagon pattern is derived from the site selection, not vice versa. The hexagon pattern derived from Figure 4.10 is shown in Figure 4.11.

Many designers proceed by starting with the hexagon pattern and selecting sites from the pattern. Where uniform terrain exists, the result can be much the same as deriving the pattern from the terrain. However, the concept of visualizing each cell as surrounded by six equal-coverage adjacent cells is easier and leads to a more flexible approach to site selection. Once a network has matured, a more valid approach is to adopt the hexagonal plan because the coverage can be determined by tailoring antenna patterns; thus the sites can be made to conform to a desired coverage. This is particularly true of the 4-cell pattern or a pattern that will evolve into a 4-cell pattern.

FILING SURVEYS

The design can be seen to proceed iteratively. Therefore, it is unwise to discard any survey result as totally unsuitable. Unused survey results can be useful in filling gaps in the current or future design. Conversely, as the design evolves, sites that once appeared ideal may become redundant. Thus, all survey results have the potential to become part of the jigsaw puzzle.

When filing surveys, remember that all surveys should be conducted using the same units (dBμV/m is recommended. See Chapter 8, "Units and Concepts of Field Strengths," for more information). This will help you avoid confusing results. In addition, all survey maps should be filed with the following data clearly marked:

- Date of survey (very important) and name of surveyor
- Location of site, owner, name, and phone number of contact person
- Survey antenna height (above ground level and above sea level)
- Cable loss (preferably measured because connector losses are difficult to calculate)
- Transmitter power
- Antenna gain (usually 6 dB)
- Correction factor used to convert actual ERP to nominal ERP (that is, all readings are corrected to the ERP of an actual cellular base station)

These details may prove essential in the future if discrepancies are detected.

MAPS AND MAP TABLES

The maps used to plot survey results should be about 1:100,000, certainly not larger than 1:250,000 or smaller than 1:50,000. Topographic projections should be used for coverage prediction. However, street maps that show street names in detail are usually best for recording individual survey results, before they are plotted at a scale suitable for system studies. Actual coverage plotted on street maps is useful for recording detail coverage of individual bases. The final result should be an easy-to-read map designed for use by the subscriber. The map shown in Figure 4.12 is the coverage map for the Extelcom system, which was prepared by the author.

A map table of at least 2m x 2m (or larger) is needed to adequately handle coverage maps, given the scale needed to ensure sufficient detail for survey results. Once a large number of maps are stored, it is essential to have a suitable cataloging system. Maps also need adequate storage and a map file (a file that can store full-size maps without folding) is best.

Transparencies can be used to completely overlay the maps to depict coverage. Felt pens used to mark the transparencies should be of the "whiteboard" erasable type. A very handy accessory is a set of map weights. Without weights, maps can be awkward to handle.

DESIGNING FOR CUSTOMER DEMAND

Continuous coverage is an essential element of any design. However, expandability for the future must also be built in. The system operator has an obligation (as well as a financial incentive) to meet the demand for services. The ultimate capacity of the system and its ability to expand should be included in the design. A good designer will produce a design that does not place constraints on future expansion. And a good designer will also know that forecasts are purely hypothetical and will build in maximum flexibility for the system. This flexibility should include the ability to expand rapidly for unexpected demand and the possible unavailability of previously selected base-station sites. A cellular system has many parameters that can be utilized to extend

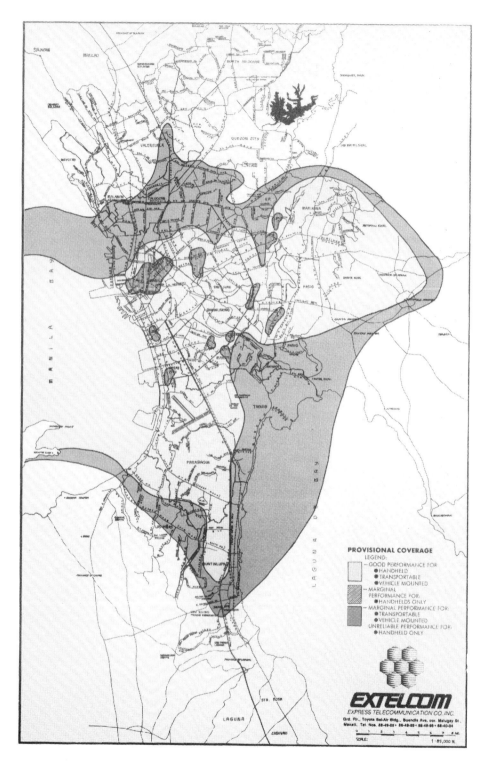

Figure 4.12   Extelcom system map

or contract base-station coverage to temporarily cover unforeseen problems, provided the capacity is available. A minimum of six months should be allowed to obtain additional channel equipment.

If a site must be moved to allow for future expansion or improved coverage, then it is possible that the design was too limited in scope. Predicting future customer demand is difficult. All good designs should allow for substantial errors in forecast demand.

DETERMINING CHANNEL CAPACITY

For the following discussion, assume an AMPS 7-cell, 666-channel A-band system. Adjustments should be made for other system types. The capacity of a fully equipped system can be determined approximately by using the first equation below. It should be noted that this equation is valid only for systems and not for individual bases, because sectored and small bases have different capacities per channel to a large omni base. However, given the usual mix of cell sizes and types in a typical system, this equation has proved valuable in practice. Where the calling rate is unknown, 30 mE can be assumed with a system capacity of 20 subscribers per channel.

$$C_A = N_C \times 20 \times \frac{30}{C_R}$$

$$= 600 \times N_C \times C_R$$

where N_C = number of speech channels
C_R = calling rate in milli-Erlangs
C_A = capacity in subscribers

The maximum value of N_C at any one site depends on the system type and the selected cell pattern. To be consistent with a particular cell plan (N = 4, 7, 12), then:

$$N_{Cmax} = \frac{Total\ number\ of\ speech\ channels}{N}$$

For example, if an operator had the A portion of the AMPS non-extended band, the total number of channels = 333-21 or 312, and if an N = 7 pattern is used, then:

$$N_{Cmax} = \frac{312}{7} = 44\ speech\ channels\ per\ site$$

If an N = 4 pattern is used, then:

$$N_{Cmax} = \frac{312}{4} = 78\ speech\ channels\ per\ site$$

It should be noted that this upper limit cannot be exceeded if the regularity of the cell pattern is to be maintained. For example, if a design called for 12 bases and used only the A band of a 666-channel (non-extended) AMPS system, and the customer-calling rate was 30 mE, the maximum channel capacity would be:

$$\frac{600}{30} \times (12 \times 44) = 10,560\ subscribers$$

(Note that the extended AMPS band has 832 channels.)

BASE CAPACITY

Channels can be redeployed from low-usage bases to high-usage bases, but base stations are fixed and expensive to move. The maximum channel capacity, as calculated above, assumes 100-percent occupancy of all bases. In a real system, the traffic is distributed so that some bases reach capacity before others and, therefore, the actual capacity is about 70 percent of the base-station capacity. In a new system (one with no traffic history), it is safer to assume 50 percent of the maximum capacity (achievable if all bases were full). This is because a mature system has less uncertainties about its traffic distribution. So the 12-base systems discussed on the previous page would have the following capacity:

$0.7 \times 10{,}560 = 7392$ *or approximately* 7400 *in a mature system*

$0.5 \times 10{,}560 = 5280$ *or approximately* 5300 *in a new system*

Of course, in this 12-base system, only the channels needed to carry the forecast subscribers would be provided.

CUSTOMER DENSITY

Customer density expressed as customers/km^2 or Erlangs/km^2 can be determined only after the system has matured. Before commissioning the system, this information is very difficult to quantify, and even the most careful estimates can be in error by factors of about 2 to 5. Thus, the designer has to work in a very uncertain environment. This means that the design should be flexible enough to be reconfigured to account for actual (as opposed to forecast) traffic demand.

Fortunately, it is very easy to relocate channels in modern cellular systems by moving channels, provided there are sufficient correctly located base stations. Base stations are usually expanded in modules of 8, 10, or 16 channels. The extra equipment needed to expand beyond the module size includes additional antennas and should be considered at major sites. The uncertainty is particularly serious in the CBD. Here the base-station configuration, not channel allocation, should allow for traffic levels of about two times the forecast traffic for new systems.

There are, however, some important factors that must be allowed for if channels are to be moved from base to base.

First, extra racks must be purchased so that channels can actually be moved. It is a good idea to purchase and install an additional spare rack for every major base station to allow for a quick expansion.

Second, the RF (radio frequency) coupling equipment, sometimes known as "plumbing," is in modules tailored for individual racks. Depending on the specific equipment layout, each connecting cable is specially designed and cut. Many of these cables are of critical lengths and it will be necessary to purchase with the spare rack cables that will be needed to interconnect the various equipment configurations.

Referring back to the example of 12 bases, if 3 of these were located in the CBD, then the CBD traffic capacity is:

$$\frac{600}{30} \text{ x 3 x 44} = 2640$$

Therefore, it would be unwise to plan this system to carry more than about 1300 customers in the CBD if it were a new system (50-percent occupancy can be assumed in the CBD).

If the CBD traffic represents 50 percent of the total traffic (this would be normal), then the actual capacity of this design is 2 x 1300 = 2600; that is, 50 percent of the "maximum" CBD traffic plus the same amount of suburban traffic (compared to 5306 without this restriction). This situation can be improved significantly by moving one or two bases to the CBD area.

DETERMINING BASE STATIONS IN THE CBD

Note: Handheld coverage will require that base stations in high-density urban areas be 2 km (or less) apart in earthquake-, hurricane-, and tornado-prone areas, or 4 km elsewhere.

To determine the number of base stations required in the CBD, perform the following calculation.

Use the estimated CBD proportion of traffic (50 percent if unknown) and define on a map the boundaries (approximate) of the CBD. Calculate the number of bases required in that area using the following equation:

$$C_A = 600 \text{ x } N_C / C_R$$

Assume a 666-channel AMPS system and N = 7, then:

$$N_{Cmax} = 44 \text{ x } N_B \text{ } (full \text{ } bases)$$

where C_A = total customers
N_C = number of channels
N_B = number of bases

$$N_B = \frac{C_A / F}{600 / C_R \text{x } 44*}$$

* Or use the maximum size if another cell pattern or system is used.

where   F = 0.5 for new designs
F = 0.7 for extensions to existing designs

OMNI CELLS

Omni cells have a significantly higher traffic capacity than an equivalent sector cell, particularly when a small number of channels (less than 20) are used. Omni cells are also simpler and cheaper to construct than sectored cells for any given number of channels. An omni-configured system thus costs significantly less per customer than a sectored installation. When frequency reuse is not a significant factor, using omni cells can be a valuable way to (temporarily) increase the capacity of a new cellular system.

In general, it will be necessary ultimately to sector the omni cells except for small service areas. The reduced handheld talk-back ability of an omni cell may result in poorer handheld performance and so may not be a good idea in central business district areas.

ANTENNAS

It is normal that two diversity antennas will each be connected to up to 64 receivers. Transmit antennas usually carry only 15 or 16 channels. To provide maximum flexibility, base stations should not be installed without reserve antenna capacity. Delivery times on antennas, cables (feeders), and couplers are about three months.

SYSTEM BALANCE

In the original cellular-system specifications, an effort was made to achieve a balanced system and, as will be shown, this balance has been achieved reasonably well. A balanced mobile system is one where the speech path between the mobile and the base, and vice versa, are of the same quality. The quality is measured as a signal-to-noise ratio. In a cellular system where all data transfers involve a "handshake" (that is, all instructions require a reply) there is little to be gained by having a better signal path in only one direction. The communicable range is limited by the loss along the least-loss-tolerant path. In general, the most vulnerable path will be talk-out (that is, from the relatively low-powered mobile to the base station).

The limiting factor for a handheld is talk-out where the radiated power is only 0.6 watts. In general, the talk-out path-loss allowance of a base station equals or exceeds that of a handheld, so any increases in base-station power will not improve handheld coverage. Vehicle-mounted mobile installations have some small advantages with increased base-station power.

Increased base-station power causes increased interference in the co-channel and adjacent-

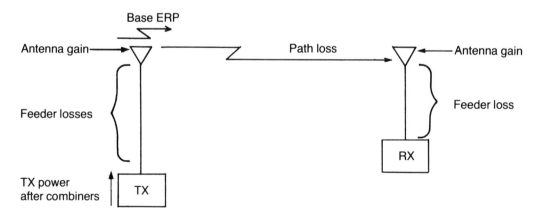

Figure 4.13 The components of a cellular-system path

channel mode, so the gains are not necessarily worthwhile, except where frequency reuse is not a consideration. When such interference occurs, data corruption is to be expected. Despite adequate available field strength, some calls cannot be made or received in these areas.

Figure 4.13 shows the parameters that contribute to overall path loss. In the direction of an omnidirectional base to the mobile, the maximum path loss is determined by base-station ERP (Effective Radiated Power). ERP is usually limited by local regulations. In this example, 50-watt ERP is assumed:

$$50 \ watts \ = \ 10 \log 50{,}000 \ dBm \ = \ + \ 47 \ dBm$$

Net allowable path loss from a base station = 148 dB

Note: Received level for S/N (at antenna) = 18 dB = -101 dBm in 50 percent of locations. Various values are used for the desired receive level, including -116 dBm (maximum sensitivity) and -95 dBm (= 39 dBμV/m). This will not affect the balance calculations, although it will lead to different allowable path losses.

	OMNI BASE	SECTORED BASE
TX Power = 3 watts = 10 log 3,000 dBm	35 dBm	35 dBm
Feeder loss	- 3 dBm	- 3 dBm
Antenna gain	+ 3 dBm	+ 3 dBm
Base receiver antenna gain	+ 9 dBm	+ 17 dBm
Base feeder loss	- 3 dBm	- 3 dBm
Base diversity gain	+ 6 dBm	+ 6 dBm
System gain	47 dBm	55 dBm
Allowable path loss for S/N = 18 dB	+ 101 dBm	+ 101 dBm
Net allowable path loss for handheld	148 dB	156 dB

Table 4.2   Allowable path loss for a vehicle-mounted mobile

	OMNI BASE	SECTORED BASE
TX Power = 0.6 watts = 10 log 600 dBm	28 dBm	28 dBm
Feeder loss	0 dBm	0 dBm
Antenna gain	0 dBm	0 dBm
Base receiver antenna gain	+ 9 dBm	+17 dBm
Base feeder loss	- 3 dBm	- 3 dBm
Base diversity gain	+ 6 dBm	+ 6 dBm
System gain	40 dBm	48 dBm
Allowable path loss for S/N = 18 dB	+ 101 dBm	+ 101 dBm
Net allowable path loss for handheld	141 dB	149 dB

Table 4.3   Allowable path loss for a handheld

The path from a vehicle-mounted mobile to omnidirectional and sectored bases will yield different allowable path losses. In this instance, there are a few additional factors as shown in Table 4.2 and Table 4.3.

As you can see, for systems where path loss becomes the limiting factor, at losses of 141 dBm the handheld will fail to provide a satisfactory path to the base. As the range is increased to a path loss of 148 dB, the vehicle-mounted unit will begin to experience an inadequate path from the base station. Increasing the base-station ERP by 3 dB (that is, to 100 watts) to balance the vehicle-mounted ERP does not improve handheld performance. Sectored antennas, however, significantly increase handheld performance and will balance the path. Using 100-watt ERP will improve mobile performance by 3 dB when sector-receive antennas are used.

SECTORED ANTENNAS AND SYSTEM BALANCE

Sectored base-station antennas have higher gains than omni directional antennas and can improve the talk-back performance of a handheld to the point of balance with the base station. This only occurs, however, in the direction of the main lobe where the gain is maximum.

The gain of a sector antenna (17 dB) is 8 dB higher than an omni antenna (9 dB), so a balanced system to handhelds can be achieved.

System balance calculations need not be done in detail for each base-station design (unless required by regulation), because the results will all be within a few dB of each other. The purpose of these calculations is to ascertain the optimum base-station TX power, which is that power just assures (or nearly assures) a balance. Because of interference, too much TX power is counterproductive.

Consider the case where omni antennas are replaced by sectored antennas with a +8-dB net gain. If the TX power is not reduced, the ERP will also increase from 50 watts by 8 dB, which means an ERP of $50 \times 10^{0.8} = 315$ watts! This ERP exceeds the legal limits and will cause serious interference.

To balance the transmission paths, the TX power should be reduced by $10^{0.8}$ (or by a factor of 6.3). In practice, it is usual to reduce the base-station TX power to between 5 and 10 watts when using high-gain sector antennas. This range is used because the actual gain of the sectored antenna varies from 17 dB to 10 dB as a function of the angle with respect to the center of the antenna.

500-WATT ERP RURAL SYSTEMS

In rural service areas (RSAs) in the US, the FCC has authorized ERPs as high as 500 watts. Although this seems to conflict with the 100-watt useful limit, in fact, it does not. The 500-watt limit can be used in rural areas that are at least 38 km (24 miles) from a metropolitan statistical area (MSA). It is envisaged that this higher power will be used along highways or other areas where elongated coverage is useful.

The system gain can be balanced by using higher-gain antennas in the receive path than in the transmit path. If the system is balanced for mobile to base at 100-watt ERP, then the imbalance at

500 watts equals 10 x log 5 = 7 dB. Provided the receiver antennas have about 7-dB more gain than the transmit antenna, the system will balance.

A typical highway coverage base station may have a 1.8-meter (6-foot) transmit antenna with a gain of 18.9 dBd, and 3.7-meter (12-foot) receiver grids with a gain of 25.4 dBd. The gain difference of 5.5 dB almost balances the system.

Chapter 5

RADIO SURVEY

A radio survey is the process of measuring the propagated radio field strength over an area of interest. It is an essential part of the cellular radio site-selection process. Many radio-survey techniques exist, but few yield consistent and satisfactory results. A radio survey is necessary as a design aid and as a maintenance tool. As a design aid, it helps determine potential coverage of a proposed base station site. As a maintenance tool, a radio survey confirms continued satisfactory coverage.

A radio survey usually uses a field-strength measuring receiver located in a vehicle to measure the field strength. Sometimes the reciprocal path (that is, the path from the mobile to the base station) is measured instead. Both measurements are mathematically equivalent.

When measuring field strength, it is important to note that what is being measured is a statistical variable and that the measurement technique must allow for this.

Three factors operate together to produce the measured field strength: path loss (free space), log normal fading, and Rayleigh fading. Figure 5.1 illustrates the free (for example, unobstructed) path loss. This loss is the most significant in microwave links, but it is only one of the losses in the mobile environment.

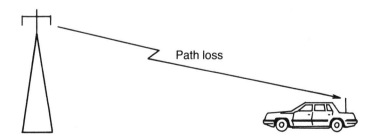

Figure 5.1   Path loss (free space)

The free space path loss P_L is given by:

$$P_L = 20 \log (42 \cdot d_{km} \cdot f_{MHz}) \, dB = 32.5 + 20 \log f_{MHz} + 20 \log d_{km} \, dB$$

where P_L = path loss in dB
             d = distance
             d_{km} = distance in kilometers
             f_{MHz} = frequency in megahertz

Figure 5.2 shows log normal fading. This process is called log normal fading because the field-strength distribution follows a curve that is a normally distributed curve, provided the field strength is measured logarithmically.

Multipath, or Rayleigh, fading is a salient feature of mobile communications and, to some significant extent, limits the coverage of mobile systems when the mobile is moving in a multipath environment. It is not such a dominant factor in handheld mobile usage but, in low-field-strength areas, it can be detected by variations in noise levels as the receiver is moved. Figure 5.3 illustrates multipath fading.

Figure 5.2   Log normal fading that is due to obstruction is known as "shadowing" or "diffraction losses."

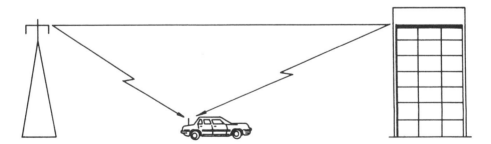

Figure 5.3   Multipath, or Rayleigh, fading is produced as a result of interference patterns between the various signal paths in a multiplan environment.

An empirical formula for the cumulative effect of these three types of fading is given in *Recommendations and Reports of the CCIR*, 1982, Volume V, Report 567-2 as:

$$P_L = 69.55 + 26.16 \log f_{MHz} - 13.82 \log h_1 - a(h_2)$$

$$+ (44.9 - 6.55 \log h_1) \log d_{km} \; dB$$

where P_L = loss in dB

f_{MHz} = frequency in megahertz

h_1 = base station antenna height in meters

h_2 = receiver antenna height in meters

$a(h_2) = (1.1 \log f - 0.7) h_2 - (1.56 \log f - 0.8)$

d_{km} = distance in kilometers

where 15 percent of the area is covered by buildings (that is, an urban area).

This formula is based on field experience. Experience dictates that in different terrains some or all of the coefficients must be recalibrated. For general use, the formula should be regarded as accurate to ± 10 dB.

It is often said that in the mobile environment, the loss is inversely proportional to distance to the fourth power. You can see that this is consistent with the formula by looking at the last term $(44.9 - 6.55 \log h_1) \log d_{km}$. This is the only term that is a function of d_{km}. If, for example, $h_1 = 30$ meters (the base-station antenna height), then this term becomes $(44.9 - 6.55 \log 30) \times \log d_{km} = 35.2 \log d_{km}$.

Because this is an expression for loss in dB, it can be rewritten in the form:

$$Loss \; \alpha \; \frac{1}{d_{km}^{3.52}}$$

This reduces the relatively complex formula to approximately d_{km}^{4}; the fourth power relationship holds exactly at an antenna elevation of 5.6 meters.

STANDING WAVE PATTERNS

In the far field, where all these loss modes are operating, the field strength varies with distance and time. At any one instance, the field strength can be shown as illustrated in Figure 5.4 on the next page.

The limiting case of standing wave patterns is one produced by a reflecting plane at right angles to the line of propagation. A standing wave produced by a wave incident on a plane reflecting surface (such as a wall) produces the familiar $\lambda/2$ standing wave pattern shown in Figure 5.5 on the next page.

Other forms of interference generally produce interference patterns with a wavelength greater than $\lambda/2$. The distance L, between the waves, is such that $\lambda/2 < L$, but L can take any larger value. In practice, however, $L \approx \lambda/2$ can be taken as the worst case. Notice that $\lambda/2 = 0.16$ m (16 cm) at 900 MHz.

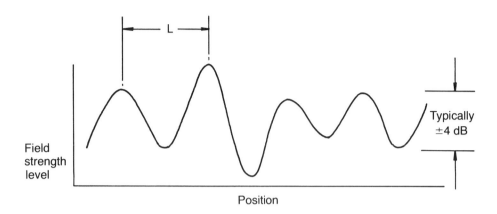

Figure 5.4   Standing wave pattern

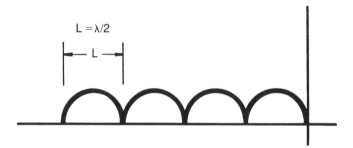

Figure 5.5   Standing wave pattern caused by reflections off a plane surface

MEASURING FIELD STRENGTH

As a mobile radio must necessarily operate over its entire service area, the field strength at a point becomes meaningless in terms of the overall performance of a mobile receiver. Consequently, individual spot readings are also meaningless.

Some operators have tried to solve this problem by using a field-strength meter in a moving vehicle and "guessing" the average level. This also yields meaningless results, as you can see from the structure of a typical field-strength meter, as shown in Figure 5.6 on the following page.

The field-strength meter consists of a receiver, which has some way to access the IF limiter or AGC (Automatic Gain Control) drive stage to measure the output of that stage via a meter. A log-law amplifier gives a usable output in dB. The meter is usually a conventional moving-coil meter.

Because these field-strength meters are designed to operate in a point-to-point environment, the smoothing capacitor C (or its mechanical equivalent) is usually incorporated to even out small fluctuations due to log normal fading, and it likely has a time constant of 0.5–2 seconds.

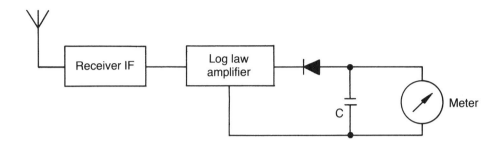

Figure 5.6   Basic field-strength meter

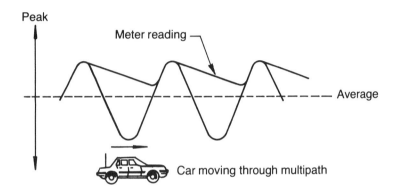

Figure 5.7   Detector response

When such a meter is confronted with a rapidly varying signal, it tends to follow the peaks, as indicated in Figure 5.7.

Because the relationship between the peak value and the average value from such a device is dependent on the depth of the fade and the multipath frequency, no firm relationship between these values can be said to exist. Thus, a sampling method that samples at a sufficiently high rate to measure the actual standing wave is necessary.

If this standing wave pattern were totally uncorrelated, the necessary sampling speed would be the Nyquist rate (2 times the pattern frequency), but, because of the existence of a correlation, in practice, it has been demonstrated that about one quarter of that rate yields results that are accurate to ±1 dB.

SAMPLING SPEED

To calculate the required sampling speed in a mobile environment, consider a vehicle moving at 100 km per hour through a standing wave pattern of a 900 MHz transmission.

The wavelength of that pattern is:

$$C / F = \frac{300,000,000}{900,000,000} = 0.333 \ meters$$

where   C = speed of light
             F = frequency

In the worst instance, the standing wave pattern is repeated every λ/2 m (as in the case of a reflection of 180 degrees from a wall).

So λ / 2 = 0.333 / 2 = 0.1665 *meters*

100 *km per hour* = 27.7 *meters / sec.*

The Nyquist sampling rate requires two samples per pattern interval, or one sample every 0.1665/2 meters. Thus, the Nyquist sampling rate is 27.7/(0.1665/2) = 332 samples/sec. As already mentioned, about one quarter of that rate (or 80 samples/sec) would suffice.

MODERN SURVEY TECHNIQUES

Modern survey equipment is based on the principles of the system illustrated in Figure 5.8. The radio receiver can be a specialized communications receiver with a wide dynamic range of both level and frequency, or it can be a special-purpose receiver such as a mobile telephone. If the receiver is not designed as a measuring receiver, some limitations will occur due to the limiter time constants.

The two main criteria for the receiver are that the RF level is a monotonic (single-valued) function of the RF level and that the limiter output has a bandpass characteristic of about 100 Hz.

Most modern receivers meet the requirement that the RF level is monotonic. Some older wide-band communications receivers with a measuring capability use switched attenuators to increase their dynamic range. Often the switching-in of these attenuators causes discontinuities in the output level, which can render them unsuitable for survey.

The response time of the limiter is deliberately damped so that the device operates only to compensate for level variations that ordinarily result from fading, but the response frequency is limited so that RTTY, MORSE, and other digital signals can be passed normally. This damping can be as elementary as a simple RC bandpass filter, or it may be more sophisticated.

Figure 5.9 shows the limiter/AGC response curve. If it was originally planned that the receiver was to operate in a fixed location, this damping may limit the response time of the limiter to the extent that it is unsuitable for survey.

The easiest way to determine suitability is to input an AM square-wave modulated carrier and compare the input with the limiter drive as the frequency is increased. Provided the output tracks fairly well up to 50 Hz, the receiver will be effective. Figure 5.10 shows the limiter response time of a survey receiver.

Most measuring receivers track only up to 50–100 Hz, although a few go a little higher than this. Notice that the inability to sample to at least 50 Hz results in damping errors of the same nature as those of a simple field-strength meter previously discussed.

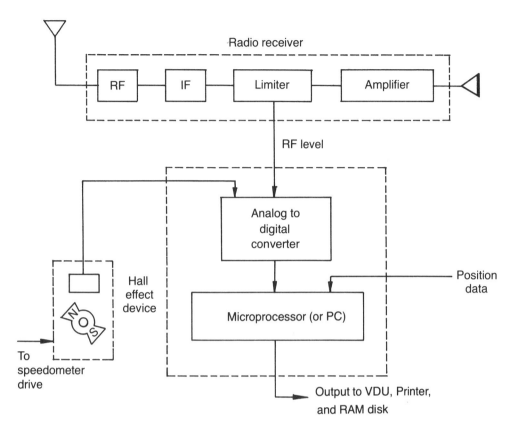

Figure 5.8   Basic survey block diagram

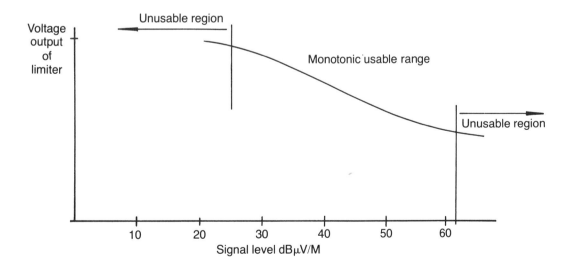

Figure 5.9   Usable portion of limiter/AGC response curve

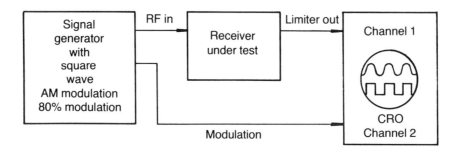

Figure 5.10   Testing the limiter response time of a survey receiver

In a digital field-strength meter, the output of the limiter is read by an analog-to-digital converter, which samples the limiter levels. As previously explained, the sample rate depends on the RF frequency and vehicle velocity. Figure 5.11 shows a typical A/D converter.

A single high speed analog-to-digital converter chip measures the limiter output voltage and outputs the level to a data bus. Usually, a multiplexer chip is used to allow multiple inlets to be sampled in turn.

Receivers not specifically designed as measuring receivers may well have a band pass characteristic similar to the one shown in Figure 5.12. You can see that the limiter drive does not respond to low-frequency signal variations. Such a receiver will probably "see" Rayleigh fading but will not respond to log normal fades. Such a receiver may also have a square-wave response, as illustrated in Figure 5.13. Notice that the behavior of the receiver in Figure 5.13 seems to be limited mainly by a simple RC network.

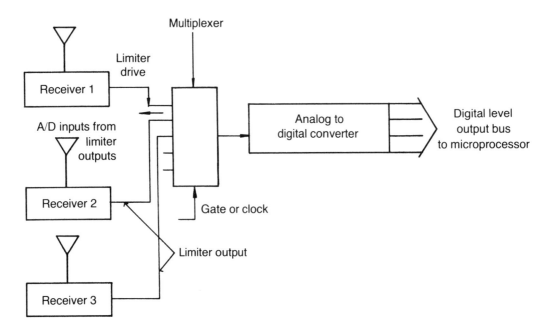

Figure 5.11   An A/D converter with multiple inputs from a bank receiver

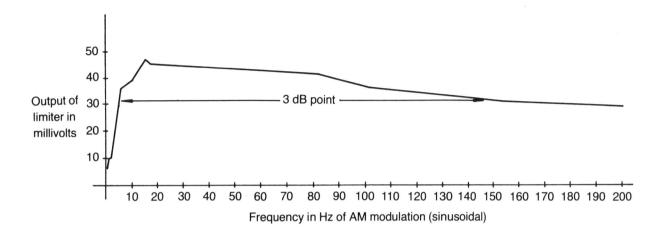

Figure 5.12   Frequency response (expressed in Hz of AM modulation-sinusoidal) of the limiter voltage with respect to a constant carrier level change. (An ideal measuring receiver has a flat response.)

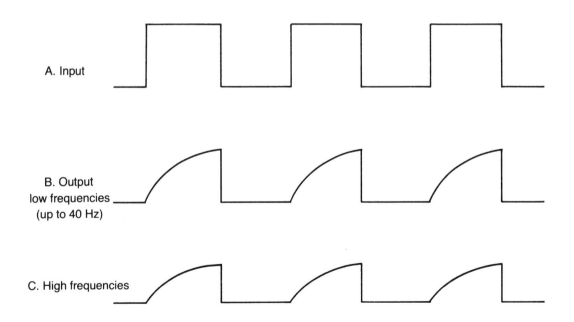

Figure 5.13   The limiter drive output response of a commercial wide-band receiver that was not designed to measure a square wave

These receivers also have a response to large-level changes that saturate very readily at excursion of about ±4–6 dB. Thus, they generally underestimate the standard deviation and the decile value.

The speedometer pulses can be used to ensure that samples are taken regularly over fixed-distance intervals. This is done to avoid the sample biases that can occur when, for example, the survey vehicle is held up at one spot (a traffic light, for example) and so records many readings at that one point, which then give a distorted average value for the area.

The speedometers in many modern vehicles are driven by a Hall effect device (see Figure 5.8) that has a sufficiently high output level to be read directly by the A/D card by using it to gate the A/D card.

Position data can also be read and recorded along with the readings.

SAMPLING INTERVAL

Within localized regions, the average field strength as measured from a distant transmitter can be shown to be approximately constant over intervals of 50 m to 500 m. Within these localized regions, the Rayleigh and log normal fades occur; the log normal mode generally dominates, although significant Rayleigh fading may occur, particularly in built-up areas.

Studies by Okumura *et al.* have shown that the local field strength in an area bisected by a 50–500 meter path can reasonably be accurately described by a single mean (average value) and a standard deviation. Figure 5.14 shows an area represented by sample measurement.

Hence, an adequate record of regional field strength can be obtained by sampling at the Nyquist rate and obtaining the mean and standard deviation over a 50–500 meter path. This is the basic approach used by most computerized field-strength measuring devices, where the sample intervals can vary from 10 meters to 1000 meters. In practice, it can easily be shown that samples over intervals of 50–500 meters yield consistent results to a few dB.

Figure 5.15 illustrates the decile method, a commercial measuring system. The illustration shows the structure of the Radio Survey Master system from Telecom Australia. The system has an external trigger (based on the speedometer pulses) that ensures that samples are taken at fixed intervals. The system records mean, standard deviation, decile value, the number of samples taken,

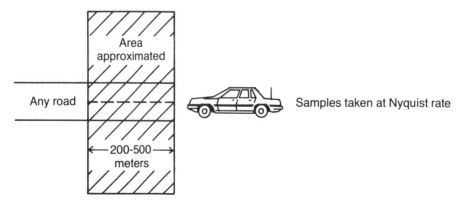

Figure 5.14   Area represented by sample measurement

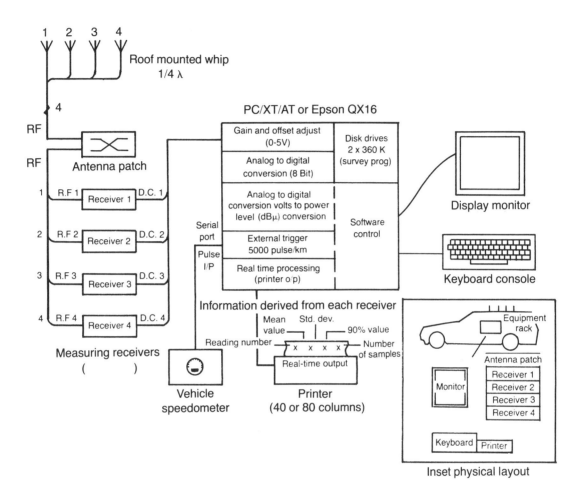

Figure 5.15   Vehicle-mounted survey equipment

and local time. The decile value is often used and favored by some because it gives a more conservative estimation of mobile performance than the average value—especially in areas of deep multipath.

In Figure 5.16 (following page), the upper 10-percent decile corresponds to T; this means that 90 percent of the readings are above the level T dBμV/m.

The 90-percent level is calculated using an iteration method. If the average is also needed from this sample, you should note that, although decile levels can be determined with equal accuracy, as an absolute level or the log of the level (dBs), the values must be converted to absolute values (μV/m) before a true average can be calculated.

In high-multipath areas with deep fades (high standard deviation), a measured mean value does not tell much about the extremities of the readings and particularly about the minimum (and hence,

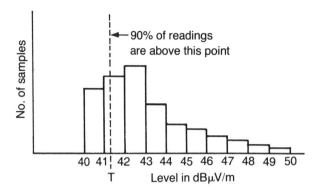

Figure 5.16   Histogram of sampled data

noisy) locations. Taken with the standard deviation, more can be extracted, but the decile method eliminates the need to know about the standard deviation and allows the use of a single standard-design field strength.

Figure 5.17 shows an example of log normal distribution of field strength.

For example, if a field strength of 39 dBμV/m average with a standard deviation of 6 dB (in a suburban area) is considered the objective, this can be equated to 39 dBμV/m − 0.87 x 6 = 34 dBμV/m for 90 percent of readings to define the boundary corresponding to 90 percent of locations having 39 dBμV/m average 90 percent of the time.

Because in lower multipath regions the 90-percent reading moves closer to the mean (and conversely, further away in high multipath regions), this method adjusts to the multipath environment in a way that an average reading cannot.

For example, if the field-strength measuring equipment was set as above (34 dB for 90 percent of readings) and it were to move from a suburban area (σ = 6 dB) to a rural area where σ = 2 dB, then the measured field strength would correspond to 34 + 0.87 x 2 = 35.7 dBμV/m average.

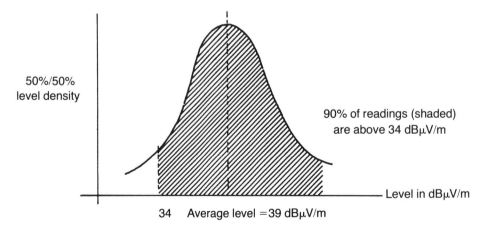

Figure 5.17   Log normal distribution of field strength in 50 percent of location and time

That is, at this lower multipath level, an adjustment is automatically made for the lower noise level.

Clearly this method gives a more effective measure of signal quality than an average reading. It is not often done, however, because it requires somewhat more "number-crunching" power than is needed to obtain an average reading.

For a more detailed discussion on these methods, see Chapter 8, "Units and Concepts of Field Strength."

REVERSE PATH SAMPLING

Some surveyors prefer to use the transmitter in the mobile and locate the receiving equipment in the base station. This has the advantage of less clutter in the vehicle, but has the disadvantage of requiring a very good position-location system that can relate readings to position.

A further disadvantage is that real-time outputs cannot normally be obtained. Thus, the mobile survey team does not have continuous direct contact with the receiving site and has no feedback on the survey's progress. This can lead to many problems, including time wasted surveying areas clearly out of range and time wasted surveying when equipment may be faulty.

When the measuring equipment is in the vehicle, the operators have a real-time measurement from which to solve these problems. Either way of measuring, however, if done effectively, gives equivalent results. Do not confuse the fact that the actual cellular transmit-and-receive hardware does not necessarily give balanced path losses with the fact that the radio paths themselves are reciprocal. Figure 5.18 shows that the paths *are* reciprocal and that it doesn't matter which way they are measured.

USING WIDE-BAND MEASURING RECEIVERS

Wide-band measuring receivers usually have poor sensitivity (typically 1–2 μV for 12 dB SINAD). Because surveys are often done on test transmitters with rather low ERPs (compared to the cellular base station), measurements are often limited by receiver performance.

It is often useful to acquire a low-noise (NF < 1dB) amplifier for the band being surveyed. Because of intermodulation susceptibility, it is best to only use such an amplifier at low signal

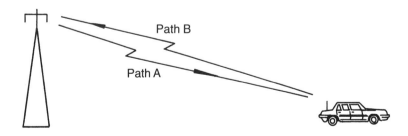

Figure 5.18   Because the signal traversing a path to the mobile uses the same path (almost) as the signal on the path from the mobile, any losses are equal. Measuring either path yields equivalent results.

levels and to switch it out when it is not needed. If the survey vehicle is equipped with a cellular phone (or other transmitting device), observe the effect on the performance of the measuring system; the transmitting device may desensitize the wide-band amplifier.

MULTIPLE RECEIVER ANTENNAS

Because of cross-coupling, if multiple-survey-receiver antennas are used, they should be spaced at least two wavelengths apart. Because most cellular sites are sectored, multiple receivers will effectively increase survey efficiency. When surveying omni sites, use a second channel as a check on the integrity of the first channel.

The best way to mount multiple antennas is on a ground plane made by suspending a sheet of aluminium between the bars of a roof-rack, as shown in Figure 5.19. This not only gives a very good ground plane, but it also avoids the necessity of drilling holes in the roof of the vehicle. The whole assembly is easily detachable when the time comes to replace the vehicle.

SURVEY TRANSMITTERS

It is preferable to have the survey transmitter on the same frequency and with the same ERP as the final cellular base. However, this is frequently not practical, particularly because transmitters of high-power outputs on 800–900 MHz are rather inefficient and are also somewhat difficult to come by.

For temporary installations (such as surveys), it is usually much more convenient to use

Figure 5.19   A survey vehicle can be fitted with a metal ground plane attached to ski bars. A number of antennas can be mounted on this good quality ground plane. Photo courtesy Extelcom, Philippines.

lower-power transmitters (especially if they are being run from batteries) and small antenna feeders that are easier to handle.

The antennas used for surveying are preferably 6 dB. This avoids the need to get the antennas accurately vertical (as would be necessary for higher-gain units) and means that they are physically reasonably small and manageable. Lower-gain antennas reduce the ERP to a point where it might be difficult to measure far-field regions.

Combining all these factors, you will probably find that the survey ERP is about a magnitude lower than a cellular base station. This can be easily corrected mathematically, but places some constraints on the receiver S/N performance because it means that the receiver must be functional at very low received-signal levels.

Figure 5.20 shows a typical configuration of survey equipment. The ERP of the transmitter is (in dB):

$$10 \log T \ (watts) - L + A$$

To convert the field strength measured by this arrangement to an equivalent cellular system level, it is necessary to correct for ERP and receiver gain. Thus, if the cell site has an ERP of E watts, the correction for the transmission system is:

$$10 \log E - 10 \log T + L - A$$

Including the passive receive gain, a correction factor of $10 \log E - 10 \log T + L - A - A_m + C$ must be applied.

Generally, the availability of transmitters on frequencies below 500 MHz is much greater than those above that frequency. It is also common for 500 MHz transmitters to have power outputs from 10–25 watts, compared to 800 MHz units, which are usually limited to about 10 watts. In most respects, frequencies from 400–500 MHz accurately enough model propagation at 800–900 MHz if a 2 dBμV/m allowance is made for slightly lower path losses at the lower frequency.

Be cautious if this is done; the 2 dB correction factor applies only to field strengths measured in dBμV/m. In other units (dB, dBμV, μV), due allowance must be made for the aperture of the antenna. (See Chapter 8, Equation A, under "Relationship Between Units of Field Strength at the Antenna Terminals.")

For survey purposes, the transmitters must be rated for continuous operation; you should note that most two-way radios with PTT (Press To Talk) switches are not so rated. Generally, it is necessary to de-rate the output power of such transmitters to 20 percent if continuous operation

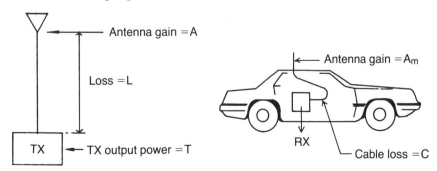

Figure 5.20   A typical survey equipment configuration

is expected. Even then, mounting a computer fan over the heat sink to provide adequate ventilation is a good idea because the difference between continuous and non-continuous rated transmitters relates to the ability of the RF power amplifiers to dissipate heat.

Notice also that continuously rated mobiles are normally duplex and have a duplex coupler to the antenna. This normally has a 3 dB loss (meaning 50 percent of the output power is lost in the duplexer), so the power efficiency is low. This can particularly be a problem when the survey station runs on batteries (as is often the case). The duplexer can be bypassed to increase ERP or save battery power as required.

When batteries are used to power the survey transmitter, it is a good idea to purchase a timer to switch on the transmitter just before work begins in the morning and switch it off in the evening. This technique usually allows three days of operation from a pair of truck batteries and saves many visits to the site. Commercially available domestic power timers with clockwork timers usually serve this purpose adequately and have the advantage of being very cheap.

AUTOMATIC POSITION-LOCATING SYSTEMS

An automatic position-location system records the position as well as the field strength each time a reading is taken. This is particularly valuable for plotting coverage areas by computer. Automatic position-locating systems can usually be described as either too inaccurate (inertial and magnetic), too expensive (omega and high-quality inertial), or too slow (satellite). Table 5.1 illustrates these systems.

Those systems that are inaccurate (inertial and magnetic) usually rely on manual correction every few kilometers. This is time-consuming and can also introduce significant errors if the correction points get misplaced.

Loran C systems can be used with just-acceptable accuracy and reasonable prices. Omega and high-quality inertial systems, which are sufficiently accurate for survey (accurate to 400 meters), are only practical where the budget is virtually unlimited. Aviation systems are suitable only for airborne applications and are not designed to work terrestrially. The satellite method is limited to periodic fixes every 20 minutes or so. In most parts of the world, however, this method holds the most promise for the future. There are plans to provide in the next few years a worldwide 24-hour satellite navigational system that will be accurate to a few hundred meters.

SYSTEM	COVERAGE	COST ($US)	TYPICAL ACCURACY (meters)
Loran C	Most parts of the world	2,000	400
Omega	Worldwide	4-20,000	6,000
Aviation	Worldwide	5,000	200
GPS	Eventually worldwide (1992)	2,000 (1992)	100
Cellular telephone	Selected areas	2,000	400
Terrestrial mobile	Selected areas	3,000	400

Table 5.1   Accuracy and costs of mobile position-location systems

This new satellite navigation system, known as GPS (Global Positioning System), works on the principle that it is possible to calculate an exact position in three dimensions if the distance to three reference points is known. The system uses the delay of the transmission from each satellite to determine the distance, and it reads the location of the satellite from the satellite broadcasts.

The range R_i as calculated in Equation A can be found for each satellite within range. Provided at least three such satellites are within range, the system, to fully determine its location, merely needs to solve the three simultaneous equations.

Equation A

$$R_i^2 = (X - X_{si})^2 + (Y - Y_{si})^2 + (Z - Z_{si})^2$$

where R_i = Range to satellite i
(determined by propagation delay)
X, Y, Z = Three-dimensional coordinates of the receiver
X_{si}, Y_{si}, Z_{si} = Three-dimensional coordinates of the satellite (as broadcast)

It is a little more complex when you realize that the range, as determined by the propagation delay, will have an error proportional to the time error of the reference clock in the mobile. Thus, if the time base of the mobile receiver is ΔT seconds in error, then the error in the range calculation will be $R_E = \Delta T \times C$ (where C = the speed of light).

To avoid the need (and expense) of all receivers carrying atomic clocks, a fourth equation containing the time error can be solved to allow for the inaccuracy of the receiver's clock.

So, the coordinate Equation A is replaced with the following equation and solved as four simultaneous equations which are now independent of the accuracy of the mobile clock.

Equation B

$$(R_i - R_E)^2 = (X - X_{si})^2 + (Y - Y_{si})^2 + (Z - Z_{si})^2$$

An accurate mobile time base is no longer required.

Since Equation B has four unknown variables, it requires a fourth equation to solve it. Thus, if reasonably priced timepieces are to be used at the receiving end, it is necessary to simultaneously obtain data from at least four satellites.

The GPS plan is for 24 satellites to be launched with 18 active units in six different orbits plus one spare in each orbit. The satellites are in an inclined orbit which takes them over any point in their path once every 12 hours.

The deployment of these satellites was held up by the delays experienced with the NASA shuttle. Full activation is now expected by 1992, although a limited number of satellites are now in service and the numbers will be gradually increased.

The satellites will use spread-spectrum techniques with a signal bandwidth of 2 MHz but an inherent base band of 100 Hz. This enables the system to work down to –163 dBm (a very low power density). The system will have a relatively inexpensive receiver (approximately $2,000) and have RS 232 output. Until the GPS arrives, manual position location or Loran C is recommended as the best compromise.

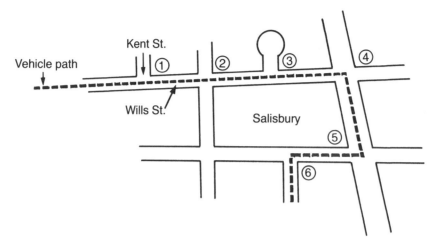

Figure 5.21   Map identification of samples

SEMI-AUTOMATIC POSITION LOCATION

This involves using the survey map with a digitizer board. When a reading is required, the surveyor touches a pen to the map location, which automatically takes a reading and records (accurately) the location coordinates.

MANUAL POSITION LOCATION

Unless a very good position-location system is used, it is necessary to use two people in a survey environment. When two people are used, the manual location does not overtax the non-driver; this is both the cheapest and most accurate method.

In normal suburban environments, it is recommended that position fixes be street intersections and that reading numbers be marked on the map in real time. Figure 5.21 illustrates the path of a typical survey run. Samples are taken at the points marked 1, 2, and so on, and the run numbers are marked on the map as the samples are taken. A conventional street map in black and white is preferred, so that the run numbers are easily seen when marked off with red pen. At some later time, the field-strength readings are transferred to the map, using another readily visible color.

PREPARATION OF RESULTS

If the results are manually collected, they should be transferred to a street map. The map should clearly show the following information:

- Date of survey and site surveyed
- ERP of transmitter
- Surveyed frequency
- Antenna gain
- Any correction factors

The run number and field strength should be marked on the map in different colors and it should be clearly indicated which is which. In order to make the results easier to visualize, it is necessary to draw the service-area contour (say, 39 dBμV/m for AMPS) and the CBD handheld contour (60 dBμV/m for AMPS).

Because frequency reuse is usually an important cellular consideration, it is necessary to survey down to the 20 dBμV/m contour (the level at which interference becomes service affecting). This defines the region where frequency reuse is not practical without sectoring. Notice that if sectoring is contemplated, an additional interference immunity of approximately 10–15 dB will be obtained, and then the interference boundary approximately equals the coverage boundary.

In order to see how various cell sites will work together, it is a good idea to transfer the coverage contour to a sheet of transparent plastic so that you can clearly see the overlapping of different cells.

Some map or computer studies will have been done before selecting sites to be surveyed. The survey results should always be compared with the predictions and discrepancies explained. A large discrepancy means either a problem in the forecasting technique or a problem with the survey procedure. The comparison can be done quickly and will frequently highlight problems.

It is advisable to have a second person check all correction factors. Use Table 8.2 in Chapter 8, to assist with this conversion. At cellular frequencies, the cable loss from the mobile antenna to the receiver port is about 3 dB and cannot be neglected.

SPECTRUM CHECK

While the survey antenna is on site, it is advisable to scan the whole band to be used from the survey site. Other services using the band will then be detected. For example, microwave links are occasionally found in the cellular band.

Notice that microwave links directed away from the site may be difficult to detect and the presence of any foreign carrier in the band is cause for concern. Spectrum analyzers are not usually sensitive enough for this check and cannot always distinguish between legitimate cellular traffic and another mobile type. The analyzer will, however, give a clear indication of the nature and extent of high-level systems such as microwave link systems, and systems with sporadic occupancy. Chart recorders can be used to examine the spectrum over prolonged periods of time. Simply listening to a wide-band receiver will sometimes give a good indication of the nature of the traffic.

CONFIRMING COVERAGE

For turnkey projects, the operator often specifies the coverage as being at a certain quality for, say, 90 percent of the area for 90 percent of the time. Suppliers often undertake to guarantee the stated coverage.

How can an operator confirm that the supplier's design meets the specification? To do this for a real city is almost impossible, but to see how it might be done, consider the case of an imaginary town, called *"Square Town,"* which is shown in Figure 5.22 on the following page.

If the operator had specified that the coverage should be such that calls could be made from 90 percent of locations for 90 percent of the time (most unwisely, as this means 10 percent of the area

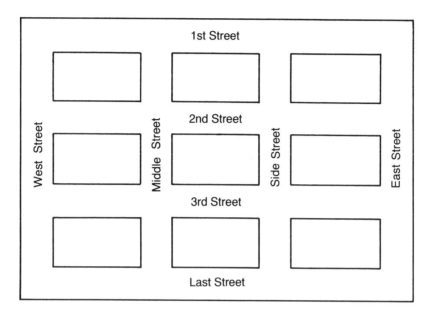

Figure 5.22   Square Town city map

is totally unserviceable), then it would be possible to proceed by attempting to make calls at various locations around the city and determining that the success rate is 90 percent or better. The results would not be conclusive, however, unless every street was sampled at the Nyquist rate (every 0.083 meters). If the town was 2 km x 2 km, then the number of samples would be 200,000, or 8 streets x 2 km/0.083 x 10^3 call attempts. Statistical sampling methods could also be used to reduce the sample size (that is explained in the next example).

If the specification called for a minimum field strength—say, F, in 90 percent of locations for 90 percent of the time—then there are two practical ways of measuring this value. First, you could calculate the corresponding mean from the *estimated* standard deviation for a field strength of F at 90 percent of the time/locations and measure the mean field strength (sampling at or about the Nyquist rate). This should be done on a street-by-street basis and, provided 90 percent of all readings are above the calculated average, then the criteria is satisfied. Or, second, you could use a measuring set that can return the field strength above which 90 percent of all readings occur.

The second method is equivalent but more accurate than the first method. The whole city can then be measured as a single entity. If the measurements on this basis yield field strength of F or higher, then the criteria is satisfied.

If a signal-to-noise ratio was specified instead of a field strength, then the measurements above, on a modulated carrier using a S/N measuring set, would be necessary. Such equipment is not commonly available.

The difficulty of performing the above measurements in a real city can be enormous. A real town, for example, might consist of 50 percent developed land and 50 percent rugged hills. An operator might find inadequate coverage in the city and complain that 20 percent of the built-up area (measured as above) is below standard coverage. The supplier could counter claim that all of the undeveloped area is covered at or above the standard level and therefore 90 percent of the city is covered at or above the specified level.

Thus, coverage guarantees for most real cities are not worth much and probably could not be effectively challenged in court. I know of no case where the guarantee has resulted in the guarantor paying any damages or costs associated with unsatisfactory coverage, and yet less-than-expected coverage is a frequent complaint.

Alternatively, the city to be tested can be divided up into a grid, and a sufficient number of coordinates can be selected to satisfy that samples taken in these regions will yield the average field strength to a given degree of confidence. This method has, in fact, been used by a number of system operators, but it has some serious limitations.

First, the sample length should be defined to be, for example, 100 meters, sampling twice every wavelength. The random selection will probably yield points that are not directly measurable (that is, a path diagonally across a city block may be chosen). In order to avoid bias (either positive or negative), it is then necessary to generate a set of unambiguous rules to select the nearest practical path—both its start and end. Next, again to avoid bias, as the interpretation of the selected path would still be quite open (to the survey team), it would be necessary to have some rules that would prevent choosing a biased sample space during measurement. One method might be to increase the sample run to the whole city block. With careful control, this method may be the most practical way of ascertaining coverage standards.

SURVEYING AS A MAINTENANCE TOOL

Very few operators fully appreciate the value of surveying as a maintenance tool. Provided the original design and acceptance-survey results are well documented, the survey equipment can be very effectively used for future maintenance.

No matter how careful the original design/surveys were, in any city of significant size it is not practical to survey every street. Soon after start-up, the operator will probably discover some areas covered that would not have been expected to be, and also some areas not covered for which good coverage was anticipated. Indirectly, therefore, some interpolation/extrapolation of measured result is needed.

When complaints come in (as they will) about coverage, the first reference is the original survey map. When, however, the area from which the complaint comes was not originally surveyed in detail, it may be necessary to look further. A survey vehicle should be sent to the area and a detailed survey of the suspect area taken. Always check nearby areas that were originally surveyed to confirm that the coverage has not deteriorated since the original acceptance survey. If there are discrepancies between the original survey and the check, then a base-station fault can be expected.

Sometimes you will find that adequate field strength is present but that calls still cannot be successfully made. This can probably be traced to co-channel or adjacent-channel interference. As a matter of routine maintenance, it is a good idea to spot check the field strength from each base station on a yearly basis (or more often if complaints warrant it).

Remember that field-strength measurements only check the transmitter path; if complaints are coming in and the field strength is adequate and interference has been eliminated (this can often be detected by the sound of the control channel—a channel suffering interference will often sound very fuzzy compared to one that is working normally), then the problem likely is with the RX path.

Where diversity is being used, it is practical to temporarily disconnect one antenna and use it to transmit a carrier that can be measured to confirm the integrity of the RX path. Diversity reception can disguise quite acute receive-antenna problems.

SOME NECESSARY PRECAUTIONS FOR RADIO SURVEY

The following considerations are important precautions for a radio survey: air-conditioning, possible errors, and equipment stability.

AIR-CONDITIONING

The survey vehicle should be air-conditioned to reduce errors due to temperature-sensitive drifts in the receivers and A/D card. Air-conditioning also helps the operators stay alert so they are more likely to pick up potential sources of errors.

Errors can occur for a large number of reasons and, where real-time output of results is available, those results should be checked for consistency as the survey progresses.

POSSIBLE ERRORS

Possible errors and their sources are considered in the following list:

- Receiver out of calibration. This should rarely occur with a good receiver but can be quite a problem when cheaper (or older) mobile radios are used as the measuring device. Frequent checks (weekly on suspect receivers and monthly on quality measuring receivers) of calibration against a good signal generator are necessary.
- Faulty antenna. The antenna should be free of visible defects and have a low VSWR. It should occasionally be checked against a few other similar antennas to confirm gain, using different feeders to the receiver.
- Faulty or damaged feeders. Use the same checks as for antennas.
- Mobile receiver de-tuned or simply not tuned to correct frequency. Double-check that the receiver is tuned to the correct frequency. The mobile can also be desensitized by other transmitting apparatus in the vehicle, particularly in regions of low field strength. Beware of transposing the TX and RX channels when setting the frequency.
- Wrong calibration table used. This often becomes a problem when the units of the signal generator output are not the same as the units used for field strength (for example, the signal generator is calibrated in dBμ and the field strength equipment is calibrated in dBμV/m). Converting units in the field often results in error. Always have a correction table handy (for example, use Table 8.2, found in Chapter 8).
- Test-receiver output is voltage/temperature dependent. Always check the test receiver for temperature and voltage sensitivity. This is more likely to be a problem with old receivers, but all sets should be checked before being placed into service.

The voltage regulation on a car battery may not be good, particularly if the receiver gets its power from a source distant from the battery where significant voltage drops may occur. Be cautious of volt drops from indicators and brake lights.

- Insufficient settling time. Even good quality measuring sets should be powered up half an hour before beginning measuring. Most of the drift occurs in the first 10 minutes after the set is switched on.
- Inaccurate records of transmitter base. Keep good records of the survey conditions, particularly for a temporary test transmitter. In particular, record the following:
  1. Power output at start (measured)
  2. Feeder loss (cable should be calibrated)
  3. Antenna gain, antenna height
  4. Frequency of test transmitter
  5. Power at end of test (remeasure to ensure no drift has occurred)
  6. Date, and TX site name

Failure to record any of these details could render the data useless in the future.

EQUIPMENT STABILITY

It is most important that all items in a moving vehicle be securely fixed. It is usually necessary to mount the receiver and other survey hardware in a rack. This rack should be padded and constructed in such a way that it does not interfere with the vision of the driver (this usually limits the rack height).

It is best if the equipment can be mounted beside the operator, where, in the event of an accident, it is unlikely to come in contact with the operator. The most dangerous mounting position is in front of the operator, where a collision will throw the operator into it. This is particularly true of a VDU.

An internal master switch for battery power also should be provided, as should a fire extinguisher (CO_2 type). For security, it is best if the rack can be completely covered, in a way to hide the hardware it contains. When it is necessary to leave the vehicle parked in the street for some time, the sight of a few expensive measuring receivers and some computer hardware is likely to attract unwanted attention.

Where possible, use laptop computers. Because they are much smaller, laptops can be placed safely on the operator's lap, requiring only an up-and-down movement of the head. An equipment rack mounted at the side of the operator places some strain on the operator's neck muscles when he or she is viewing the keyboard or screen. The operator must a have clear field of vision forward to allow for proper street identification and to minimize the effect of car-sickness. The motion sickness that results from operating survey equipment is similar to that caused by reading in a moving vehicle. It is such a problem for some people that they are unable to do this task effectively.

Chapter 6

CELLULAR RADIO INTERFERENCE

In cellular radio systems, which have frequency reuse, some interference is inevitable. Equipment designers have allowed for interference and have incorporated many elaborate countermeasures into this environment. The system designer also must be aware of the nature, causes, and control of interference to achieve effective frequency reuse.

FREQUENCY REUSE INTERFERENCE

Interference can occur in many ways, but the most significant interference in cellular radio is from a mobile unit to a distant cell on the same frequency. Figure 6.1 shows this form of interference.

This interference is not usually noticed by users, but it can temporarily block the channel being

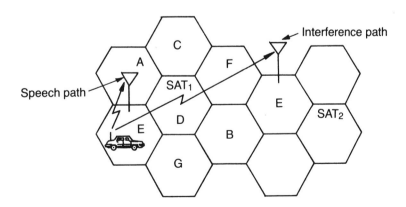

Figure 6.1   The most significant interference is from a mobile unit to a distant cell on the same frequency.

interfered with or cause audible interference in some instances. In the AMPS system, adjacent cells are provided with different SAT (Supervisory Audio Tones) tones (around 6 KHz) that are used to identify foreign carriers. These tones minimize the probability of erroneous control decisions due to co-channel or adjacent-channel interference.

Interference is normally assumed to become objectionable at carrier-to-interferer ratios of 18 dB or less on analog systems. In low multipath environments (such as stationary handhelds) it becomes a problem only at much lower levels, especially in high-deviation systems. This level allows successful operation in a more severe interference environment. To avoid interference problems in AMPS systems, the detection of foreign SAT codes (implying a carrier from a mobile unit is directed at another cell) temporarily blocks the channel (if not in use). If the channel is in use at the time, then a handoff occurs. In both cases, the channel experiencing interference becomes unavailable for traffic, and the system capacity is reduced when the interference reaches a system-defined level. In systems with multiple frequency reuse, the interference level at which the channel is blocked can be reduced to increase the traffic capacity but only at the expense of more interference.

Cross talk and call dropouts can result from co-channel (same channel) interference. These types of interference are usually associated with high sites and handheld use in high-rise buildings or mobiles operating from hilltops. For the same reason, operation of handhelds in aircraft is not encouraged.

Increasing frequency reuse initially increases system capacity, but as the problems of interference increase, further frequency reuse decreases the system capacity. The objective is to maximize frequency reuse with minimum interference.

ADJACENT-CHANNEL INTERFERENCE

Other forms of interference can also occur, notably adjacent-channel (a channel either one channel space above or below the reference channel) interference. Adjacent-channel interference on the control channel corrupts data and causes call failure (particularly in AMPS and TACS).

Adjacent-channel interference can be managed by the frequency planning techniques discussed in Chapter 7, "Cell Plans," on cell planning. This problem often requires reassignment of frequency of one or more cells. The problem occurs from interference between base stations on adjacent channels in certain areas, and it is recognized by high incoming-call (to the mobile) failure rates in certain areas (where the field strengths are sufficiently close to cause data corruption). Except where the problem is most severe, once a call is established, it usually proceeds normally, although some blocking may be noticed.

The two most effective ways to limit adjacent-channel interference are antenna downtilt and antenna height reduction. These methods can restrict the range without drastically reducing near-end power (and hence building penetration). Note that the NMT900 systems not using interleaving is essentially free of this type of interference because of its wide-channel bandwidth-to-deviation ratio. In AMPS and TACS however, there is no guard-band between channels and so it is not possible to fully filter adjacent channels.

INTERFERENCE FROM OTHER SYSTEMS

Cellular systems are designed to operate in an environment of interference. One way the system reduces interference is by instructing the mobiles to reduce power when they are sufficiently close to the base station to perform adequately at lower power levels. For AMPS and TACS, the power output of the mobile can vary 28 dB during a conversation. A similar facility is available on the NMT systems. Table 6.1 shows the power reduction levels in an AMPS system.

It is not unusual in new systems that the cellular band reserved for the system has previously established users in it. Where these can be identified, the cellular operator must make the necessary arrangements to move the offending systems to other frequencies. This is ordinarily done at the cellular operator's expense (except where the occupiers are illegal). This has largely occurred because CCIR recommended spectrum allocations do not include spectrum for cellular systems.

Some interesting effects occur when unidentified "other users" remain on the band. Most systems in this frequency band are point-to-point microwave links, and therefore operate continuously and are highly directional, thus limiting their area of interference.

Figure 6.2 illustrates an interesting problem that occurred with an AMPS system located near a microwave link (also owned by the cellular operator) that used mobile transmit frequencies corresponding to those used at a neighboring cell. In this example, the problem occurred when a vehicle at cell B requested a handoff attempt. If the vehicle was traveling away from cell B when a handoff attempt occurred, the signal-measuring receiver at cell A would be asked to measure the

LEVEL	POWER REDUCTION (dB)
1	4
2	8
3	12
4	16
5	20
6	24
7	28

Table 6.1   Power reduction levels in an AMPS system

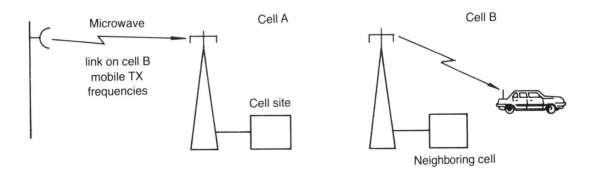

Figure 6.2   A false handoff report can occur if interference on the same frequency as the mobile is detected during a handoff attempt. In this example, cell A measures a signal from a microwave system on the same frequency as the scanned mobile, wrongly reporting ideal handoff conditions.

field strength of the vehicle. It would report a strong signal (because only the RF carrier level is measured), when the signal was in fact from the microwave system. The switch would then request a handoff attempt to cell A. The mobile would attempt to handshake with cell A on the specified channel and would find no carrier. The call would then fail. If the mobile was roaming towards cell A, an effective handoff could perhaps occur. The solution was to change the microwave link to a new frequency.

INTERMITTENT AND MOBILE INTERFERENCE

This type of interference is the most troublesome because it is very difficult to locate the offending source. There is really no effective way of tracking an offending mobile in the urban environment, except with a great deal of patience and elaborate tracking equipment.

The intermittent nature of these transmissions means that you must spend a good deal of time waiting. Once transmission commences, the most effective method is to use two vehicles, each equipped with direction-finding and two-way communication equipment, to locate the target, preferably by homing in from different directions and using the intersection of the directional vectors to pinpoint the target. Remember, though, that the propagation mode is such that accurate fixes are nearly impossible and that it is necessary that the target transmit for some considerable time (usually hours).

Doppler-effect devices are available that use four receivers and give the bearing of an interferer directly via a digital readout.

Less difficult to locate, but still not easy, are intermittently operated links or repeaters. The first step is to measure the field strength from a number of different cellular bases and then triangulate the position. The next step is to obtain a portable field-strength measuring receiver and, equipped with a Yagi antenna, set out in search of the target. Unless the sources are very close and hence easy to locate, you should allow at least a week to locate an interferer.

IMPROVING FREQUENCY REUSE

Cellular systems have from 180 to about 1000 channels. As the number of users rises, it is necessary to reuse frequencies in order to accommodate additional traffic.

In small cities and rural areas, frequency reuse will probably never be necessary, so prominent high sites can be used to maximize coverage. In urban areas, however, every effort must be made to maximize frequency reuse. In large cities it is usual to design for between 4 and 15 times reuse of each frequency.

For example, when the UK initiated service in London in January 1985, the extent of the future demand was not fully appreciated. The original service was based on 2-km radius cells, and these evolved quickly into 1-km radius cells. The smaller cells meant more interference problems, and London is now operating in what is probably the world's most hostile interference environment.

One solution to the cell-density problem is to go from a 7-cell to a 4-cell pattern. Cell splitting below 1-km radius cells has been achieved with fair success in a number of cities, but the constraints placed on the narrow search areas for sites and the difficulty in obtaining suitable sites is formidable.

What are the upper limits of frequency reuse? Some cities—Los Angeles and Chicago, for example—have virtually reached saturation with their AMPS systems at about 200,000 customers on 666 channels (that is, frequency reuse of 15 times). Note, however, that these are big, sprawling cities compared to places like Hong Kong, where 40,000 customers on 300 channels (frequency reuse of 5 times) appears to be the limit. (Hong Kong is especially difficult. It is built along foothills and it cannot spread inland, and it is separated by a harbor which is not wide enough to be a barrier but rather serves as a giant signal reflector.)

BLOCKING

The easiest way to determine the magnitude of frequency reuse problems is to look at channel blocking information (channels out of service due to interference). This information is normally available from the system housekeeping data; it should be kept below 5 percent.

You may choose a trade-off between blocking (which may mean lost calls) and increased interference (which occurs as cross-talk and spurious noise). This trade-off is controlled by adjusting the C/I limit for blocking, which is a user-definable system parameter.

USE OF TERRAIN AND CLUTTER

The primary aim of a cell site is to provide coverage for its service area. With careful planning, it is possible to achieve this goal while minimizing cell coverage outside the intended service area. Using local terrain and clutter can be an effective way of containing a cell. Effectively using local terrain and building obstructions can greatly enhance frequency reuse, if base sites are placed so that far-field propagation is reduced. Placing a base station behind a hill greatly increases the chances of successful frequency reuse. Similarly, placing the base station so that buildings form a natural shield can also be very effective (see Figure 6.3).

In very high-density areas, effective frequency reuse is more critical than coverage.

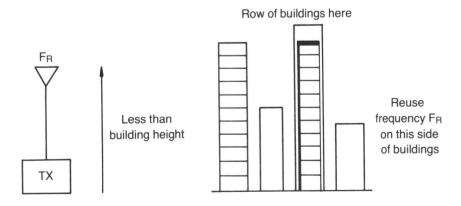

Figure 6.3   Use of obstructions to enhance frequency reuse

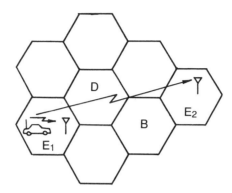

Figure 6.4   Omnidirectional interference mode

USE OF SECTOR ANTENNA

Sector antennas are important tools in minimizing interference. Figure 6.4 illustrates the basic principle of sector antennas. In this example, consider a mobile roaming from cell E_1 to E_2, both of which use the same frequencies. If the cell sites are omnidirectional and E_1 is relatively high, the mobile might roam well into cell D before a handoff occurs. All the time it is getting closer and closer to E_2 and the probability of interference is increasing.

As shown in Figure 6.5, sectoring E_1 and E_2 can improve things considerably. When the car begins its journey (at the maximum distance from E_2), the situation is the same as for the unsectored site. However, when the mobile passes the base station on the way to E_2, a handoff is initiated to cell E_B. The mobile is now using cell E_B at site E_1 but is behind the antenna for cell E_B at site E_2. This gives the added protection of the effective front-to-back ratio of the antenna (usually 6–15 dB) against interference. Note that the effective front-to-back ratio is almost always less than the value quoted by the antenna supplier (which is true in free space only). When, finally, the mobile hands off to cell E_2, it uses cell E_A and again interferes with cell E_1 only via the back of the antenna.

Interference protection can thus be increased by use of sectored antennas and the degree and type of sectors. Antenna sectors as narrow as 60 degrees can be used effectively.

4-CELL PATTERNS

It can be shown theoretically that significantly higher cellular densities can be obtained using a 4-cell pattern rather than a 7- or 12-cell pattern. In practice, the gains are not as large as they are in theory, but there is evidence that 4-cell patterns are measurably more efficient than 7-cell patterns. The trade-offs are that even though a base-site capacity is higher, the individual cells are smaller. This occurs because, for any given number of channels, the 4-cell pattern divides the channels into 24 channel groups and the 7-cell pattern divides the pattern into 21 channel groups.

The 4-cell pattern requires more critical placement of cells sites; its efficiency is reduced if this

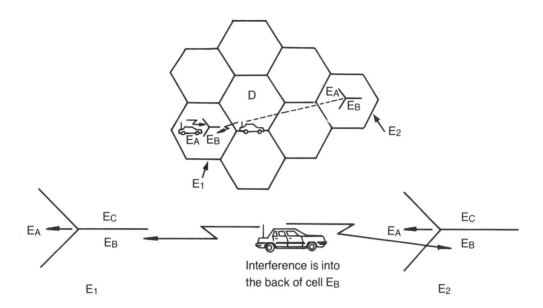

Figure 6.5   Sectoring to enhance frequency reuse. Antennas are orientated in the same direction for maximum interference protection.

cannot be attained. Despite these limitations, it is generally more efficient overall in Erlangs/km to use 4-cell patterns.

CHANNEL BORROWING

This channel-borrowing technique allows dynamic channel reassignment of a channel in a cell that has some of its "normal" channels blocked due to interference. It assumes that channels can be retuned remotely on command. Channel borrowing may increase traffic densities by 10 to 20 percent, but it is costly to implement and uses a good deal of processing power. The technique is sometimes called "Dynamic Channel Assignment."

ANTENNA HEIGHT

Simply mounting the antenna at a lower level is probably the most effective way to decrease the effective cell radius and increase the chances of frequency reuse. When cells are located on buildings, it is generally very difficult to reduce the antenna height; you should consider this when selecting building sites.

Tower-mounted antennas, on the other hand, have a wide scope for varying the level of the antenna and even of locating different sectors at different levels. The conventional triangular mounting structure for cellular antennas presuppose all sectors mounted at the same level. This is often appropriate, but in some cases (particularly where the local terrain is relatively flat) it can restrict the degrees of freedom available to the designer.

As discussed earlier, downtilt is effective against co-channel and adjacent-channel interference, but it is not particularly effective against mobile-to-base station problems. For more information, see "Effective Use of Downtilt," later in this chapter.

ANTENNA TYPES USED IN CELLULAR RADIO

The most commonly used antenna patterns are illustrated in Figure 6.6.

HANDHELD BENEFITS

Using sectored antennas can improve handheld coverage by overcoming, to some extent, the limited talkout power of a handheld. This is due to the additional antenna gains achieved with sectored antennas.

LEAKY CABLES

Using leaky cables to feed buildings, tunnels, and large complexes from inside rather than from an outside antenna provides better-quality service and improved frequency reuse, because the

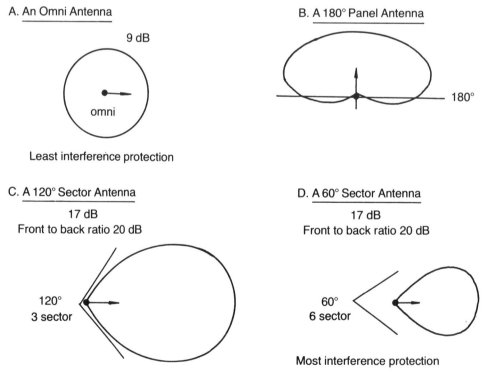

Figure 6.6   Antenna patterns used in cellular radio

building itself is an effective shield against interference. This technique is more attractive as cells get smaller.

SYSTEM PARAMETERS

Cellular systems have a number of system parameters that can be modified to alter the performance of a cell. The following parameters can improve frequency reuse:

- Signal level for handoff (variable over 64 dB). Raising this level reduces the effective size of the cell.
- Signal-to-noise level before handoff is attempted. Increasing this level causes handoff to occur earlier.
- Signal-to-noise level before releasing a call. This can be used to extend or contract the effective range.
- Level at which mobile is instructed to increase/decrease power. This can be effective in small cells. Causing earlier power reductions reduces interference.
- Signal-blocking level is the level at which interference blocks channels. Increasing this level can reduce congestion due to channel blocking, but if it is adjusted too low, it can lead to data corruption and hence lost calls.

These parameters are loaded via the switch to the base stations and can be adjusted by a simple data entry. Each base should be studied separately to determine the usefulness of changes in these levels. Those most likely to benefit are those in congested areas, those experiencing interference, those requiring extended range, and those from which interference originates.

EFFECTIVE USE OF DOWNTILT

Downtilt is one of the most frequently used techniques for producing small cells and achieving good frequency reuse. It works well if used only on small cells (about 500 meters radius) and with carefully calculated downtilts. It is not effective for large cells.

Downtilt effectively reduces:

- Cell coverage to provide for higher traffic densities
- Co-channel and adjacent-channel interference from other cell sites
- Channel blocking from co-channel mobiles

However, downtilt does not improve handheld coverage and may in fact reduce the effectiveness of handhelds in the normal service area of the cell.

Small and new systems generally should not use downtilt until it is proven necessary because of the attendant range reduction of the cell and reduced handheld performance even at moderate ranges.

Large angles (greater than 5 degrees) of downtilt are rarely justified and will usually not be effective. Before downtilting an antenna, determine its original coverage (by survey) and after applying the calculated downtilt, resurvey to be sure that patchy coverage has not resulted from the change. Downtilt will decrease the far-field strength dramatically; the designer must be aware of the consequences when employing this technique.

Inadvertent downtilting of high-gain omni-antennas (greater than 6 dBd) of more than a few degrees can cause significant shortfalls in performance. This can occur when the installer fails to properly align the antenna, or when the installation is poor and the antenna tilts mechanically with time. Storms and other external influences can also cause such problems. Those familiar with lower-gain antennas (6 dBd and less) are aware that this problem is not very pronounced with such antennas. Caution is needed with any antenna that has a gain greater than 6 dBd and is tilted.

DOWNTILT AND HOW IT WORKS

For the purposes of illustration, consider an idealized corner reflector with a pattern as shown in Figure 6.7. Figure 6.8 shows the radiation pattern that occurs when this antenna is downtilted by B. From Figure 6.8 you can see that the far-field ERP (that is, the horizontal ERP) is now H. H is normally expressed as a fraction of the field strength measured as the E or H components with respect to the value along the main axis (that is, $0 \le H \le 1$). Thus, the far-field attenuation by downtilt is 20 log H dB.

The near-field (and the far-field) field strength in practice is highly dependent on local clutter, particularly in high-density environments. For the purposes of this idealized study, we must, for the time being, assume zero clutter.

The downtilted antenna will always have a relative gain over a vertically mounted antenna in the middle field. This can be seen in Figure 6.9.

Now, consider the gain of the downtilted antenna compared to the vertical one. As a function of the Angle M (the angle from the antenna to the receiver):

The gain at $M = 0°$

Vertical antenna $= 1$

Tilted antenna $= H$

Relative gain $= 20 \log H - 20 \log 1 = 20 \log H$

Notice that $H \le 1$, so the relative gain of the downtilted antenna is negative or zero at M = O. At any other angle M, the difference in gain equals:

$20 \log G_V - 20 \log G_T$

You can see in Figures 6.10 and 6.11 that this difference (the relative gain of the downtilted antenna) increases as M increases and equals zero at the value of M where the two patterns intersect (point I). This angle is such that $M = B/2$. The downtilted antenna has a relative positive gain which peaks and finally equals zero at the angle $B + T$, where T = the angle to the lobe tangent.

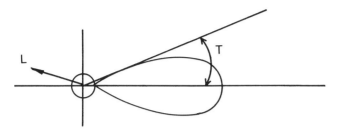

T is the angle of the tangent to the main lobe.

L is the gain outside the main lobe.

Figure 6.7   Vertical radiation pattern of a gain antenna

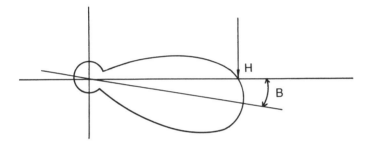

Figure 6.8   Pattern of a downtilted antenna

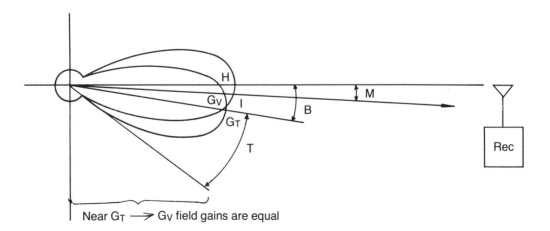

Near $G_T \longrightarrow G_V$ field gains are equal

Figure 6.9   The pattern of the far field changes with downtilt.

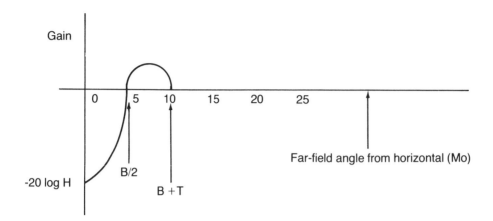

Figure 6.10   Relative gain of downtilted antenna compared to a vertical antenna (dB) as a function of angle of elevation to receiver

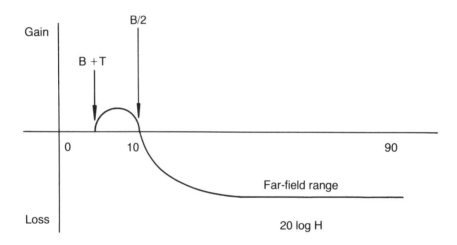

Figure 6.11   The gain (dB) of a downtilted antenna compared with that of a vertical antenna (dB) as a function of distance from the antenna

The important points to note are:

1. Far-field loss = 20 log H (and hence interference protection factor = 20 log H).
2. Handheld gains are only realized in the angular range B/2 to B+T; that is, at an angular range of h/tan (B/2) (where h = base station height). Outside that range the field strength decreases. The problem of downtilt is now apparent.

If reasonable handheld range is sought, B/2 (B = downtilt angle) must be small. However, a small B/2 gives a low far-field loss (and hence poor interference immunity) unless the beam width is very narrow.

Thus, the interrelated parameters are height and downtilt. Table 6.2 shows the effect of height and downtilt on small cell operation. Because of these conflicting factors, where frequency reuse is important, it is better to use lower antenna heights and less downtilt than higher antennas and greater downtilt. Remember also that the interference received from the back of the antenna

	HEIGHT AND DOWNTILT EFFECT ON SMALL CELL OPERATION	
PARAMETER	PROS	CONS
Height	Increased height will increase far-field losses by use of greater downtilt (for same range).	Increased height will decrease far-field interferer losses by height gain. Increased height will also reduce interference immunity of base-station receivers.
Angle	Increased downtilt will increase far-field losses.	Increased downtilt will reduce handheld range within the service area.

Table 6.2   Effect of height and downtilt on small cell operation

increases with height gain and is virtually independent of downtilt. This is a big problem in high-density areas.

To estimate the trade-off, note that:

height gain α 20 x log *height (meters)*

and far-field loss can be calculated by first determining the handheld range required, then using (B/2) = arctan (height/range) (where B = downtilt). From B, use the radiation pattern to find H = far-field loss.

EXAMPLE

Consider the antenna pattern in Figure 6.12. The objective is to achieve good handheld coverage at 1 km from a 30-meter site. As shown in Figure 6.13 on the next page, you can see that:

$B / 2 = arctan (30 / 1000) = 1.7°$

$B = 3.4°$

The far-field strength reduction available from this antenna would be:

$20 x log H = 1.1 \, dB$

which is hardly worth having.

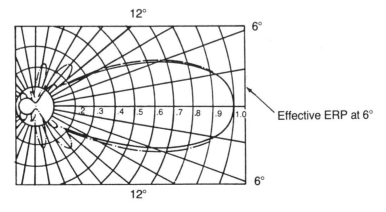

Figure 6.12 Radiation pattern of typical 17-dB unidirectional antenna

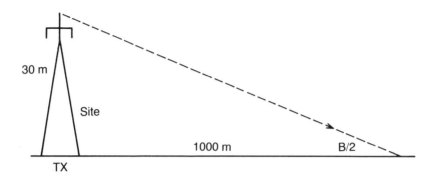

Figure 6.13   Range calculation for a downtilted antenna

SAME ANTENNA AT 60 METERS

The same antenna at 60 meters can be shown to require a downtilt angle of 6.8 degrees (to achieve the same 1-km range) and a far-field loss of 8.6 dB. The height gain is 6 dB, and the net far-field loss is 2.6 dB.

Notice, however, that the 6-dB height gain is also added to the gain of the antenna in the backward direction. Thus, the antenna is significantly more vulnerable to receiver interference and hence to channel blocking in high-density areas.

Although larger cells from high sites do not really benefit much from downtilt, smaller cells from moderate elevations do. Consider the field-strength plot of the original 30-meter site with 6-degree downtilt, as shown in Figure 6.14. You can see that the two systems have an equal field strength at a distance of only 572 meters (B/2 = 3°). At smaller distances, the downtilted antenna has a relative gain, but except for very marginal places (like elevator shafts and basements), most

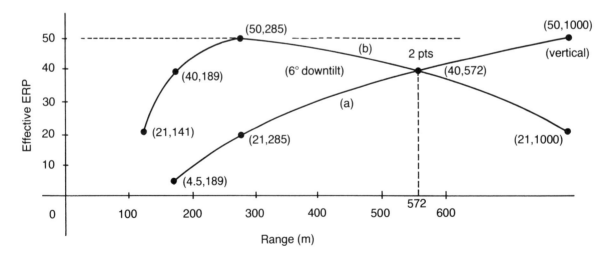

Figure 6.14   Effective ERP versus distance from 30-meter transmitter for curve (a) vertical antenna and (b) antenna at 6° downtilt

areas within 570 meters are likely to be well covered without downtilt. Any improvement in coverage close to the antenna and attributable to downtilt is marginal.

The most useful effect of downtilt is the relative reduction in the far-field strength (in this case, to 21 watts ERP or 3.8 dB, which is still not substantial). Notice, however, that one more degree downtilt (to 7 degrees) increases the far-field loss to 9.1 dB, while decreasing the equal field-strength range to 490 meters.

Downtilt is most effective for very small cells (of about 500-meters radius), when used in combination with downtilt angles that approach T (that is, half the beam width of the vertical propagation pattern). Downtilt provides greater opportunity for frequency reuse but not improved handheld coverage, as is often mistakenly believed. *Use it with care!*

The final way of improving frequency reuse is by careful cell and frequency planning. This topic is discussed in Chapter 7, "Cell Plans."

Chapter 7

CELL PLANS

The objective of a cell plan is to cover the service area as economically as possible, while allowing for maximum flexibility for future frequency reuse. As an aid to visualizing frequency patterns, the hexagonal cell plan has been devised. In the real world, however, radio waves do not propagate to hexagonal boundaries, and cells often overlap. Indeed, cell overlap is sometimes an intentional part of the pattern design. This real-world untidiness means that some conceptual mental gymnastics may be necessary to relate the theory to practice.

BASIC CONSIDERATIONS

The most common cell plans involve splitting the available spectrum into 3, 4, 7, or 12 frequency cells, each using a hexagonal grid to define the relative positions of the base sites. Such cell plans define a "tidy" and systematic frequency-reuse pattern, although most cellular switches do not *require* the use of such patterns. It is possible to devise a workable system that is virtually "non-cellular"; that is, a system that uses large numbers of channels at one or two sites or that is cellular but has very irregular channel configurations. Some designers plan a "non-cellular" system to meet short-term objectives such as lower cost, but these systems are not recommended in areas where even moderate levels of frequency reuse is contemplated. As discussed later, in "Non-Reuse Plans," this technique should be reserved for small remote service areas.

Because most cellular systems are prone to adjacent-channel interference, the first requirement of a system is that adjacent channels, as far as possible, should not be located in neighboring cells (for example, those to which a handoff can be attempted).

Cell-site selection should not be unduly influenced by something as hypothetical as the hexagonal cell plan (see Figure 7.1). Use the cell plan to help determine the most suitable frequency to be used at a site selected by taking into account real-world constraints. (An exception is the 4-cell pattern where very little flexibility is possible.)

Theoretically, cell plans with the smallest number of cells can achieve the highest subscriber densities but are more sensitive to cell-site location and design. For example, a 4-cell plan cannot

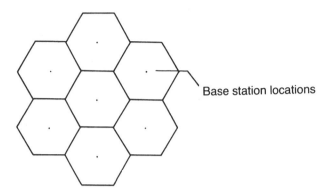

Figure 7.1   The hexagonal grid where each cell (hexagon) contains a group of frequencies

avoid adjacent-channel interference without using sector antennas (if frequency reuse is contemplated). For this reason, it is not unusual, even where a 4-cell plan is the ultimate goal, to start with a 7-cell plan, which is more flexible and can even incorporate some omni bases.

A 7-cell plan has some cells that are both physically adjacent and have adjacent channels. The adjacency problem can be confined to three cells only, and sectoring these three may be necessary. In smaller systems and in areas of irregular terrain it is often possible to avoid sectoring altogether. The 12-cell plan is designed to use omnidirectional cells and does not have the immediate adjacency problem.

Because each of these cell plans is based on a hexagonal grid, it is possible to change from one cell plan to another by simply rearranging frequency assignments. Doing so, of course, may not be so simple on a working system.

THE 7-CELL PATTERN

The most common cell plan uses 7 cells, which can be further subdivided to form a 21-cell (sectored) plan. Figure 7.2 shows a commonly adopted 7-cell pattern. As you can see from Figure 7.2, despite an effort to separate adjacent channels, the pairs D,C and D,E are still adjacent.

Careful frequency planning can substantially reduce the cost of a new installation. For example, the adjacent channel pairs D,C and D,E will probably ultimately require that the sites be sectored for separation. For relatively small systems, however, sectoring in the early stages can be avoided. These sites can be installed as omnidirectional sites provided that lower frequencies are assigned to the D cell and higher frequencies are assigned to the C and E cells (or vice versa). Thus, if each cell had seven channels, they could be allocated as shown in Table 7.1. Such frequency planning works only for bases that are at less than 50 percent of maximum channel capacity, and it can delay the need for expensive sectoring for a few years.

Some base stations can be programmed to preferentially choose odd or even channel numbers so that adjacency problems are unlikely to occur whenever less than half the channels are in use. These two techniques—cell subdivision and use of odd or even channels—can be combined.

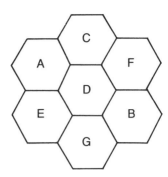

Figure 7.2   Cell distribution. Cells that are alphabetically adjacent are also adjacent in frequency (note that A is also adjacent to F).

CELL	D1	C1	E1
CHANNELS	4	192	194
	25	213	215
	46	234	236
	67	255	257
	88	276	278
	109	297	299
	130		
CONTROL CHANNEL	316	315	317

Table 7.1   Channel allocation of low channel numbers for the D cell and high channel numbers for the C and E cell will avoid adjacency problems for small systems.

Sectoring further subdivides channel groups and prevents the worst effects of adjacent-channel interference. Tables 7.2 and 7.3 show the channel groups subdivision. As shown in Figure 7.3, once the site is sectored, the adjacency problem is lessened by the antenna patterns.

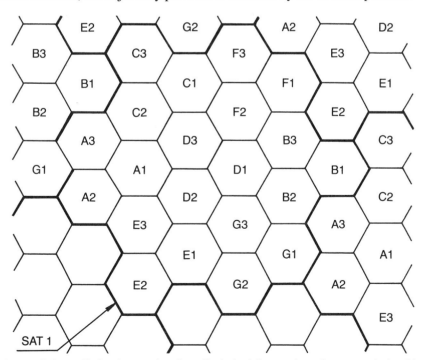

Figure 7.3   Sectored sites cell plan (sectored at the cell edge). Adjacent channels are now isolated by the antenna patterns. The orientation and separation provide the interference immunity required.

	A₁	B₁	C₁	D₁	E₁	F₁	G₁	A₂	B₂	C₂	D₂	E₂	F₂	G₂	A₃	B₃	C₃	D₃	E₃	F₃	G₃
Control Ch	313	314	315	316	317	318	319	320	321	322	323	324	325	326	327	328	329	330	331	332	333
Voice Ch	1	2	3	4	5	6	7	8	9	10	11	12	13	14	15	16	17	18	19	20	21
	22	23	24	25	26	27	28	29	30	31	32	33	34	35	36	37	38	39	40	41	42
	43	44	45	46	47	48	49	50	51	52	53	54	55	56	57	58	59	60	61	62	63
	64	65	66	67	68	69	70	71	72	73	74	75	76	77	78	79	80	81	82	83	84
	85	86	87	88	89	90	91	92	93	94	95	96	97	98	99	100	101	102	103	104	105
	106	107	108	109	110	111	112	113	114	115	116	117	118	119	120	121	122	123	124	125	126
	127	128	129	130	131	132	133	134	135	136	137	138	139	140	141	142	143	144	145	146	147
	148	149	150	151	152	153	154	155	156	157	158	159	160	161	162	163	164	165	166	167	168
	169	170	171	172	173	174	175	176	177	178	179	180	181	182	183	184	185	186	187	188	189
	190	191	192	193	194	195	196	197	198	199	200	201	202	203	204	205	206	207	208	209	210
	211	212	213	214	215	216	217	218	219	220	221	222	223	224	225	226	227	228	229	230	231
	232	233	234	235	236	237	238	239	240	241	242	243	244	245	246	247	248	249	250	251	252
	253	254	255	256	257	258	259	260	261	262	263	264	265	266	267	268	269	270	271	272	273
	274	275	276	277	278	279	280	281	282	283	284	285	286	287	288	289	290	291	292	293	294
	670	671	672	673	674	675	676	677	678	679	680	681	682	683	684	685	686	687	688	689	690
	691	692	693	694	695	696	697	698	699	700	701	702	703	704	705	706	707	708	709	710	711
	712	713	714	715	716	717															
							991	992	993	994	995	996	997	998	999	1000	1001	1002	1003	1004	1005
	1006	1007	1008	1009	1010	1011	1012	1013	1014	1015	1016	1017	1018	1019	1020	1021	1022	1023			

Table 7.2   AMPS A band

	A₁	B₁	C₁	D₁	E₁	F₁	G₁	A₂	B₂	C₂	D₂	E₂	F₂	G₂	A₃	B₃	C₃	D₃	E₃	F₃	G₃
Group	1	2	3	4	5	6	7	8	9	10	11	12	13	14	15	16	17	18	19	20	21
Control Ch	334	335	336	337	338	339	340	341	342	343	344	345	346	347	348	349	350	351	352	353	354
Voice Ch	355	356	357	358	359	360	361	362	363	364	365	366	367	368	369	370	371	372	373	374	375
	376	377	378	379	380	381	382	383	384	385	386	387	388	389	390	391	392	393	394	395	396
	397	398	399	400	401	402	403	404	405	406	407	408	409	410	411	412	413	414	415	416	417
	418	419	420	421	422	423	424	425	426	427	428	429	430	431	432	433	434	435	436	437	438
	439	440	441	442	443	444	445	446	447	448	449	450	451	452	453	454	455	456	457	458	459
	460	461	462	463	464	465	466	467	468	469	470	471	472	473	474	475	476	477	478	479	480
	481	482	483	484	485	486	487	488	489	490	491	492	493	494	495	496	497	498	499	500	501
	502	503	504	505	506	507	508	509	510	511	512	513	514	515	516	517	518	519	520	521	522
	523	524	525	526	527	528	529	530	531	532	533	534	535	536	537	538	539	540	541	542	543
	544	545	546	547	548	549	550	551	552	553	554	555	556	557	558	559	560	561	562	563	564
	565	566	567	568	569	570	571	572	573	574	575	576	577	578	579	580	581	582	583	584	585
	586	587	588	589	590	591	592	593	594	595	596	597	598	599	600	601	602	603	604	605	606
	607	608	609	610	611	612	613	614	615	616	617	618	619	620	621	622	623	624	625	626	627
	628	629	630	631	632	633	634	635	636	637	638	639	640	641	642	643	644	645	646	647	648
	649	650	651	652	653	654	655	656	657	658	659	660	661	662	663	664	665	666			
						717	718	719	720	721	722	723	724	725	726	727	728	729	730	731	732
	733	734	735	736	737	738	739	740	741	742	743	744	745	746	747	748	749	750	751	752	753
	754	755	756	757	758	759	760	761	762	763	764	765	766	767	768	769	770	771	772	773	774
	775	776	777	778	779	780	781	782	783	784	785	786	787	788	789	790	791	792	793	794	795
	796	797	798	799																	

Table 7.3   AMPS B band

Sectoring can be visualized to occur either at the cell edge (as shown in Figure 7.3) or at the cell center (as shown in Figure 7.4). Figure 7.5 shows a cell split based on site D by maintaining the central cell as a D cell while rotating the rest of the pattern counterclockwise by 120 degrees. The new cells have sides one-half the size of the original cells.

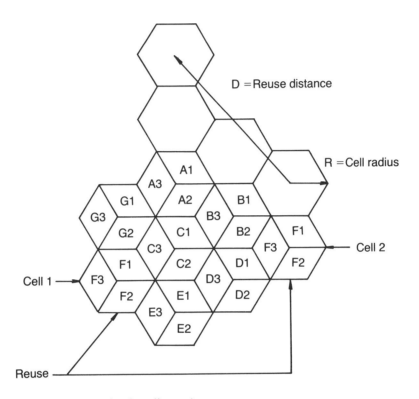

Figure 7.4   7- cell pattern (sectored at the cell center)

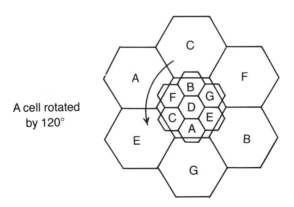

Figure 7.5   Cell orientation during a cell split can be seen as a 120-degree counterclockwise rotation. Note how the small cell C has moved with respect to the larger cell C.

SAT CODES

An AMPS system has SAT (Supervisory Audio Tones) codes (6 KHz tones of frequencies 5910, 6000, and 6030) that are transmitted on the speech channels. These tones are used to identify the local cell traffic from adjacent interfering traffic. The SAT is generated by the base station and looped back via the mobile circuitry.

Allocating SAT codes is very simple, following the rule that adjacent cell clusters have different SATs. There are only three SATs: SAT 0, 1, and 2. As Figure 7.6 shows, each cluster intersects only two other cells.

At the cell junctions (marked with triple lines) shown in Figure 7.6, the condition that all adjacent cells must have a different SAT is sufficient to determine all SAT assignments once any two have been designated.

SAT CODES AFTER CELL SPLITTING

Allocating SAT codes with cell splitting is also quite simple. As Figure 7.7 shows, cell splitting occurs around the central cell. The new cell takes on the SAT of the old cell but what happens to the SAT of the old cell? To answer this question, it is necessary to consider splitting the center cell of the adjacent cluster. If the new clusters at the center of the two original clusters retain their old SAT codes, then the new cluster between them must be the third SAT code.

Figure 7.8 shows the new SAT codes that result from replacing the SAT codes of the old cells with the SAT code at the center of the overlay split cell.

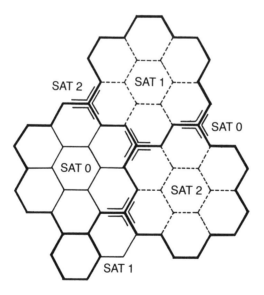

Figure 7.6   SAT are allocated so that any three adjacent cell groups are different.

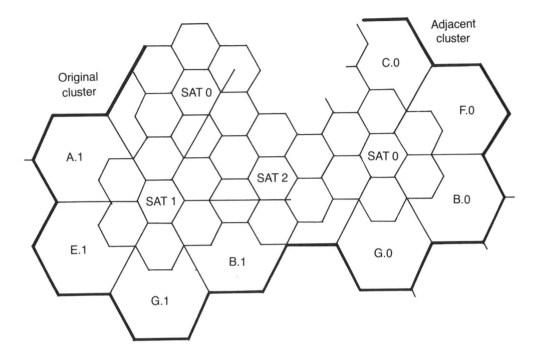

Figure 7.7   Cell split of two adjacent clusters

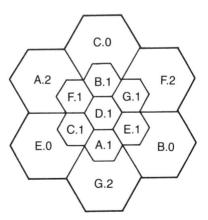

Figure 7.8   New SAT codes for the old cells

DIGITAL COLOR CODES

The Digital Color Code (DCC) is a similar identification tag to SAT, except that DCC applies to control channels. There are four DCC codes: 0, 1, 2, and 3. As shown in Figure 7.9, randomly allocating three of the codes to adjacent cell clusters determines the pattern for the rest.

Each cell can be visualized as being at the center of a 7-cell cluster (that is, surrounded by six other cells). If the D cell is first considered to be at the center of a cell cluster (cluster 1) and DCCs are randomly assigned, then $D=3$, $C=0$, $F=1$, and $B=2$. By repeating the pattern around the D cell, it follows that $A=2$, $E=1$, and $G=0$.

This method can now be extended to all other cells. If the B cell is now regarded as being at the center of a new 7-cell cluster (cluster N), by applying the same pattern, the assignments for cells X, Y, and Z can be found. This is continued until all DCCs are allocated ($X=1$, $Y=3$, and $Z=0$).

Cell expansion proceeds in much the same way as described earlier, except that instead of the center cells taking on the DCC of the "old" cell, all center cells take on the same DCC. This means that the reorganization of the first cell can be random but that future cells are locked into the first one. Again, the old cells take on the DCC of the overlay control cell.

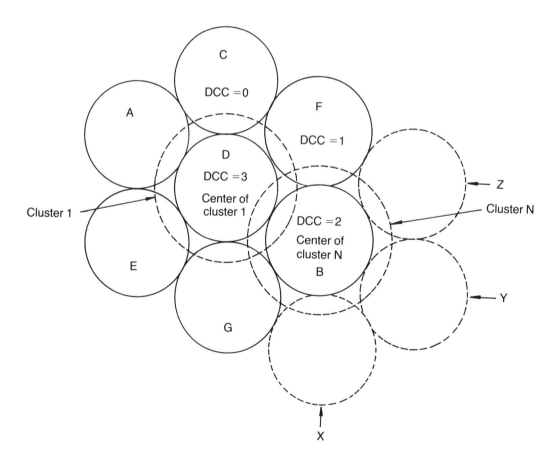

Figure 7.9   Allocation of digital color code

4-CELL PATTERN

The 4-cell pattern is really a 4/12 or 4/24 pattern. That is, the cells are grouped into four frequency plans with either 12 or 24 sectors. The 4-cell sector pattern uses either six 60-degree antennas or three 120-degree antennas at a common site. This arrangement produces the 24-sector (6 sectors by 4 cells) and 12-sector (3 sectors by 4 cells) patterns, respectively. Figure 7.10 shows the 24-cell pattern.

12-CELL PATTERN

Less commonly used is the 12-cell pattern, which has good carrier-to-interference ratios even when the sites are omnidirectional. The individual cells in this configuration are very small, however, and the traffic efficiency is correspondingly low. Figure 7.11 shows the 12-cell pattern.

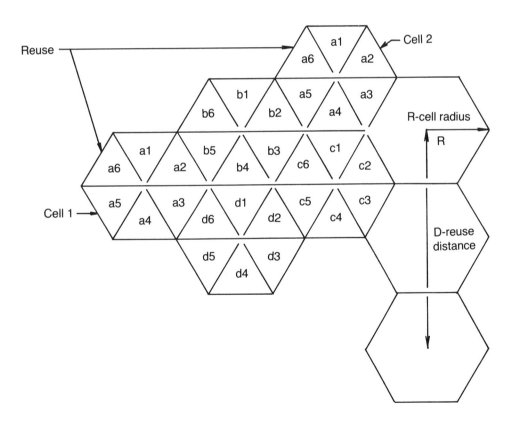

Figure 7.10   4-cell sector pattern (24 cells)

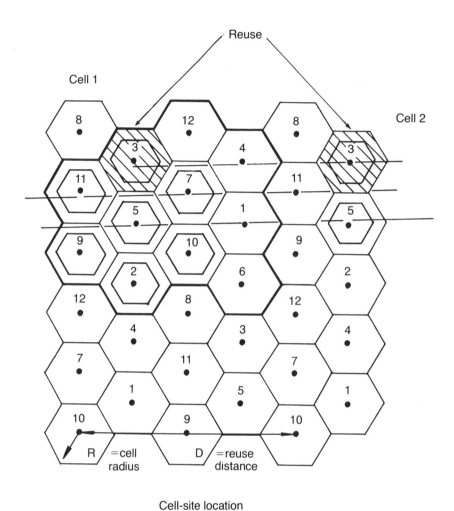

Cell-site location

Figure 7.11   12-cell pattern

THE STOCKHOLM RING MODEL

The Stockholm Ring model has been used on NMT450 systems, originally in Stockholm, and later in Jakarta, Malaysia, and Bangkok. This technique was designed to respond to high traffic densities in central Stockholm. Using quite different principles from the conventional hexagonal pattern, the Stockholm Ring attempts to increase the CBD capacity by using all available frequencies at a central site. NMT450 has 180 frequencies and the plan divides those channels into 6 by 60 degree co-sited sectors, as shown in Figure 7.12. Co-siting all frequencies implies that good adjacent-channel separation is possible. This requirement, of course, limits this technique to the low-deviation systems, NMT450 and NMT900 (without channel interleaving only).

Figure 7.13 shows the next expansion outward, using another concentric ring with all the frequencies rotated 120 degrees. As cell expansion occurs away from the CBD, wider antenna patterns accommodate the lower densities efficiently.

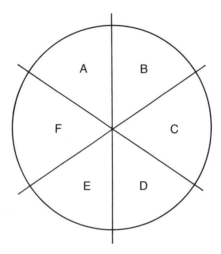

Figure 7.12   Inner cell shown with 6 by 60 degree sector cells using all 180 degree frequencies

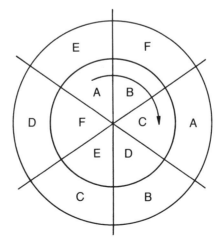

Figure 7.13   First expansion pattern

NON-REUSE PLANS

An analog cellular system frequency plan can be said to be "non-cellular" when there are less than four cells in the pattern. In certain rural areas, however, there may be no need for frequency reuse, making it economical to place a large number of channels at one location—along the lines of the central cell of the Stockholm Ring.

In AMPS and TACS, non-adjacent channels can be co-located so that a cell about one-half the total spectrum at one site is conceivable. Nothing short of a major base station rearrangement, however, will permit future efficient reuse of this system, so this plan should be used only in small communities.

MIXED PLANS

When a 4-cell plan is used to obtain high density, it can be economical to design the network to gradually taper to a 7-cell plan at the city boundaries, and maybe evolve into a 12-cell plan farther away. There are no standard techniques for carrying out this type of graduated plan, but careful frequency allocation and a few fortuitous hills can make this technique a little less daunting.

Chapter 8

UNITS AND CONCEPTS OF FIELD STRENGTH

There are a diversity of units of field strength in use in RF engineering today. Mobile engineers tend to prefer the unit dBμV/m, but some still use units like dBm which have their origins in the land-line network. This chapter seeks to clarify the usage of the different units and the concepts behind the measurement of field strength.

In free space, the energy of an electromagnetic wave propagates through space, and, according to an inverse-square law, with distance. The energy is dispersed over the surface of an ever-increasing sphere, the area of which is R (where R is the sphere radius). The total energy is constant because there is no loss in free space. The energy measured in watts/square meter (or any other units) will be constant in total and so the energy per unit area will vary as $1/R^2$. Notice that this equation holds for all frequencies. This concept is illustrated in Figure 8.1.

The total energy within a solid angle is a constant at any radial distance from the origin.

In a mobile-radio environment, the signal is attenuated much more rapidly than in free space and follows approximately an inverse fourth power law with distance. There is a common misconception that this attenuation increases rapidly with frequency. It will be shown later in this chapter that, although the attenuation is an increasing function of distance, it is not a very strongly frequency-dependent function.

However, because the capture area, or aperture, of an antenna decreases directly with frequency, the energy captured by the antenna is directly a function of frequency. Thus, a quarter-wave antenna at 450 MHz is twice as long as a 900-MHz antenna and so can capture more energy from a field with the same intensity. This difference in capture area or aperture is what mainly accounts for the better long-range performance of lower-frequency systems.

Of course, this discussion must be limited to a frequency band where the propagation mode is similar. If the range 150-1000 MHz is considered, then the assumptions will hold.

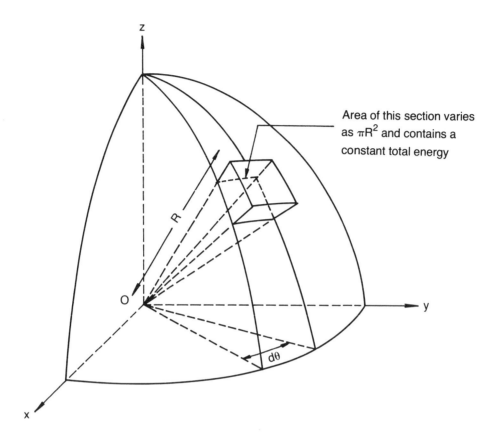

Figure 8.1   The originating energy from the origin, "O," is dispersed uniformly over a spherical surface as it propagates in space.

Mathematically, the effective aperture or capture area A, of an antenna is:

$$A_{eff} = \frac{\lambda^2 G}{4\pi}$$

where λ = wavelength
          G = antenna gain
          A_{eff} = effective aperture

Therefore, if the frequency is increased by a factor of 2, the effective capture area will increase by a factor of 4. The energy received will also increase by a factor of 4. Thus, the power collected by a 450-MHz dipole antenna compared to a 900-MHz dipole antenna is 10 log 4 = 6 dB higher in a field of the same intensity. The same result can be derived by visualizing the antenna as immersed in an electric field of V volts per meter. A longer antenna is swept by more lines of field strength and so induces a higher voltage.

If a 900-MHz antenna is compared to a 450-MHz antenna of the same type, the 450-MHz antenna is twice as long and will intercept twice the electric field potential. The resulting increase in received field strength is 20 log 2 or 6 dB.

The effective aperture of an antenna of the same gain (for example, 3 dB) will also slightly effect the relative performances of 800-MHz and 900-MHz systems.

The AMPS system has a center or mid-frequency of approximately 867.5 MHz and the TACS/NMT900 systems have a central frequency of 927.5 MHz (although both of these frequencies are outside the respective system frequencies). By comparing the aperture we can obtain the relative gain, as shown here:

$$\textit{The relative aperture is then} \left[\frac{927.5}{867.5}\right]^2$$

$$\textit{or a relative gain of } 10\log\left[\frac{927.5}{867.5}\right]^2$$

= 0.6 dB relative gain for the AMPS system

Field strength is a measure of power density at any given point. The main units of field strength are dBμV/m, dBμV, μV, and dBμ. The dBμV/m unit measures power density of radio waves. The other units measure received energy levels. Most mobile-radio engineers prefer the dBμV/m unit, while microwave engineers prefer dBm, and bench technicians use μV. This preference is partly historical and partly the practicality of the units in the different fields.

To visualize how these units relate, consider Figure 8.2, which shows a test dipole in the plane at right angles to the direction of propagation. It measures the electric field strength in the direction of the antenna.

The units dBμV, dBm, and μV measure the power or voltage received by a dipole antenna in

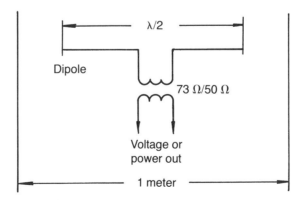

Figure 8.2   Measurement of field strength

the field. Voltages in mobile equipment are usually measured at 50 ohms, but any impedance can be used. These units are defined as:

$$dB\mu V = 20 \log \frac{\text{voltage at transformer output (terminated in 50 ohms)}}{1 \text{ microvolt}}$$

$$dBm = 10 \log \frac{\text{power from transformer}}{1 \text{ milliwatt}} \text{ (impedance independent)}$$

μV = voltage at transformer terminal into 50 ohms load in microvolts

The unit $dB\mu V/m$ is the voltage potential difference over 1 meter of space, measured in the plane at right angles to the direction of propagation and in the direction of the test antenna. It is defined as follows:

$$dB\mu V / m = 20 \log \frac{\text{voltage potential over 1 meter}}{1 \text{ microvolt}}$$

To see how all these units relate to each other in a real environment, a case study based on the work of Okumura et al. (a classic work on mobile propagation) will help.

Consider a typical base station that has a reference height of 200 meters and a transmitter ERP of 100 watts. Table 8.1 lists the received field strength for various transmitter frequencies in the far field; for example, 10 km.

From the table, it should be clear that the actual receiver signal varies enormously with frequency, even though the transmitter power, site, and antenna height remain fixed. Therefore, if the units μV, dBm, or $dB\mu V$ are selected, the results of a survey of one frequency cannot easily be translated to frequencies that are significantly different. However, if $dB\mu V/m$ is selected, only a few dB separate the readings. Furthermore, if additional sites are studied (that is, different heights for the transmitter, and different distances), this relationship is retained. Thus, any field strength measured in $dB\mu V/m$ measures energy density at a given point and is dependent on frequency, only to the extent that atmospheric and clutter attenuation is dependent on frequency. Therefore, it is possible to use results from a survey done at one frequency to draw conclusions about another if an allowance of 2 dB per octave is used. Some caution should be exercised when using these broad generalizations. However, they can be most useful approximations of coverage.

FREQ	dBμV/m	μV	dBm	dBμV
150 MHz	49	71	-70	+37
450 MHz	47	17	-82	+25
900 MHz	45	8	-89	+18

Table 8.1   Field strength at 10 km for a 100-watt TX at 20 meters in an urban environment. (Note: This table was derived from a paper by Okumura et al. entitled "Field Strength and Its Variability in VHF and UHF Land-Mobile Radio Service," *Review of the Electrical Communication Laboratory*, Vol. 16, Nos. 9, 10, Sept-Oct 1968, pp. 835-873.)

RELATIONSHIP BETWEEN UNITS OF FIELD STRENGTH AT THE ANTENNA TERMINALS

Assuming a 50-ohm termination, a dipole receiving antenna (unity gain), and a zero-loss feeder, the relationship between units of field strength at the antenna terminals is as follows:

$$E(\mu V / m) = \frac{\mu V}{39.3924} \; x \; FREQ \; (MHz)$$

Equation A. Starting with dBm all into 50 Ω

$$\mu V = 2.236 \; x \; 10^5 \; x \; 10 \, \frac{dBm}{20}$$

$$dB\mu V / m = 20 \log (5.676 \; x \; 10^3 \; x \; FREQ \; (MHz) \; x \; 10 \, \frac{dBm}{20})$$

$$dB\mu V = 20 \log (2.236 \; x \; 10^5 \; x \; 10 \, \frac{dBm}{20})$$

Equation B. Starting with μV all into 50 Ω

$$dBm = 20 \log \frac{\mu V}{2.236 \; x \; 10^5}$$

$$dB\mu V = 20 \log \mu V$$

$$dB\mu V / m = 20 \log (\mu V \; x \; FREQ \; (MHz) \, / 39.3924)$$

Equation C. Starting with dBμV/m all into 50 Ω

$$dB\mu V = dB\mu V / m - 20 \log \frac{FREQ(MHz)}{39.3924}$$

$$dBm = 20 \log \left[\cdot 1.76 \; x \; 10^{-4} \; x \; \frac{10 \, \frac{dB\mu V / m}{20}}{FREQ \; (MHz)} \right]$$

$$\mu V = \frac{39.3924 \; x \; 10 \, \frac{dB\mu V / m}{20}}{FREQ \; (MHz)}$$

Equation D. Starting with dBμV all into 50 Ω

$$\mu V = 10^{dB\mu V / 20}$$

$$dBm = 20 \log \frac{10^{dB\mu V / 20}}{2.236 \; x \; 10^5}$$

$$dB\mu V / m = dB\mu V + 20 \log \frac{FREQ \; (MHz)}{39.3924}$$

CONVERSION TABLES

Because it is very easy to make a mistake when applying the formulas to translate between units, the conversion table in Table 8.2 can be very helpful. This table is useful for calculating the relationship between the variables illustrated in Figure 8.2. If a different antenna impedance is considered (a 300-ohm folded dipole antenna, for example), then the results cannot be used directly. Similarly, in a real-life environment, the signal levels will probably be measured at the receiver output, as shown in Figure 8.3 below, and the necessary corrections must be applied.

The relationship between the variables must be adjusted by the antenna gain minus the cable loss. At 900 MHz, using a 3-dB antenna, the cable loss is about 3 dB and thus the correction factor approaches zero. At other frequencies, this approximate relationship will not hold.

STATISTICAL MEASUREMENTS OF FIELD STRENGTH

In point-to-point radio, field strength is a one-dimensional variable of time. Most of the time variance is due to log-normal fading, and the nature of the signal variability is well documented. Because it is a simple function of time, field strength in point-to-point radio can be easily understood and measured. The situation is somewhat more complex in the mobile RF environment.

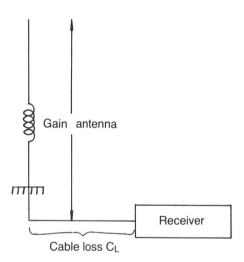

Figure 8.3   Actual field strength measurement

			FREQ.MHz (dBµV/m)		
dBm	µV	dBµV	150	450	900
10	707945	117	129	138	144
1	251188	108	120	129	135
-1	199526	106	118	127	133
-2	177827	105	117	126	132
-3	158489	104	116	125	131
-4	141253	103	115	124	130
-5	125892	102	114	123	129
-6	112201	101	113	122	128
-7	100000	100	112	121	127
-8	89125	99	111	120	126
-9	79432	98	110	119	125
-10	70794	97	109	118	124
-11	63095	96	108	117	123
-12	56234	95	107	116	122
-13	50118	94	106	115	121
-14	44668	93	105	114	120
-15	39810	92	104	113	119
-16	35481	91	103	112	118
-17	31622	90	102	111	117
-18	28183	89	101	110	116
-19	25118	88	100	109	115
-20	22387	87	99	108	114
-21	19952	86	98	107	113
-22	17782	85	97	106	112
-23	15848	84	96	105	111
-24	14125	83	95	104	110
-25	12589	82	94	103	109
-26	11220	81	93	102	108
-27	10000	80	92	101	107
-28	8912	79	91	100	106
-29	7943	78	90	99	105
-30	7079	77	89	98	104
-31	6309	76	88	97	103
-32	5623	75	87	96	102
-33	5011	74	86	95	101
-34	4466	73	85	94	100
-35	3981	72	84	93	99
-36	3548	71	83	92	98
-37	3162	70	82	91	97
-38	2818	69	81	90	96
-39	2511	68	80	89	95
-40	2238	67	79	88	94
-41	1995	66	78	87	93
-42	1778	65	77	86	92
-43	1584	64	76	85	91
-44	1412	63	75	84	90
-45	1258	62	74	83	89

Table 8.2   Conversion tables for a 50 Ω dipole antenna (for 150 Mhz, 450 Mhz, and 900 Mhz) (continued)

Table 8.2. *continued*

dBm	μV	dBμV	FREQ.MHz (dBμV/m)		
			150	450	900
-46	1122	61	73	82	88
-47	1000	60	72	81	87
-48	891	59	71	80	86
-49	794	58	70	79	85
-50	707	57	69	78	84
-51	630	56	68	77	83
-52	562	55	67	76	82
-53	501	54	66	75	81
-54	446	53	65	74	80
-55	398	52	64	73	79
-56	354	51	63	72	78
-57	316	50	62	71	77
-58	281	49	61	70	76
-59	251	48	60	69	75
-60	223	47	59	68	74
-61	199	46	58	67	73
-62	177	45	57	66	72
-63	156	44	56	65	71
-64	141	43	55	64	70
-65	125	42	54	63	69
-66	112	41	53	62	68
-67	100	40	52	61	67
-68	89	39	51	60	66
-69	79	38	50	59	65
-70	70	37	49	58	64
-71	63	36	48	57	63
-72	56	35	47	56	62
-73	50	34	46	55	61
-74	45	33	45	54	60
-75	39	32	44	53	59
-76	36	31	43	52	58
-77	32	30	42	51	57
-78	28	29	41	50	56
-79	25	28	40	49	55
-80	22	27	39	48	54
-81	20	26	38	47	53
-82	18	25	37	46	52
-83	16	24	36	45	51
-84	14	23	35	44	50
-85	13	22	34	43	49
-86	11	21	33	42	48
-87	10	20	32	41	47
-88	9	19	31	40	46
-89	8	18	30	39	45
-90	7	17	29	38	44

(continued)

Table 8.2. *continued*

dBm	V	dBV	FREQ.MHz (dBμV/m)		
			150	450	900
-91	6.3	16	28	37	43
-92	5.6	15	27	36	42
-93	5	14	26	35	41
-94	4.5	13	25	34	40
-95	4	12	24	33	39
-96	3.5	11	23	32	38
-97	3.2	10	22	31	37
-98	2.8	9	21	30	36
-99	2.5	8	20	29	35
-100	2.2	7	19	28	34
-101	2	6	18	27	33
-102	1.8	5	17	26	32
-103	1.6	4	16	25	31
-104	1.4	3	15	24	30
-105	1.3	2	14	23	29
-106	1.1	1	13	22	28
-107	1	0	12	21	27
-108	0.89	-1	11	20	26
-109	0.79	-2	10	19	25
-110	0.71	-3	9	18	24
-111	0.63	-4	8	17	23
-112	0.56	-5	7	16	22
-113	0.5	-6	6	15	21
-114	0.47	-7	5	14	20
-115	0.4	-8	4	13	19
-116	0.35	-9	3	12	18
-117	0.32	-10	2	11	17
-118	0.28	-11	1	10	16
-119	0.25	-12	0	9	15
-120	0.22	-13	-1	8	14
-121	0.2	-14	-2	7	13
-122	0.17	-15	-3	6	12
-123	0.16	-16	-4	5	11
-124	0.14	-17	-5	4	10
-125	0.126	-18	-6	3	9
-126	0.122	-19	-7	2	8
-127	0.1	-20	-8	1	7
-128	0.089	-21	-9	0	6
-129	0.079	-22	-10	-1	5
-130	0.071	-23	-11	-2	4

Thus the field strength in the point-to-point environment is a simple function of time as shown here:

$$F = f(t)$$

In the mobile environment, the field strength also varies with location (space), and so the measured value is a four-dimensional statistical variable:

$$F = f\ (t,x,y,z)$$

or, if $(x,y,z) = L = $ location, then:

$$F = f\ (t,L)$$

The real-life measurement of the field strength of an area is the collective result of a number of measurements made in different points in space and time, as illustrated in Figure 8.4. Thus, if a measure of field strength is required to typify an area, then its statistical nature dictates that the actual result depends on when and where the measurement is made. Let's assume a measurement of the average field strength along a 500-meter section of a road is required. If all samples are taken at the Nyquist rate (in space at locations $L_0, L_1, \ldots L_m$) and at one instant (T_0), the result would be an average value at a particular time, T_0.

Mathematically this can be expressed as:

$$F_0 = \frac{\sum\limits_{i=0}^{m} f\ (T_0, L_i)}{m + 1}$$

($m + 1 = $ the number of readings taken at positions $L_0, L_1, \ldots L_m$)

In practice making simultaneous measurements is difficult and the individual space measurements would be made at different times ($T_1 \ldots Tn$). The average value in time and space of the field strength in that region is then given by F as shown here:

$$\frac{\sum\limits_{K=0}^{n} F_K}{n}$$

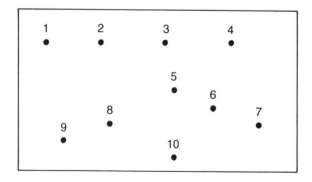

Figure 8.4   The measurement of field strength over an area is made at unique points in space and time. Actual readings vary with time (due to log-normal fading) and space (due to log-normal and Rayleigh fading). Any statement about the field strength in this area is a statement about a variable of space and time.

Notice that since Nyquist-rate space samples were taken, it is not necessary to use different locations for samples in space; but this could be done. Virtually all real measurements are an average (or some other statistical measure) of samples in different locations in space and time.

Other statistical measures are widely used. For example, all readings can be recorded and separated into two equal groups depending on level. Thus, it is possible to obtain the value above (or below) which 50 percent of the recorded samples occur. This is known as the median value (not the same as the average value). If all readings were taken at a single instant, the median value F_{50} can be found and the result would be at time T_0:

$$F_{50} = f \text{ median } \{f(T_0, L_i)\} \text{ for i = 0 to m}$$

where f_{50} = the level at which 50 percent of the samples 0 to m lie below

In practical terms, as survey readings are taken at different points in space and time for a given region, the average of a set of readings in a region is the true time/space averaged measurement. Because time fluctuations of field strength are usually fairly fast relative to measurement periods, this assumption holds fairly well. Log-normal fades produce field-strength variations with a periodicity of a few seconds, while a measurement is usually taken over 1- to 5-minute intervals. Therefore, if measurements taken from a moving vehicle in one survey record F_{50} field strength, it is considered to be the field strength for 50 percent of locations and 50 percent of the time.

As the number of samples in space equals the number of samples in time, it is also reasonable to use the same technique to obtain the 70 percent/70 percent or 90 percent/90 percent values. However, because F_{50} is a function of space and a different one of time, a simple survey technique cannot be used to directly obtain the value for 90 percent of locations for 50 percent of the time. To obtain this value it would be necessary to obtain the standard deviation of the time-dependent variation independently (approximately the standard deviation of the log-normal fade).

Interpreting a cellular requirement for the field strength to be, for example, 32 dBμV/m for 90 percent of locations and 90 percent of the time is a difficult task because the concept has never been adequately defined. In particular, the concept originally applied to small regions of space and its application to service areas is vague. It might mean that within the service area only 10 percent of all readings will be below 32 dBμV/m and this is how most people think of it. However, the goal of a system designer is not to have 10 percent of the service area substandard. Probably what is really meant by this is that 10-percent substandard coverage is a worst-case scenario and is an upper limit rather than a design parameter. One would not expect, for example, a batch of resistors with 10-percent tolerance to be *designed* to be 10 percent out of tolerance. The 10-percent limit merely means the manufacturer's tolerance or worst case is 10 percent.

In a multicell system, 90 percent of time could be interpreted to include channel availability for a given number of customers; that is, service is not available or is of poor quality for 90 percent of the time because of cell congestion. The 90 percent/90 percent are two separate criteria and the service is now satisfactory for only 81 percent of the time/space points bounded by the coverage area. This is difficult to achieve in a design, as it requires accurate knowledge of customer distributions. However, it would also be undesirable, as it results in 19 percent of the service area being unsatisfactory.

CONCLUSION

It is much simpler and less ambiguous to define the field strength of the service area boundaries as 39 dBμV/m average or 32 dBμV/m in 90 percent of locations and times. Some European administrations define the boundary as the 20 dB SINAD contour. At the edge of this boundary, it is only just possible to have an intelligible conversation. (At the 39dBμV/m level, the line quality at the boundary is noticeably below normal telephone service but is still of quite an acceptable standard.) This contour defines a much noisier service area than the 39 dBμV/m FCC specification and is easier to define.

Any specification wherein the time and space variables are not equal is impossible to realize in practice (that is, 90 percent of places 95 percent of the time is physically meaningless unless what is meant by each parameter and how it can be measured are clearly defined).

Chapter 9

RADIO BASE STATIONS

Radio base stations are effectively radio concentrators that serve a large number of customers using a few channels. Typically, 20–30 customers can be handled on each channel (depending on the calling rate). The base station is always modular in construction. Figure 9.1 shows the basic components of a base station. These elements of the base station shown in Figure 9.1 are discussed in the following sections.

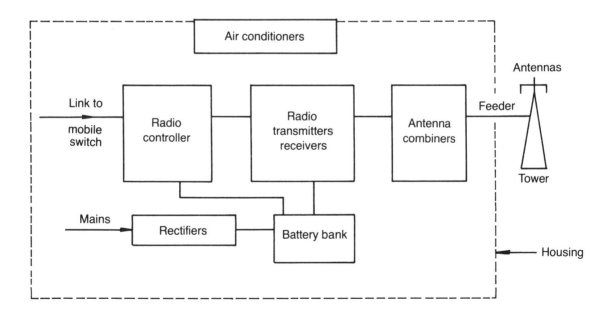

Figure 9.1   Diagram of a typical base station

RADIO CONTROLLER

The radio controller is the interface between the mobile switch and the base station and operates similarly to a remote-subscribers switch. Under the control of the mobile switch, the radio controller selects (switches) the radio channels as required. It also supervises various system parameters, including alarm conditions and radio field strength.

NMT systems do not have a radio controller at the base site; most of the radio controller functions are performed by the switch, making NMT base stations smaller and significantly simpler than TACS or AMPS bases, as they are virtually only remote-controlled transceivers with an associated scanning receiver. A small NMT base (about five channels) can fit completely in one rack.

RADIO TRANSMITTERS/RECEIVERS

The earliest bases had separate active and passive RF components, which had the disadvantage that a minimum-sized installation (a few channels) still needed at least two racks. Figure 9.2 shows an early NMT450 base station in Jakarta, Indonesia. The two outside racks contain the receivers and transmitters; the combiners are housed in the middle.

The transceivers are the links to the mobiles. They are sometimes fully frequency-agile (that is, they are software-programmed to a particular channel, but older versions may need mechanical tuning). Once installed they ordinarily act as a fixed frequency transceiver. In some systems from some suppliers, a voice channel can be designated as a standby control channel and may change frequency automatically when a control channel failure is detected. In other systems, the transmitters and combiners can be retuned remotely so that "channel borrowing" or the reallocation of the channel frequency can occur during periods of blocking on the regular frequency. This is used in some high-density systems.

Because it is difficult to predict the traffic distribution, it is advantageous to have channel equipment that can be moved easily from site to site. One of the earliest attempts at this was in the NMT450 system in Kuala Lumpur, Malaysia, in which each of the four racks was a self-contained set of four channels. As equipment became smaller, eight channels per rack became standard.

Figure 9.2   This NMT450 system in Jakarta, Indonesia has two active RF racks and two separate multicoupler racks. Photo courtesy L.M. Ericsson.

AMPS BASE STATION SPECIFICATION

Table 9.1 on the next page shows a typical AMPS base-station transceiver rack specification.

ANTENNA COMBINERS AND SPLITTERS

Because many channels ordinarily operate at one base, it is necessary to use antenna combiners. Antenna combiners combine transmitter outputs into a small number of antennas. The frequencies used at any one base have been chosen so that their spacing ensures good isolation between the channels.

PARAMETER	SPECIFICATION
Frequency stability	±1 PPM
RF sensitivity	- 116 dBm @ 12 dB SINAD
Expander	1:2 attack time 3 ms recovery 13.5 ms ±20% (per EIA PN1377)
Audio de-emphasis	- 6 dB for 300 Hz to 3000 Hz (per EIA PN 1377)
Audio distortion	2.5% at telephone interface
TX spurious	To meet EIA PN 1377
Compressor	2:1 attack time 3 ms recovery 13.5 ms ±20% (per EIA PN1377)
DC power	24 volt nominal (neg. earth)
Power/channel	8 amps
Equipment floor load	800 kg/m^2
Operational temperature range	-4°C to 45°C

Table 9.1   A typical AMPS base station transceiver rack specification

Figure 9.3 shows a typical antenna combiner. The whole combining system is known as a multicoupler or multiplexer. Because of confusion that can arise in discussing the RF coupling device and the MUX (also known as a multiplexer), the term multicoupler is preferred.

The following is the specification for a typical cavity:

*   Resonant circuit
*   Impedance: 50 ohms
*   Insertion loss: 2 dB
*   Input isolation: 14 dB
*   Maximum input power: 50 watts
*   Minimum frequency separation: 630 KHz

This is the specification for a typical isolator:

*   Impedance: 50 ohms
*   Insertion loss: 0.6 dB
*   Reverse isolation: 50 dB or more
*   Maximum input power: 60 watts

This is the specification for a typical junction:

*   Impedance: 50 ohms
*   Maximum input power: 600 watts
*   Intermodulation of 16 TX carriers of +45 dBm shall not produce more than -105 dBm of 5th order intermodulation products

Isolaters allow the TX signal to pass with only 0.5 dB loss in the forward direction but attenuate any reflected signal by at least 50 dB. These devices are known as circulators and are generally wide-band devices.

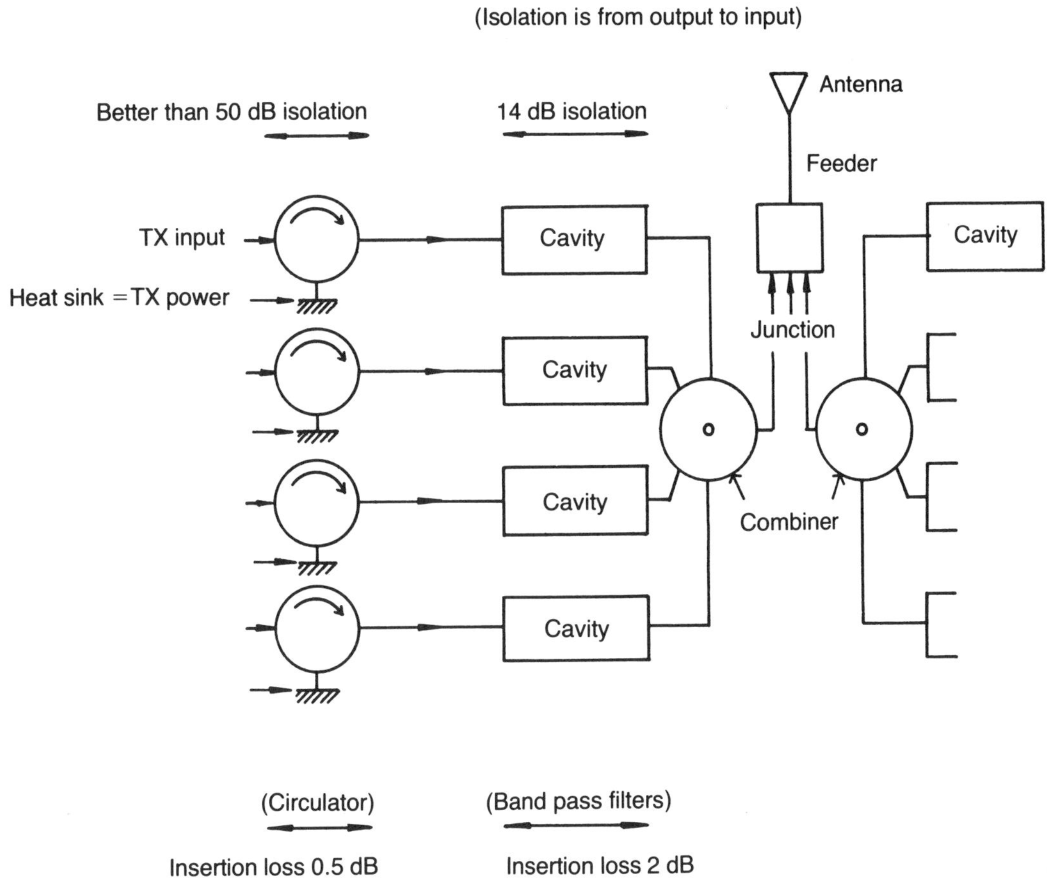

(Isolation is from output to input)

Antenna

Better than 50 dB isolation   14 dB isolation

Feeder

TX input →   Cavity   Cavity

Heat sink = TX power →   Junction

                                   Cavity

                                                                   Combiner

                                   Cavity

                                   Cavity

(Circulator)   (Band pass filters)

Insertion loss 0.5 dB   Insertion loss 2 dB

Figure 9.3   Typical combiner assembly

The cavities are resonant filters. They have insertion losses of about 2 dB and an input isolation of 14 dB, resulting in about 2.5 dB overall loss of the combiners. Ordinarily, up to 16 transmitters are combined into one antenna.

For cellular systems, 630 KHz separation (minimum) is generally allowed between channels being coupled together. Channel separations less than this are possible only if greater insertion losses are tolerated.

Receivers are multicoupled to one antenna, usually in multiples of 64. The most important part of the receiver multicoupler is a good low-noise preamplifier that will determine the ultimate S/N ratios and hence the base station receiver performance. Figure 9.4 on the next page shows a six-antenna multicoupler. Figure 9.5 shows a typical splitter.

In most cellular bases, diversity reception is used. This means that two antennas are used for each receiver that has two RF inputs and a diversity combiner. These antennas are physically separated by about 3–4 meters so that their received signals are not correlated. When one antenna receives a multipath fade, the other antenna probably will not.

Figure 9.4   This Motorola receiver multicoupler is contained in its own housing. The six "black boxes" at the top are
the low-noise preamplifiers with preselectors for a 3-sector site (with diversity). Each of the preamplifiers then feeds a
four-way splitter (the six light -colored boxes below). Additional splitters are found in the receiver racks. Photo courtesy
Extelcom, Philippines.

The diversity receivers come in two types: the diversity-combining receiver and the switched
diversity receiver. The diversity combining receiver aligns the phases of the incoming signals and
then adds them. The switched diversity receiver chooses the best of the two signal paths and
switches to that path. A gain of 6 dB can be obtained in the first instance, and a gain of 3 dB in
the second. Figures 9.6 and 9.7 illustrate these configurations.

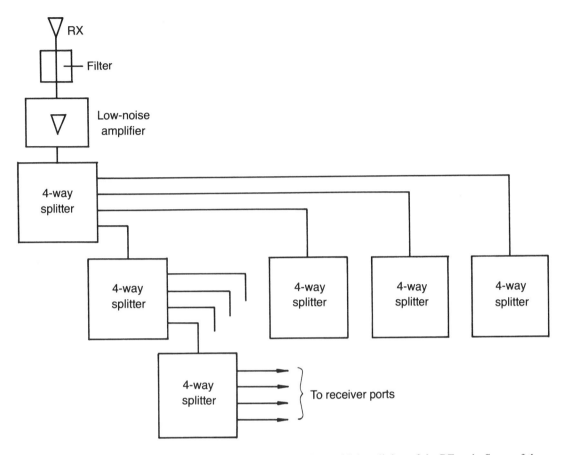

Figure 9.5   A total of 64 receivers can share a single antenna by multiple splitting of the RF path. Some of these splitters may be active but the S/N performance is largely determined by the first low-noise amplifier.

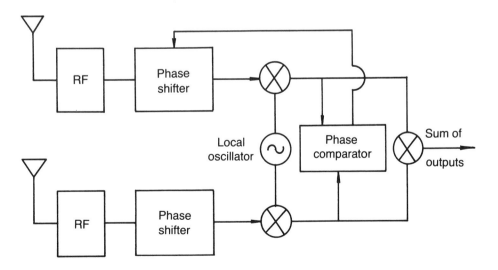

Figure 9.6   A diversity combining receiver continuously adjusts the phase of one of the received signals to produce a phase consistent output that can then be added. Such a receiver configuration can improve S/N by about 6 dB.

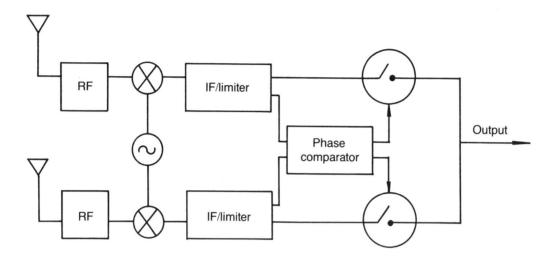

Figure 9.7   A switched diversity (also known as selection diversity) receiver chooses the best of the two signal paths.

Antenna duplexers are sometimes used to connect transmitter-and-receiver ports to the antenna, as shown in Figure 9.8. Antenna duplexers are used mainly when antenna space is at a premium. The duplexer consists of a number of resonant cavities that isolate the transmitter output from the receiver multicoupler input by at least 80 dB. Duplexers are used extensively in mobile equipment (allowing the use of one antenna only) but are only occasionally employed in cellular base stations because they are quite lossy (2 to 3 dB) and impair the overall sensitivity of the base-station receiver.

CAVITIES

Figure 9.9 shows a simple resonant cavity used in UHF applications. This simple cavity has a resonant frequency that is dependent on the physical dimensions of the inner conductor. Such cavities are not often used because they are very temperature-dependent and cannot be adjusted for different frequencies.

More commonly, a few simple enhancements are made to improve the frequency stability and to give some degree of control over the resonant frequency.

A practical cavity should be constructed so that thermal expansion effects have automatic compensation. Figure 9.10 shows a cross-section of a practical cavity. In this figure, you can see that by suitably dimensioning the components, it is possible to neutralize the effects of thermal expansion. You can adjust the dimension of the internal conductor (and hence the resonant frequency) by turning the adjustment screw at the top.

Cavities are usually made of low electrical-loss, high thermal-conductivity metals such as copper, brass, or aluminum. The thermal dissipation of the cavity depends on its loss, but a 3 dB loss cavity connected to a 100 watt transmitter can dissipate 50 watts—quite a lot of heat. Cavities are designed to dissipate about 30 mW per square centimeter.

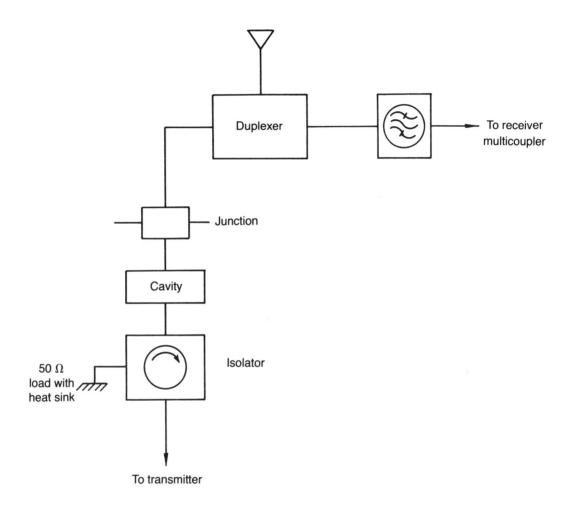

Figure 9.8   When receivers and transmitters are coupled into one antenna, a duplexer is needed to isolate the transmitter output from the receiver input.

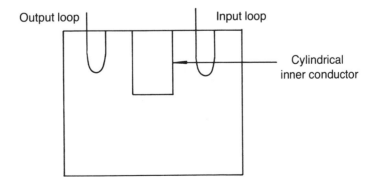

Figure 9.9   The interior of a simple cavity

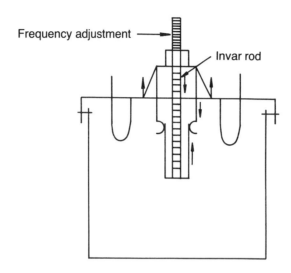

Figure 9.10   A practical cavity with temperature compensation and some flexibility in resonant frequency. The arrows indicate the direction of expansion of the components when heated.

ANTENNA CABLE FEEDERS

It is important that the cables connecting the base to the antennas (known as feeders) are low loss, particularly for receiver cables where any loss results in degraded performance. This factor is less significant for transmit cables because losses can be accommodated by adjusting the final P.A. gain. (Of course, there are limits to the compensation that can be achieved by brute power.)

The connecting cables are usually 7/8-inch, low-loss foam cables that have a loss of about 4 dB/100 meters. For short cable lengths (less than 30 m), it is practical to use 1/2 inch cable.

Generally, cellular systems have N-type connectors. Use only high-quality N-type connectors; cheaper units often have poorly mating pins, which can cause problems, especially when used for high-power transmit functions.

Because large cables are very awkward to handle, particularly when they are required to be connected directly to antennas or equipment, it is advisable to provide "tails" at both ends of the cable. Special connectors are made to join large cables into smaller ones via N-type connectors. Transition connectors are available in all sizes. To enable connections to be made easily, it is typical to provide approximately 2–3 meter "tails" at each end of the large feeders. Figure 9.11 shows how connecting "tails" are used.

You should use feeder cable clamps (on the tower), and you should use ground straps to directly ground the main cable outer conductor to the top and bottom of the tower as well as to the cable window. Seal all connectors used outside with a suitable sealing compound and then wrap them in insulation tape to seal out water (as an alternative, use a self-amalgamating tape). Tighten the cable window glands onto the sheath to prevent water entering the building.

Tighten connectors to firm-hand-tightness only. Both overtightening and undertightening can

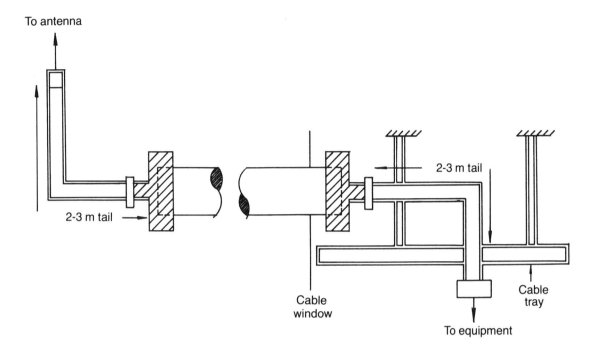

Figure 9.11   Small-diameter cables ("tails") can be used to connect equipment conveniently.

cause degraded connector performance, resulting in increased losses or even intermodulation problems.

ANTENNAS

Antennas are typically either 9 dB omnidirectional, or 17 dB or 14 dB, 120-degree or 60-degree, sector antennas.

Because diversity reception is frequently used, antennas should be mounted as shown in Figure 9.12 on the following page. Diversity results in an effective 6 dB improvement in the receive path where diversity combiners are used and 3 dB where switching diversity is used. The mounting arrangement shown in Figure 9.12 ensures acceptable isolation and diversity reception.

AIR CONDITIONERS

Air conditioners are an essential part of any mobile-telephone base-station installation. The heat output from the equipment is so intense that the equipment would soon overheat without

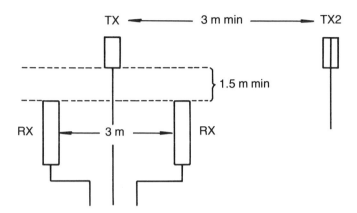

Figure 9.12   Antenna mounting

air-conditioning in all but the coldest climates. In very cold climates, reverse cycle units may be needed.

An air conditioner in its simplest form is a heat pump, that pumps heat energy from one region (the equipment rooms) to another region (the outside). Figure 9.13 shows the basic operation of an air conditioner.

In planning for reserve battery power, you should note that unless the air conditioners can also be run off the emergency power, the base station power amplifiers may exceed the base's operating temperature (usually around 60 degrees C) and protection circuits will close down the base during power failures.

Table 9.2 shows the typical heat loads for 25 watt bases.

The air-conditioning load is the sum of the components shown in Table 9.2. Notice that the use of two air conditioners (in case one fails) is a good idea.

Domestic air conditioners are ordinarily too small for cellular base station operation and units typically two to four times more powerful are required. Because of the high power consumption (typically 4 kilowatts), the air conditioner motors are often three phase. This can be a problem in some areas, so a single-phase alternative should be available.

To conserve space, wall-mounted air-conditioner units are preferred, but these can be difficult to find, particularly if a single-phase motor is required.

BASE TYPE	HEAT LOAD
Radio channels	180 watts/channel
Radio controllers	400 watts each
Rectifiers (100 A)	500 watts each
Exhaust fan	40 watts each
Microwave link	200 watts

Table 9.2   Heat loads for 25-watt bases. (Increase radio-channel load proportionally for higher outputs.)

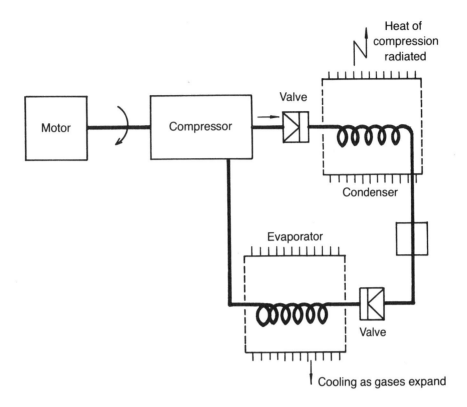

Figure 9.13 In a typical air conditioner, the compressor pumps coolant into a compression chamber, where the generated heat is radiated to the atmosphere. Upon release from the chamber, the coolant expands to produce cooling.

RECTIFIERS AND BATTERIES

Ordinarily, two or more rectifiers are provided on a load-sharing basis, with provision to load share in the event of a rectifier failure.

Batteries are needed to keep the equipment functional during power failures. Sealed batteries are popular because of their low maintenance cost and flexibility in mounting arrangement. You can save space with sealed batteries because the need to partition the battery room is eliminated. Although some installers are content to place wet cells in an equipment room, the practice is not universally accepted. The initial capital cost of sealed batteries is, however, somewhat higher.

Some operators insist on having two battery banks in parallel (with half the reserve capacity each) to ensure against base station failure in the event of the failure of one battery bank. A single cell or even a fuse can cause battery bank failure.

You can calculate the rectifier load and battery load from heat loads:

> Rectifier load = sum of heat the air-conditioning loads

> Battery load = rectifier load + air-conditioning load (if included)

> *where*   the air conditioners also run from a battery of approximate load = 2 x rectifier load (that is, most electrical energy is dissipated as heat so the air-conditioning load is approximately equal to the rectifier load).

Some manufacturers produce equipment to temperature specifications that are sufficiently high so that air-conditioning is not essential during power failures. This results in significant savings in both batteries and emergency plant costs. It, however, results in a substantial temperature cycle for the equipment, which has been shown convincingly to increase MTBF.

Battery ampere-hour ratings are usually quoted for a 10-hour discharge. A mobile base is ordinarily equipped with 2–3 hours battery reserve; at these higher discharge rates, the battery ampere-hour capacity is reduced by 10 to 20 percent.

AC power can be provided either as single phase or three phase. If a three-phase supply is used, the load can be distributed (as equally as possible over the three phases). This must be done carefully because most base-station hardware is single phase.

Rectifiers can be supplied in rack sizes compatible with the cellular equipment. Most equipment on the market, however, is not rack compatible. They usually come in modules of 25, 50, 100, or 200 watts (single phase), and somewhat larger in three phase.

Because rectifiers are relatively bulky (in terms of weight and floor space), plan ahead for the floor space needed. It should be possible to get about 200 amps single phase in one 600 mm rack, although this depends on the rack height. Plan your space carefully—for example, avoid buying rectifiers of about 100 amps that use just over half a rack height (" wasting" the rest).

EMERGENCY PLANT

It is often necessary to provide a diesel generator set to back up the base station. When such a generator plant is provided, it is possible to reduce the battery capacity to the point where it merely ensures continuous operation from the time of power failure until the generator starts. Ordinarily, the generator starts automatically when a power failure occurs.

Because it is costly to move a generator plant, it is best to purchase a unit that can run a fully equipped base station from the start. However, diesel plants do not perform well at partial load, and, unless expansions are foreseen within a few years, it may be necessary to plan to upgrade the generator plant at a future time.

Generators are usually rated in KVA. Unless the power factor is known, a figure of 0.7 should be used. Thus the generator rating, in KVA, is:

$$\frac{\text{WATTS}}{1000 \times \text{PF} \times E_f}$$

> *where*   WATTS = the total power consumed by the base station
> PF = power factor
> E_f = rectifier efficiency (typically 70 - 80 percent)

Such a generator consumes about 0.3 liters fuel/KVA.

You should provide a fuel tank sufficiently large to provide a one-week backup (this can be tailored according to the reliability of the local power supply).

Diesel fuel does not keep indefinitely; do not store it longer than six months.

Gravity feed tanks can be very dangerous in the event of a fuel-line failure, which could flood the equipment room, and are therefore not recommended. The preferred delivery method is via a pump. To ensure that sufficient fuel is available for starting the generator, however, place a small gravity feed tank in sequence with the main fuel tank. The main fuel tank is best placed underground with a pump for fuel delivery. Dual pumps with manual change-over are a good idea.

Because the base stations consume hundreds of amps of current, the DC distribution system must be properly designed. In particular, it is important to provide switches that can isolate each piece of equipment used. This isolation of the batteries, rectifiers, and equipment bays is imperative. Heavy-duty copper cables, carrying around 60–100 amps each, are usually used for power distribution, with each RF rack being individually supplied via a separate fused path equipped with a circuit breaker.

Use cables for rack wiring that are sufficient to carry the current safely. Table 9.3 on the following page shows the current capacity of various wire gauges. As the ambient temperature increases, a derating factor must be applied to the cable. Table 9.4 gives the appropriate derating. Table 9.4 assumes that not more than three separate conductors are placed in one cable or raceway. When more than three cables are bundled together, a further reduction in capacity occurs, as shown in Table 9.5.

AWG	MAX. CURRENT IN AMPS		
26	1		0.04
22	5		0.064
18	10	1.16	1.016
14	17	1.84	1.63
12	23	2.32	2.05
10	33	2.95	2.59
8	45	3.7	3.26
6	60	4.67	4.13
4	80	5.9	5.2
2	100	7.42	6.54
1	125	8.43	7.35
0	150	9.47	8.25
00	175	10.06	9.27
000	200	11.9	10.04
0000	225	13.4	11.68

Table 9.3   Copper cables and their dimensions and current carrying capacity at DC continuous rating at room temperature (30 degrees C)

TEMPERATURE (C)	DERATING FACTOR
40	0.82
45	0.71
50	0.58
55	0.41

Table 9.4   Correction factors for higher temperatures

CONDUCTORS IN ONE CABLE OR RACEWAY	DERATING FACTOR
4 - 6	0.8
7 - 24	0.7

Table 9.5   Derating factor for multiple bunched cables

Chapter 10

BASE STATION CONTROL AND SIGNALING

The air interface (or the interface between the base station and the mobile) is specified precisely for all systems. This precise specification allows different manufacturers to produce compatible systems. All systems use "handshake signaling," meaning that all instructions must be acknowledged. Most systems use data blocks and some error-correcting code.

Base stations have a control, or signaling, channel for signaling purposes. In some instances, the control channel is also available as a speech channel (when not used for signaling). Two such instances are NMT systems, which use a voice channel for signaling (control), and some AMPS/TACS systems, which use the redundant control channel as a voice channel when it is not required for control. Although some limited signaling is done on the speech channel, this is usually by out-of-band (or inaudible) tones. Control channels usually send instructions in a digital format, because this is faster than using analog tones and is more immune to simulcast-type interference.

The following section discusses an AMPS system for mobile-originated and terminated calls together, with handoff.

CALL TO MOBILE STATION

A call to a mobile station requires sending a page to the mobile, a response from the mobile, sending a ringing tone, and finally, connecting the voice circuit. This process involves a number of distinct steps, as illustrated in Figures 10.1. through 10.9.

First, the land party dials the mobile number. The PSTN forwards the number to the cellular switch for verification and forwarding to the mobile. The PSTN switch generates call progress tones. All mobiles receive the page call. See Figure 10.1.

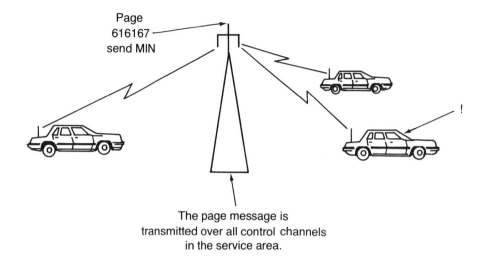

Figure 10.1   A paging call is sent.

Only the desired mobile responds by sending an identification that acknowledges it received the page. See Figure 10.2.

Next, the base station determines an appropriate free channel, turns on that channel, and transmits its SAT tone. See Figure 10.3.

The mobile now has to be directed to the appropriate channel and told which SAT tone to expect. See Figure 10.4.

When the mobile switches to the voice channel it automatically loops the SAT tone, which informs the base that the connection is complete. See Figure 10.5.

An alert order (a 10 KHz tone) tells the mobile that a call is on line. See Figure 10.6.

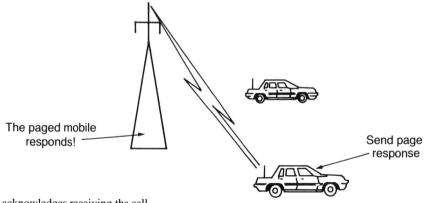

Figure 10.2   The mobile acknowledges receiving the call.

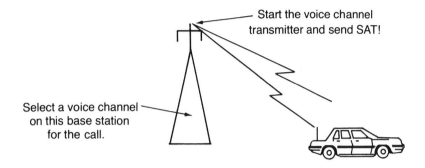

Figure 10.3   The base station selects a call (voice) channel.

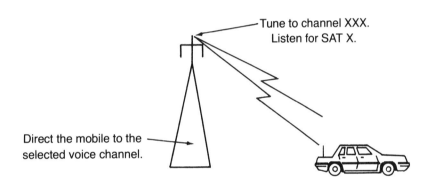

Figure 10.4   The mobile switches to the appropriate voice channel. The base informs the switch of the selected channel.

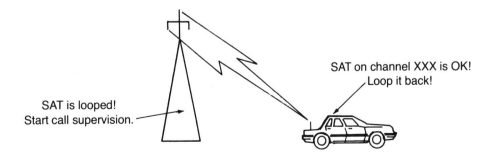

Figure 10.5   The looping of the SAT tone, which is generated by the base, received by the mobile, then sent back to the base, confirms that the mobile has arrived on the call channel. This process is time-supervised.

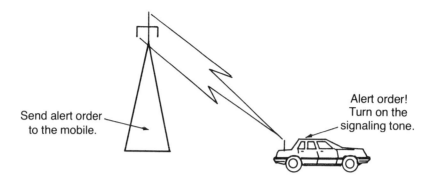

Figure 10.6 The ringing (call ready to proceed) tone is generated for the land subscriber by the switch and locally in the mobile for the vehicle unit.

The ringing tone is then generated locally at the mobile and sent back to the network by the mobile switch. See Figure 10.7.

When the handset is lifted, the mobile loops the SAT tone, telling the base to stop the ringing tone and that a normal call is in progress. See Figure 10.8.

During the conversation, call supervision occurs for the entire duration of the connection to ensure quality and check the continuity of the call. See Figure 10.9.

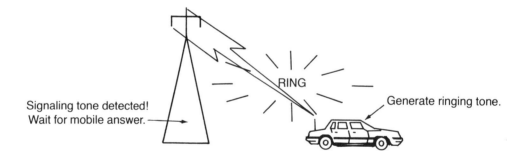

Figure 10.7 Signaling is done via a 10 KHz tone burst from the mobile to signify that its ring signal has been started.

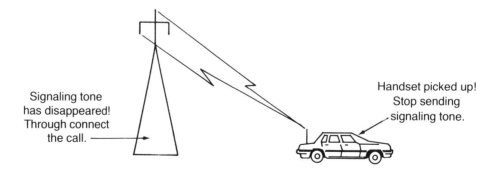

Figure 10.8 The looping of the SAT tone indicates that the mobile has answered.

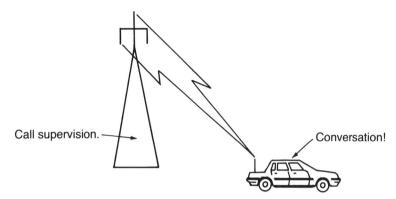

Figire 10.9   The call undergoes constant supervision.

MOBILE-ORIGINATED CALL

A mobile-originated call proceeds much the same as a terminated call, except that this time the mobile calls the base. The mobile knows if it is out of range and does not attempt to send if there is no control channel present, although it continues to scan for the presence of a control channel.

Periodically, the base station transmits overhead information that, among other things, defines what the mobile should send for identification. Particularly where encryption is used, it may not be desirable to send all the identification information because some of it is used as part of the encryption "seed."

Because all mobiles use the same control channel to request a call, the possibility of two mobiles attempting to call simultaneously must be allowed for. Should an originated-call attempt clash with another mobile (and cause data corruption), a built-in algorithm ensures a retry attempt at a time that (hopefully) is different from that of the conflicting mobile.

The steps involved in a mobile-originated call are as follows:

The mobile subscriber enters the desired number and then presses the send button. The call can then be sent almost instantly on the correct channel because, in idle mode, the mobile regularly scans for the best control channel. See Figure 10.10.

At first, the mobile sends its identity (MIN2 plus, if requested, MIN1 and its serial number). The MIN2 is associated with the subscriber's telephone number, and it may, in fact, be the

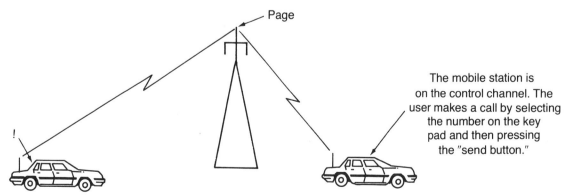

Figure 10.10 The mobile is in "listen mode" when idle; it listens out on the best control channel. The subscriber initiates a call by pressing the send button.

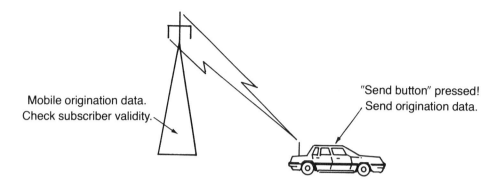

Figure 10.11 The base station communicates the subscriber's information to the switch, which checks the validity of the subscriber. The base station assigns a free voice channel.

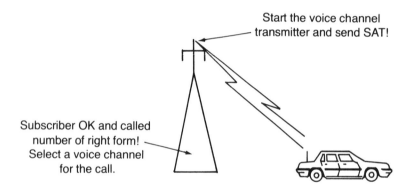

Figure 10.12 The SAT tone is generated locally at the base station to distinguish the base from others using the same frequencies.

subscriber's number (but not necessarily). This information is checked for validity at the switch. The called number is also sent. See Figure 10.11.

Next, the base selects a voice channel and sends a SAT tone, as in the case of a terminated call. See Figure 10.12.

The mobile is then instructed to switch to the designated voice channel. See Figure 10.13.

The looping of the SAT tone confirms that the mobile is on frequency. See Figure 10.14.

The call is connected and the network ringing tone is received until the call is answered. Then call supervision takes place. See Figure 10.15.

CALL SUPERVISION

Call supervision lasts for the full duration of all calls. The call quality is "sampled" by a scanning-signal measuring receiver, which samples each of the active channels. This process takes

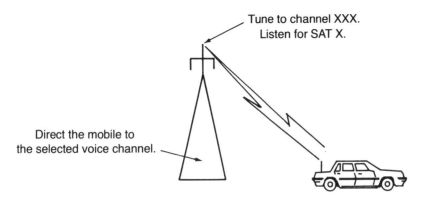

Figure 10.13 The mobile switches to the speech channel

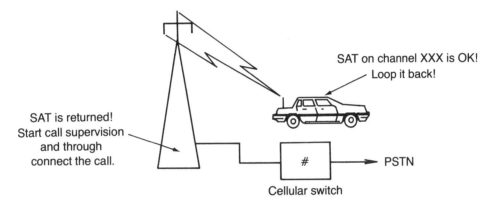

Figure 10.14 .The mobile loops the SAT tone, confirming its presence.

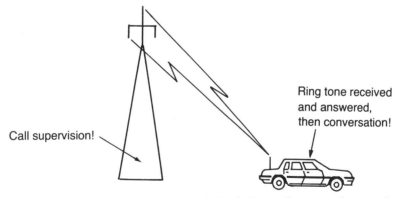

Figure 10.15 Tones to the mobile are generated by the switch to indicate call progress (for example, ringing tone, busy tone, and number unobtainable).

about 50 milliseconds per channel. The channel itself is monitored for interference by constantly monitoring the presence of foreign SAT tones.

If necessary, a handoff request is made. Figure 10.16 on the following page shows the call-supervision process.

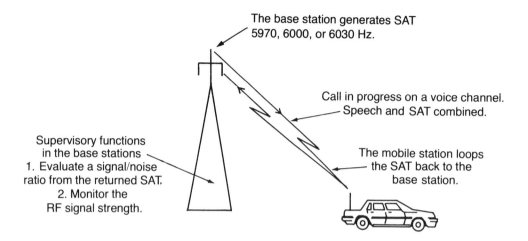

Figure 10.16 Active subscribers are logged on to a lookup table at the base station. A dedicated scanning receiver monitors for S/N performance, RF field strength, and the presence of the correct SAT tone. When the subscriber hangs up, a 10-KHz tone burst from the mobile signifies release of the call.

SETTING UP A CALL BETWEEN TWO CARS

The call-setup procedure for calls between cars is very similar to that for mobile-to-land-line subscribers where an external telephone exchange is involved, except that the connections are made via the incoming/outgoing interface. That is, the call is completely handled by the mobile switch. Note that the switching process described in this section is for an AMPS system, but the basic concepts are the same for all systems.

A call request from the car in cell A is sent via the access channel (a radio channel dedicated to setting up calls from mobiles) to the terminating circuit. This circuit separates out control signals (which are sent to the controller) from voice signals (which are handled by the switch).

The terminating circuit sends the request for a channel to the controller, which examines the A number (the calling party's telephone number) to determine the nature of the caller.

The controller checks the following:

• Is the caller local or a roamer?
• What is the category (barred international calls, barred outgoing calls, and so on)?
• Are there any special facilities (call forwarding, call answering, and so on)?

Assuming the call is legitimate, it is marked as busy to prevent conflict with other calls that might be attempted to the A party.

The B party (or number called) is then analyzed, usually only to the first three or four digits to determine the charging, number length, and routing. The B party category is also analyzed to confirm that the called party is currently a valid subscriber.

Next, a check is made to see if the B party is free; if it is, the line is marked as busy.

The B party is then called over the control channels (from all bases, because the location has not yet been determined) via the terminating circuit.

When the B party mobile acknowledges, the most appropriate cell is selected. (The B party has already designated the best cell by choosing the strongest control channel.)

The B party is then assigned a voice channel and instructed to switch to it.

The B mobile switches to the voice channel and acknowledges its presence. This acknowledgement is sent to the controllers.

Similarly, the A party is assigned a voice channel. The controller then instructs the switch to connect the two voice channels.

Call supervision now takes place via the terminating circuit to the controller. Call-quality assessment, as judged by S/N, absolute signal strength, and detection of interference occurs throughout the call; necessary handoffs to adjacent cells are attempted if problems occur.

At the same time, the system is monitoring for possible cleardown signals (end of call indications) from the A or B party (which, of course, terminates the connection).

HANDOFFS

Handoffs are initiated when the signal, as monitored at the base station by the signal-level scanning receiver, is deemed to be below a certain quality. This quality may be judged simply on signal strength or it may involve S/N measurements, or both.

At the base station, in all systems, there is a receiver, the sole function of which is to scan all mobiles in use and to monitor the signal from each. The scanning time is usually about 50 milliseconds per channel. This scanning receiver can be a standard voice channel transceiver or a dedicated scanning receiver. A dedicated unit is usually cheaper than a transceiver (which will have an unused transmit section), but it also requires dedicated spare parts for backup.

When a mobile is judged to be below a predefined standard (and this is a user-definable parameter), the base-station controller requests that the switch attempt a handoff.

The switch refers to its lookup table of sites that are adjacent (or neighboring) to the site requesting the handoff. The switch then asks each of these sites to scan the transmit frequency of the mobile in question and report its field strength (or S/N, or other quality). The returned parameters are then compared to those of the existing base; if they are better by a defined amount (usually 3 dB), then a handoff to a free channel on the preferred base is initiated.

The mobile then has a set period (typically 50 milliseconds) to report on the new channel.

The scanning channel measuring receiver first detects a substandard signal path and then requests a handoff from the switch, as shown in Figure 10.17 on the next page.

The switch now refers to its lookup table to determine which base stations are adjacent to base

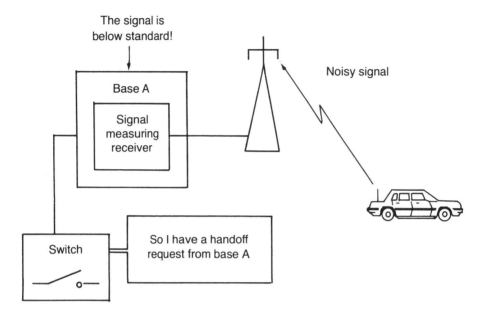

Figure 10.17 A noisy path is detected by the scanning signal measuring receiver and the switch is informed. A handoff is requested.

A and therefore candidates for a handoff. Because this process consumes considerable processor time, it is typical to limit the "neighboring" bases to only immediately adjacent ones.

The neighboring bases are then asked to tune their measuring receivers (which can scan any channel) to the channel used by the mobile, and then to report back on the signal level. Figure 10.18 shows neighboring bases reporting signal levels.

Each base station then sends back a report on the received level; the switch attempts a handoff if the next best base is better than the original signal by a predefined level (a system parameter). Figure 10.19 illustrates this process.

If one of the bases reports a sufficiently high level, then a handoff is attempted and the mobile is requested to report to the new base on a new channel. In Figure 10.20, base C has the highest field strength and so is the station chosen for handoff.

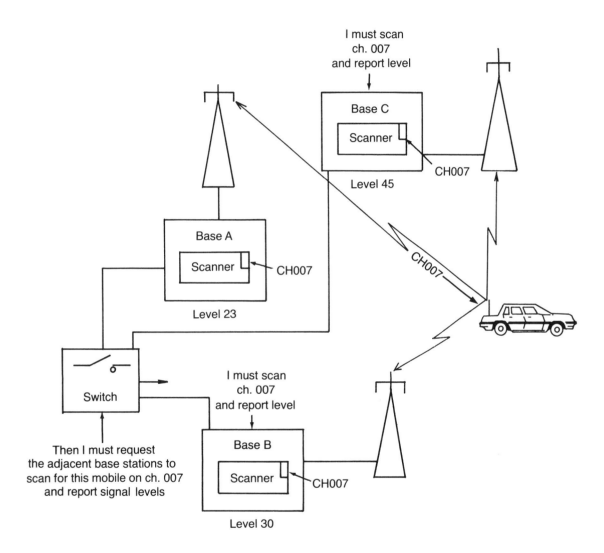

Figure 10.18 Neighboring bases are requested to tune to the mobile frequency and report on the received signal level.

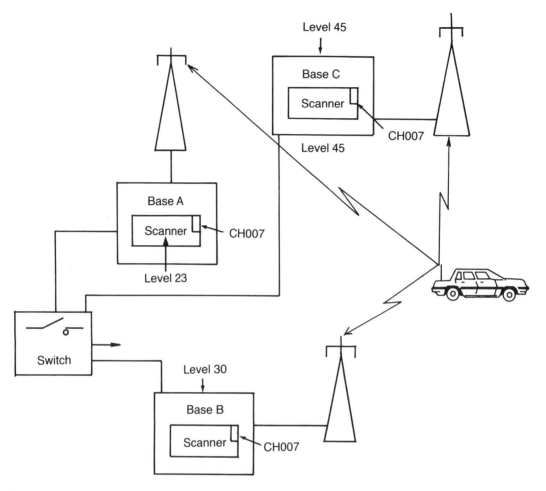

Figure 10.19 The bases send back a report on the received signal level from Channel 007.

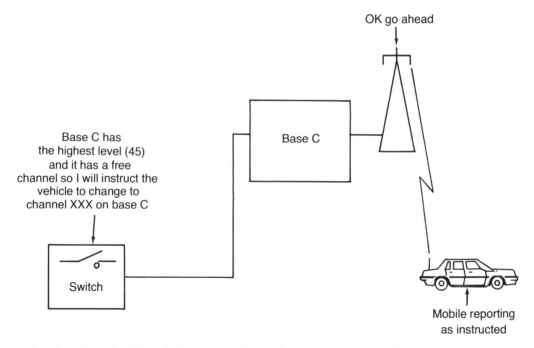

Figure 10.20 Base C has the highest field strength so the mobile is instructed to handoff to that base. After reporting, the conversation continues as before.

AMPS SIGNALING FORMAT

The signaling on the air interface is achieved by sending binary-coded-hexadecimal code in bit streams called words. A word sent to the mobile consists of 28 bits; a word from the mobile contains 36 bits. An additional 12 bits of error correction code is added, resulting in word lengths of 40 and 48 bits, respectively. For security, each word is repeated a number of times.

A word train begins with dotting, a term used to describe the signal sent to synchronize the mobiles. Dotting consists of the digits 1010..10. This is followed by a word synchronization code of 11100010010, and then by two words.

A mobile can request a connection to a given number by sending the information shown in Figure 10.21.

The message is preceded by the dotting and word synchronization and the whole message is repeated five times. A 16-digit number is assumed.

A forward control-channel stream (a message from the base station to the mobile) takes much the same form. The general form of such a message is shown in Figure 10.22 on the following page. The word is repeated five times to ensure the correct transfer of information that might otherwise be lost in a multipath environment.

Figure 10.23, also on the following page, shows a typical word sent by a base station for the case of a voice-channel assignment.

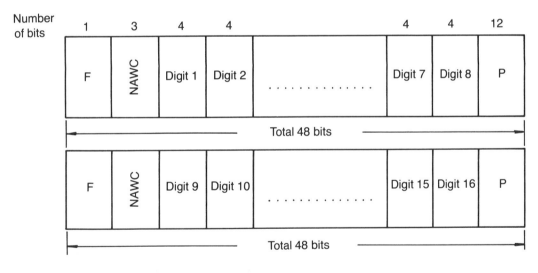

where   F = first word in the field
        NAWC = number of additional words coming
        P = parity bits

Figure 10.21 The format of the message from a mobile requesting a network number is two words of 48 digits each, describing a 16-digit number. This information is repeated five times.

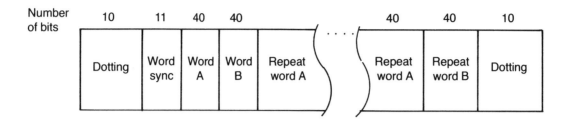

Figure 10.22 The general form of a forward message includes the dotting and word synchronization followed by the word repeated five times.

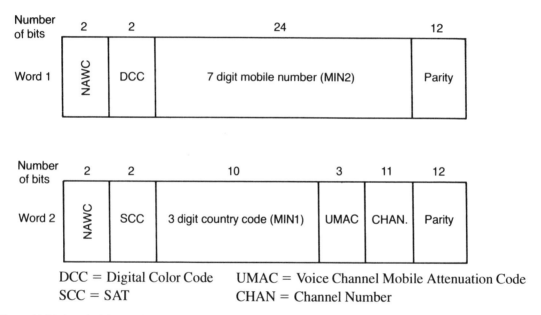

DCC = Digital Color Code   UMAC = Voice Channel Mobile Attenuation Code
SCC = SAT   CHAN = Channel Number

Figure 10.23 A typical forward control-channel message

SIGNAL STRENGTH PARAMETERS

There are a number of user-defined system parameters that determine the behavior of a base station. Table 10.1 shows some of the most important system parameters and typical values for them.

Other parameters that can be set include SSHY, which sets the hysteresis for a handoff attempt (usually about 3 dB, which means the alternate base must be at least 3 dB better before a handoff is attempted), and SUH, which is the supervision time for a call before the handoff is declared unsuccessful.

Many systems also measure signal-to-noise (S/N). Two other parameters SNH (signal-to-noise

SIGNAL STRENGTH	PARAMETER
- 72 dBm	SSD - mobile power decrease
- 82 dBm	SSI - mobile power increase
- 95 dBm	SSH - mobile handoff level
-110 dBm	SSB - block channel level

Table 10.1   A number of user-defined levels are available to customize the coverage and behavior of a base station.

before attempting a handoff, usually around 25–30 dB) and SNR (signal-to-noise before the call is released, usually around 12–15 dB) are also very useful.

Chapter 11

CELLULAR REPEATERS

Because cellular base stations are expensive, repeaters are sometimes used to fill in small areas not properly covered by the base-station network. Cellular repeaters are cheaper than base stations and can be very useful, but their limitations must to be understood. A number of cellular-radio repeater systems are now available. They are useful in low-density isolated areas. This chapter discusses typical applications for repeater systems.

There are basically two types of repeater: the broad-band repeater (also called the cell-extender repeater) and the cell-replacement repeater. These repeaters are shown in Figure 11.1. The broad-band repeater amplifies all the channels in, for example, the A or B band. To have reasonable total power consumption and isolation, this repeater must have low power per channel.

The high-power cell-replacement repeater amplifies a few channels only. Frequency translation occurs so that the repeated cells are on a different frequency from the main base. Cell-replacement

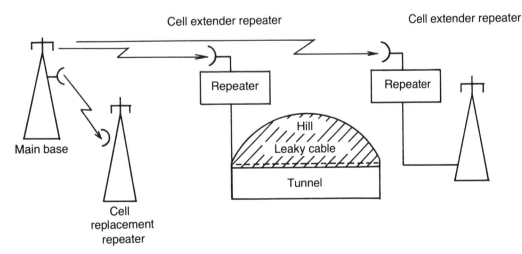

Figure 11.1   A cell with a cell replacement repeater and two fill-in cell extenders, one behind a hill and another feeding a leaky cable in a tunnel

repeaters are used along highways where the extended coverage is not expected to attract high traffic volumes.

CELL-EXTENDER REPEATERS

The cell-extender repeater allows the coverage of a cell to be extended cheaply in areas adjacent to the main cell that may not otherwise be covered. No modification is required at the original cell site. This extender is designed to cover relatively small areas, typically about 10–30 km.

As shown in Figure 11.2, this type of repeater is simply a wide-band amplifier that amplifies and repeats the host base-station channels. The cell extender typically costs only about 10 percent of the cost of a new cell site, but it has some serious limitations.

Using a wide-band amplifier makes it difficult to have a really sharp cutoff at the edge of the band. Figure 11.3 shows that the decision to place the control channels for the A and B bands adjacent to each other clearly was done without regard for this type of repeater limitation.

In Figure 11.3, you can see that with practical filters, either the gain of the system over the control-channel band is not consistent or the adjacent control channels are also amplified.

A filter that causes differential control-channel amplification severely restricts the cell-pattern flexibility because only cells with control channels, the frequencies of which are close to the voice channels, can be repeated.

If a filter that amplifies all the control channels equally is used, then problems can occur in the

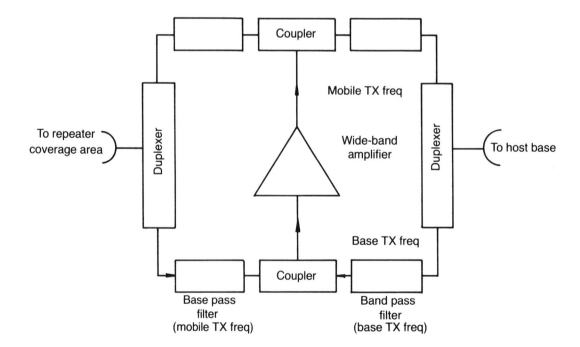

Figure 11.2   A simple wide-band repeater (cell extender)

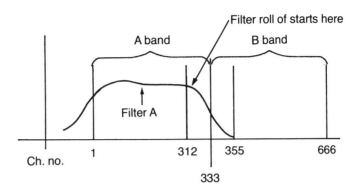

Figure 11.3   Repeater filters have cutoffs asymmetrically across the control channels. This band-pass filter cuts off at the edge of the speech channels.

adjacent band. For example, if such a filter is used on the A band, as shown in Figure 11.4, then the adjacent B band control channels are amplified to some extent. Amplifying the A band control channels causes many unsuccessful call attempts to occur on the A band system because the repeaters repeat the control channels but not the voice channels. (This applies only to incoming calls that use the control channels for paging.)

Another problem in a simple repeater system is multipath cancellation on the control channels. Figure 11.5 on the next page illustrates this problem. Destructive interference that renders the area unserviceable can occur over the area where the direct path from the host base and the repeated transmission have field strengths within ±3 dB. This is a problem mainly on the data channels; it is recognized by the existence of areas where calls cannot be reliably established or received, but where voice communications, set up elsewhere, are available.

Because the repeater amplifies its own input, it is a potential oscillator, as shown in Figure 11.6 on the next page. In practice, the isolation between the receive and transmit antennas must exceed the gain by a margin of at least 10 dB, to ensure stable operation. That is, the total isolation between the input and output must be 10 dB greater than the gain. Unity gain or 0 dB isolation is needed for oscillation.

This isolation is usually provided by high-gain antennas with good front-to-back ratios, plus

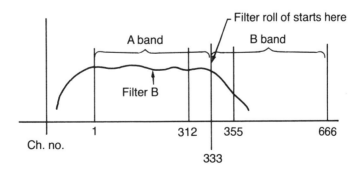

Figure 11.4   Filters can give a flat response over the desired control channels. This band-pass filter is flat over the band of operation, but significantly amplifies the adjacent control channels.

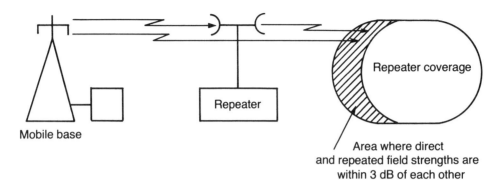

Figure 11.5   Destructive interference of control channels can occur in cell extender repeater systems.

vertical and horizontal antenna separation. Repeaters usually have a variable gain, adjustable over a wide range.

In order to get reasonable power levels from the repeater, it is necessary to operate fairly close to the maximum-gain margin (10 dB). Gains of about 60 dB are typically used. Instability can be caused by the external environment, however, when such things as passing aircraft or wind in the trees alter the feedback levels.

You can calculate the antenna isolation using the following formula:

$$\textit{Vertical isolation} = 25 + 40 \log{(2.8\ d)}\,dB$$

$$\textit{Horizontal isolation} = 22 + 20 \log{(2.8\ d)}\,dB$$

where   d = spacing in meters

Figure 11.7 shows the antenna locations and the way in which the spacing (d) is measured.

Usually, although not necessarily, the repeated cell covers a relatively small region that can be served by a directional antenna. Where an omnidirectional antenna must be used, the isolation is decreased and the maximum power at the repeated site is limited (due to the oscillation margin).

The maximum gain that can be used at the repeater is as follows:

$$\textit{Maximum Gain} = \textit{Isolation} - \textit{Oscillation Margin}$$

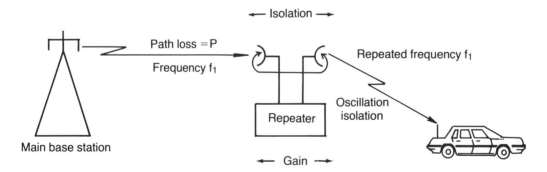

Figure 11.6   A repeater is a potential oscillator. Cell extenders are potential oscillators as the input signals are amplified and retransmitted without frequency translation.

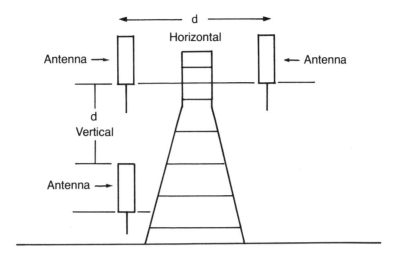

Figure 11.7 Vertical isolation is measured as the smallest distance between the two antennas. The distances are measured as indicated.

As mentioned earlier, 10 dB is a practical value for oscillation margin.

To calculate the radiated power, start with the ERP at the base; for example:

$$50 \; watts = 47 \; dBm$$

$$Subtract \; path \; loss \; P = -P$$

$$Received \; signal = 47 - P$$

$$Maximum \; repeater \; gain = Isolation - 10 \; dBm$$

$$Maximum \; ERP \; at \; repeater = 47 - P + Isolation - 10 \; dBm$$

This isolation is mainly determined by the antenna selection and location.

Intermodulation is always a problem with wide-band amplifiers and a well-designed repeater minimizes intermodulation products. Because of the wide separation between the send-and-receive frequencies of cellular systems (usually 45 MHz), using separate amplifiers in the send-and-receive directions can contribute significantly to lower intermodulation.

Because of the bandwidth, gains, and number of channels, the output power of a cellular repeater is typically limited to about 0.25 watts per channel. This is adequate for many repeater applications.

ENHANCED CELL EXTENDERS

Because most of the limitations of the simple repeater adversely affect the control-channel operation more than the voice channels, an obvious refinement is to separate out the control channel and, using a narrow-band amplifier, process it separately. The enhanced-cell repeater is a refinement of the simple repeater. Using frequency translation on the control channel only can eliminate the interference problem previously discussed. This is done in modern repeaters.

Figure 11.8 on the following page shows an enhanced-cell repeater.

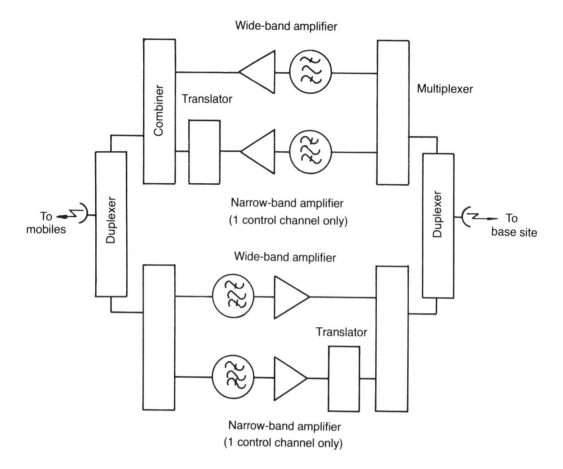

Figure 11.8   In an enhanced repeater, the control channel can be filtered and translated in frequency. The send-and-receive channels are independently translated.

TRAFFIC CAPACITY OF THE SIMPLE REPEATERS

Because this type of repeater repeats only those channels required for calls, the base station cannot distinguish between direct traffic and repeater traffic. Thus, the marginal traffic capacity of the repeated cell is added arithmetically to the local traffic. In this way, the base station is equipped with enough channels to carry the cell traffic only. It is not so simple with extended cells.

CELL-REPLACEMENT REPEATERS

The cell-replacement repeater operates quite differently from a simple repeater and requires a hardware and software interface with the main base and switch.

Figure 11.9 shows the cell-replacement repeater. In this instance, the main base can act as a normal base or, under the control of the extender controller, a particular channel can be used as a link channel to the repeater base. Seen from the view point of the mobile switch, the repeated channel is frequency agile and can change from the normal frequency to the repeated frequency.

This system avoids interference from the main base to the extended base because different channels (frequency translation) are used at each base. The repeater can therefore use high-power omnidirectional antennas. Typically, the repeater channels are 10 watts ERP, as shown in Figure 11.9 below.

The cell-replacement repeater is useful when additional coverage of a large area with few subscribers is anticipated.

The repeater base consists of two back-to-back repeaters; it performs much the same as a normal base station, but it needs no link system and no controller. This is cheaper than a full base station only for small cells, as you can see from comparing the equipment required in Figure 11.10 and 11.11 on the following page.

For small numbers of channels to be repeated, the cost of an extra transceiver is less than the cost of a link-plus controller. However, the cost per channel of the repeater (requiring two transceivers/channel) rises faster with additional channels than does the cost of adding channels at a base station.

For example, suppose you have the following costs:

- A transceiver costs $8000 (including combiners and antennas).
- A link costs $9000.
- A controller costs $30,000.
- An extender-controller costs $10,000.
- Assume other infrastructure costs are equal (batteries, building, towers, and so on).

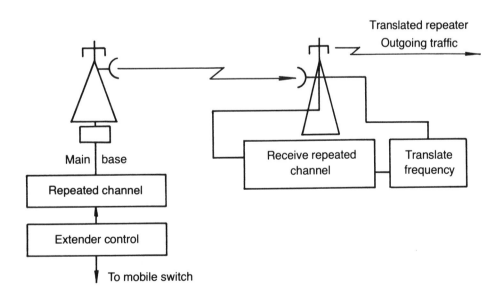

Figure 11.9   Cell-replacement repeaters are useful for repeater operations over wide areas.

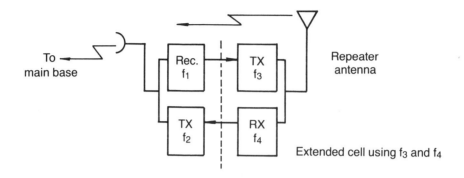

Figure 11.10 A cell-replacement repeater is basically two back-to-back transceivers that allow frequency translation. No site controller or link system is necessary.

The cost of the base station is:

$$90 + 30 + 8 \times N = 120 + 8\,N \text{ \$1000}$$

where   N = number of channels

The cost of a repeater is:

$$(8 + 16) \times N \ (2 \ repeaters \ / \ channel + original \ channel) + \$10,000$$

The costs are equal when:

$$120 + 8\,N = 24\,N + 10$$

(that is, N = 7 channels)

Because dedicated repeaters are usually available in modules of seven channels (at about 10 watts each), it is reasonable to assume that this is the upper limit of the economic viability of this type of repeater. However, there are many applications where seven channels is quite adequate.

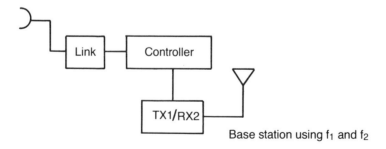

Figure 11.11 A base station consists of a transceiver, a site controller, and links to the switch. The controller sends different channel assignment instructions to the mobile depending on whether the channel f_1/f_2 is directly accessed or is accessed through the repeater.

TRAFFIC CAPACITY OF CELL-REPLACEMENT REPEATERS

The traffic capacity of the cell-replacement system presents some interesting problems. Imagine channels that are extended as last-choice routes (channels that are extended and used only for local traffic when all non-extended channels are fully occupied). Figure 11.12 illustrates this situation.

The optimum number of channels, K, can now be found using the conventional traffic theory for a main route with one overflow path. The optimum number of channels, K, may exceed the traffic capacity required just to serve the extended base mobiles. This depends strongly on the relative costs of the alternative routes, but in 1989, little or no overflow could be justified. Be sure to calculate this extra capacity when comparing the cost of a repeater with the cost of an additional base.

Figure 11.13 on the following page shows a cell extender as an alternate routing system. (For more information about routing, see Chapter 16, "Traffic Engineering Concepts.") From Figure 11.13, it can be seen that:

Cost of the main route $= \$ 8,000 \, / \, channel = C_1$

Cost of the overflow route $= \$ 24,000 \, / \, channel = C_2$

Occupancy of the main circuits $\approx 0.7 = B_1 \, Erlangs \, / \, circuit$

Occupancy of the overflow circuits $\approx 0.5 = B_2 \, Erlangs \, / \, circuit$

$$Cost \, factor \, H \approx C_1 \left[\cdot \frac{B_2}{C_1 + C_2} \right] = \frac{8 \, x \, 0.5}{(8 + 24)} = 0.125$$

A cost factor of 0.125 suggests that almost no overflow is justified, so it would be just as economical to dedicate the repeater channels at the base to repeater use only. If the cost per channel of the repeater were lower, however, this might change.

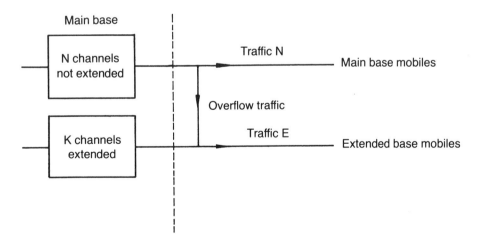

Figure 11.12 A system that uses extended channels only for local traffic when all non-extended channels are used

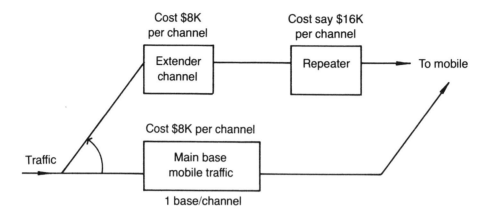

Figure 11.13 A cell extender as an alternate routing system

TUNNELS

In the case of a leaky cable feeding a tunnel or building, the isolation is very high, and therefore very high-gain (high-power) repeaters can be used.

Chapter 12

ANTENNAS

Antennas play an important role in shaping cell patterns and in determining reuse patterns. There are two basic antenna types. The first type has an omnidirectional pattern; the second has a sector, or unidirectional, pattern.

OMNIDIRECTIONAL ANTENNAS

An omnidirectional antenna is usually a collinear dipole (a number of dipoles in a line with a phasing harness). Figure 12.1 on the following page shows an omnidirectional antenna.

In an omnidirectional antenna, a power divider may be required to phase a number of dipoles within the one gain antenna or to connect two antennas to the one feed-line. The quarter-wave transformer shown in Figure 12.2 is a simple power divider. Because the power divider is fed inside the antenna, the internal wiring harness is quite complex. A cellular omnidirectional antenna is usually of this kind, inasmuch as it has a good wide-band performance. The matching section of the quarter-wave transformer has a characteristic impedance of $Z_o = \sqrt{Z_{IN} \cdot Z_i}$

The unit shown in Figure 12.1 is an 8.5 dB collinear antenna constructed with a fiberglass radome. Extreme care must be taken with the choice of materials used in the antenna, as dissimilar metals can cause corrosion, which in turn leads to intermodulation problems.

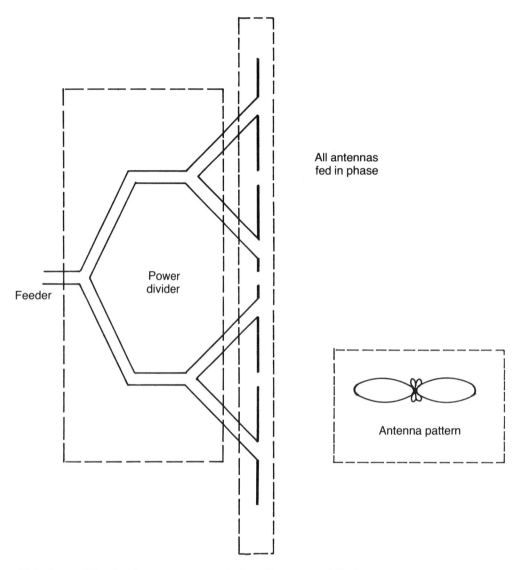

Figure 12.1   An omnidirectional antenna constructed of a collinear array of dipoles

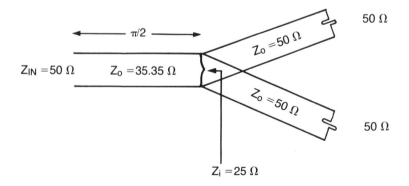

Figure 12.2   A 50-Ω power divider

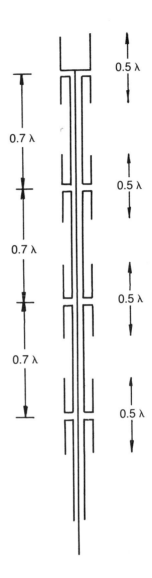

Figure 12.3   A collinear dipole with passive radiators. It is possible to drive only the top radiator and to use induced currents to derive the required pattern. This principle is similar to that of the familiar Yagi.

Simpler, more compact collinear antennas can use passive radiators, as shown in Figure 12.3. Such antennas are widely used in two-way radios because they are much cheaper to construct than the type illustrated in Figure 12.1. However, they operate only on a narrow bandwidth.

SECTOR ANTENNAS

Sector antennas usually have higher gain than omnidirectional antennas and are typically 14 dBd or 17 dBd. Sector antennas combine the power gains obtained by using phased arrays with the additional gains obtained by using reflectors. Figure 12.4 shows a sector antenna with 17 dBd gain. Such an antenna is, in fact, constructed with the omnidirectional radiator (see Figure 12.1) mounted in front of a reflector. This configuration has the advantage of needing only one set of

Figure 12.4   This sector antenna has 17 dBd gain and an overall length of 3.76 meters. Photo courtesy DELTEC.

spares (an omnidirectional radiator) to service both omnidirectional and sector antennas. Table 12.1 lists the specification for the antenna shown in Figure 12.4.

Figure 12.5 shows that the VSWR varies considerably over the rated bandwidth but always is less than 1.5. The maximum gain, however, is vertically flat at 17 dBd over the rated frequency ranges. The radiation pattern illustrated in Figure 12.5 was derived from a computer simulation.

Type number	MTA860-8-UN
Frequency	825-896 MHz (to 860 MHz)
Bandwidth	71 MHz
Input impedance	50 ohms
VSWR	Less than 1.5:1 over 71 MHz
Gain	17.0 dBd
Vertical beamwidth	6.5 degrees
Horizontal beamwidth	60 degrees
Maximum power	500 watts
Polarization	Vertical
Termination	'N' type plug fitted to 50 cm of PTFE coaxial cable
Reflector screen	All-welded alloy finished Alocrom 100 pretreated and coated in white polyester

Table 12.1   Specification for 17 dB sector antenna

(continued)

Table 12.1   *(continued)*

Mounting section	300 mm x 48.5 mm x 7 swg aluminium grade 6082 finished Alocrom 100
Overall length	3.76 meters
Weight (incl. mounting clamps)	18.2 kg
Wind area	0.734 square meters
Wind loading	107.5 kgf at 160 kph
Mounting clamps	2x9099 galvanized steel parallel clamps
Packing	Case with timber ends and sides; hardboard top and bottom panels

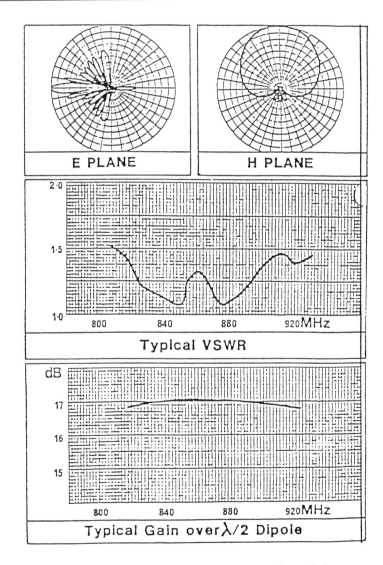

Figure 12.5   Radiation patterns, VSWR, and gain of the antenna shown in Figure 12.4

Figure 12.6   A simple corner reflector can provide gains of about 8 dBd in a very compact unit. Photo courtesy DELTEC.

When measured on the test range, real antennas show anomalies in their patterns that are not suggested by the simulations.

In places where high gain is not necessary, a smaller and simpler reflector can be used. Figure 12.6 shows a simple corner reflector. Such reflectors are made from unity gain radiators mounted in front of a radiator. The main application for this type of antenna is covering small localized regions such as tunnels and parking garages.

POLARIZATION

Electromagnetic waves propagate with the polarization defined by the relationship between the electric and magnetic field vectors. Cellular radio (and all VHF/UHF land mobile systems) uses vertical polarization, which means that the electric vector (E-plane) is vertical, as is the orientation of the antenna. Figure 12.7 illustrates the polarization of cellular systems. The magnetic field (H) oscillates in the same plane as the direction of propagation.

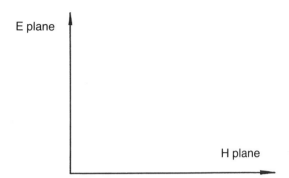

Figure 12.7   Polarization of cellular systems

ANTENNA MATERIALS

Because antennas are exposed to the elements (sun, rain, ice, smog), the choice of materials is critical. All metals must be electrolytically compatible or else local corrosion cells will form. Joints with corrosion are potential intermodulation sites. Cellular omnidirectional antennas usually have about 9 dBd gain and are of collinear construction, encapsulated in a fiberglass radome.

The radome must also be of high-grade fiberglass because water leakage can lead to corrosion of the elements and the ultimate failure of the antenna. Many fiberglass products, however, have metallic additives that make them unsuitable for antenna construction.

MOUNTING

Cellular antennas must be properly mounted to function effectively. Omnidirectional antennas should be mounted on top of the tower or building, because side mounting can seriously distort the original omni pattern.

Because cellular antennas are high-gain devices with a very narrow E-plane pattern, even small deviations from the vertical can very seriously effect coverage. The antennas should be vertical to $\pm 1°$.This variance is usually much less critical in land mobiles, where base-station antennas rarely have gains greater than 6 dB.

Unidirectional antennas can be side mounted. Side mounting can improve the front-to-back ratio by using the mounting structure to decrease sensitivity to interference from the backward direction. In most big cities it is a good idea to mount unidirectional antennas on a mounting that allows for the future downtilt of the antenna. Figure 12.8 shows such a mounting.

The connection between the main feeder and the antenna should include a tail of 2–3 meters, as shown in Figure 12.9. The tail makes installation easier and is important for future maintenance, when it may be essential to replace an antenna with one of a different type.

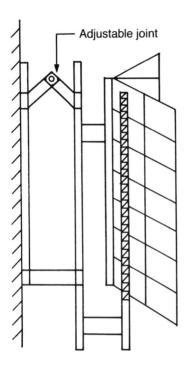

Figure 12.8   A mounting that allows for antenna downtilt

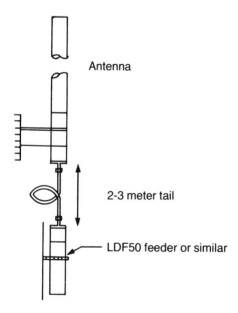

Figure 12.9   The connection between the feeder and antenna should be by a 2–3 meter tail.

Frequency	825–896 MHz
Bandwidth	71 MHz
Input impedance	50 ohms
Gain	8.5 dBd
Maximum power	500 watts
Half Power bandwidth	6.5°(E plane)
Wind area	0.175 square metres

Table 12.2   Specifications for an omnidirectional antenna

Table 12.2 lists the typical specifications of an omnidirectional cellular antenna.

DRAINAGE

Drain holes must be clearly identified and correctly oriented. Because even the best-constructed antennas will suffer some water leakage, antennas are usually fitted with small drain holes. These holes are located at the lowest points of the antennas. This is an important factor when considering mounting the antenna upside down (as is often done in cellular installations). If the antenna is not designed to be mounted this way, there will be a drain hole only in what was the bottom (and is now the top). In these cases, it is necessary to seal the existing drain hole and carefully drill a new hole in what was the top (and is now the bottom).

INTERMODULATION

When two signals of a different frequency mix in a non-linear device (for example, a rusty wire, loose joints on towers, and rusty fences), the result is intermodulation. Intermodulation can be a problem at any site that has two or more transmitters. Cell sites, which often have co-sited paging, trunked radio, and other mobile services, are likely to experience intermodulation. This problem appears as interference to other (nearby) mobile service users, or it can introduce interference into the cellular operator's equipment and cause blocking.

Finding the intermodulation source can be very time-consuming and usually proceeds by eliminating likely offenders by trial and error. Because the problem is often intermittent, it can be very frustrating to track down.

Rusty bolts have long been blamed as the major culprits, but recent tests have revealed that the worst offenders are galvanized mild-steel rope, mild-steel chains (the very worst offenders), and mild-steel wire fences.

Loose joints with small areas of contact are the main intermodulation points. Large areas of corrosion, such as on decking or galvanized iron sheets, may not produce high levels of intermodulation.

MEASURING VSWR

Antenna VSWR's should be in the range of 1 to 1.7, and each antenna should be checked upon installation. Most base stations have a built-in VSWR meter, which can be switched to each channel.

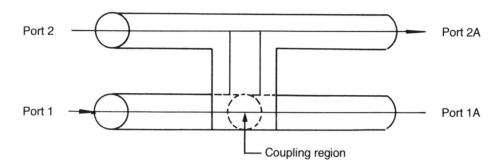

Figure 12.10 A simple directional coupler

To measure VSWR, it is necessary to distinguish between forward signals and reflected signals. In a VSWR meter, a directional coupler is used to selectively read the signal in either direction. If two couplers are used, it will be possible to get a direct reading of VSWR by using some simple circuitry to derive the ratio of the forward and reflected signals. Figure 12.10 shows a simple directional coupler.

A directional coupler relies on placing the sensing loop so that the induced currents from the electric and magnetic fields are equal. The induced currents resulting from the electric field are indifferent to the direction of that field. The magnetic fields, however, are of opposite phase in the forward and reflected waves and will thus cancel the electric field in one direction. By specifying port 1 or port 2 as the sensor, a directional coupler can be made to read either the forward current only or the reflected current only.

Chapter 13

MICROWAVE LINKS

In its simplest form, a cellular system can be thought of as a switch connected to a number of base stations. The connections between the cellular radio bases and the switch (or the switch and the PSTN) are an essential part of the network. Cellular operators usually think of microwave, fiber optic, or lines rented from the wireline service as the means of linking up the network. Often a mix of these options is the best choice.

The short lead times, low installation costs, and often the availability of existing towers (particularly for non-wireline operators) make microwave links particularly attractive for cellular operators. Base-stations towers can also be used for microwave links.

Microwave systems are so called because they operate in the "microwave band," which is the UHF band (300–30,000 MHz). Typically, they have capacities ranging from a few voice channels to many thousands.

A number of software programs are available that can quickly and reliably calculate microwave path performances. These programs can also give the path performance as a function of tower height and determine the effect of potential interferers. Cellular operators who use microwave extensively should obtain this software and become familiar with it. This chapter looks mainly at the concepts involved in the selection and design of microwave links. It is not intended, however, that the calculations discussed should be used for path design. This is particularly true for a cellular operator who uses microwave only for cell-site links. Because the link system allows for only part-time application for microwave design, the value of using a well-designed software package is obvious.

Microwave systems suitable for cellular radio will be in the band 2–23 GHz. Frequencies lower than 2 GHz have been used in the past but are not likely to be generally available for cellular applications. Table 13.1 shows the bands and their suitability.

FREQUENCY GHz	APPLICATIONS
2	Widely used for cellular, low and medium capacity
6	Widely used for cellular, medium and high capacity
10	Only recently utilized for small capacity
11	Medium capacity, medium-haul
18	Short-haul, moderate to low rainfall only
23	Short-haul, moderate to low rainfall only

Table 13.1   Microwave frequencies

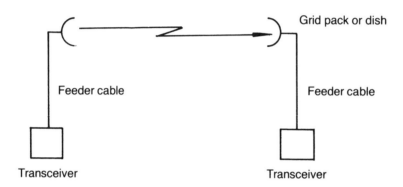

Figure 13.1   A typical microwave link

Modern microwave links use digital modulation microwave (earlier systems used analog links). Figure 13.1 shows a typical microwave link. The performance of a digital link is measured in BER (Bit-Error Rates). The threshold level for a digital link is often defined to be the level at which the BER equals 10^{-3}, although other BERs can be used. The threshold level is approached very rapidly from much higher BERs as the S/N level of the carrier approaches the threshold. This can be seen in Table 13.2. Terminal equipment often has the threshold level defined for a 10^{-3} BER (or similar BER) as a specified parameter. Table 13.3 lists typical threshold values.

Thus, the BER is a function of the input signal level, which in turn depends on the transmission loss or the loss between the two ends of the link. You can calculate the transmission loss for Figure 13.1 from the following formula:

$$\text{Transmission loss} = \text{Free-space path loss} + \text{Terrain losses} + \text{Antenna feeder losses}$$
$$+ \text{Antenna branching loss} - \text{Antenna gains}$$

	CARRIER S/N		
BER	2 LEVEL	4 LEVEL	16 LEVEL
10^{-3}	11	17	28
10^{-4}	12	18	29
10^{-5}	13	19	30
10^{-6}	14	20	31

Table 13.2   BER as a function of carrier S/N and the number of states in the digital code of a typical link

	SYSTEM SIZE		
	2 Mbit/s	**2 + 2 Mbit/s**	**8 Mbit/s**
10^{-3} BER THRESHOLD	− 95 dBm	−92 dBm	−89 dBm

Table 13.3   BER threshold levels for typical 2, 4, and 8 Mbit/s systems

The free-space path loss is calculated as:

Equation A

$$L_f = 32.5 + 20 \text{ x } \log d + 20 \text{ x } \log f$$

where   d = hop length km
         f = frequency in GHz
         L_f = path loss in dB

FRESNEL ZONE

In a microwave link, the radio transmission exhibits wavelike characteristics, and the zone where wavelike interference can affect the propagation path can be approximated by the Fresnel Zone. The Fresnel Zone is widest in the middle of the link and can be calculated from the formula:

Equation B

$$R_{FZ} = 17.3 \text{ x } \sqrt{(d_1 \text{ x } d_2) / (d \text{ x } F)}$$

where R_{FZ} = Fresnel Zone Radius
         d_1 = distance to zone from base 1 (km)
         d_2 = distance to zone from base 2 (km)
         d = d_1 + d_2 or the length of the hop
         f = frequency in GHz

Figure 13.2 shows the calculation of the first Fresnel Zone radius.

Microwaves do not normally propagate within the atmosphere in straight lines; they ordinarily travel in curved paths (usually curved downward) due to atmospheric refraction. The amount of curvature is usually defined with respect to the earth's curvature, which is designated as K, where K x R (R = the earth's actual radius) gives the effective radius of the earth as seen by the microwave path.

If the Fresnel Zone is obstructed, some additional path losses will occur. When there are no obstacles within 50 percent of the Fresnel Zone radius for K = 4/3 (the most usual value that approaches a "flat earth"), then the obstacle generally causes negligible loss. When, however, an obstacle protrudes into the path of the link by more than 50 percent of the first Fresnel Zone, an adjustment must be made for the additional losses incurred.

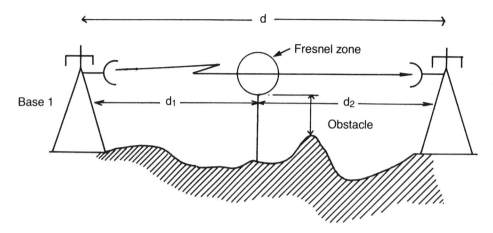

Figure 13.2   Calculation of the first Fresnel Zone radius

The terrain loss L_{TR} (in dB) can be calculated as:

Equation C

$$L_{TR} = 10 - (C / R_{FZ}) \times 20$$

> *where*   C = the clearance in meters of the obstacle from the Fresnel Zone (as shown in Figure 13.2)
> R_{FZ} = Fresnel Zone radius

Notice that:

- C can be negative if it protrudes into the Fresnel Zone.
- This approximation is valid only for $-1.5 \leq C/R_{FZ} \leq +0.5$

Because of changes in the refractive index of the atmosphere, the effective value of K varies with time. Smaller values of K increase the attenuation due to obstructions, particularly on longer path lengths. You should check to ensure that potential variations in K will not degrade the service.

The change in clearance (C_C) for changes in K can be approximated by:

Equation D

$$C_C = 0.078 \times d_1 \times d_2 \times \{0.75 - (1 / K)\} \ meters$$

The limiting values of K are:

> K = 1 for wet climates
> K = 0.8 for temperate climates
> K = 0.6 for desert climates

It is normal to check the path profile for the extremes of K = 4/3 to K = 0.8.

FADING DEPTH

The correlation between BER and fading on a digital link is not very high, and the fade margin has less meaning than it does in analog systems. The flat-fade margin can be calculated from the empirical formula. The formulas used are tailored to geographic regions so that the constants vary a little from place to place. The main variables are temperature, humidity, and rainfall. For Western Europe, Equation E applies.

Equation E

$$F_D = 35 \times \log d + 10 \times \log f - 10 \times \log p - 58.5 + 10 \times \log C_f + 10 \log Q_f$$

where   p = the percentage of time of outage
C_f = the climatic factor
Q_f = the terrain factor
F_D = fading depth

and Q_f = 0.5 for hilly or mountainous paths
  = 1 for rolling terrain
  = 2 for very smooth terrain (or over water)

and   C = 0.5 for dry climate
  = 1 for temperate climate
  = 2 for humid climate or coastal area
  = 4 for coastal area in hot humid climate

LOSSES IN ANTENNA COUPLING

Table 13.4 on the next page shows typical coupling losses. The antenna coupling losses are dependent on the actual coupler used. Use Table 13.4 as a guideline only; consult the manufacturers specifications.

The multicoupling used on a microwave system varies according to configuration. A typical protected system is illustrated in Figure 13.3, also on the next page.

CALCULATION OF OUTAGE TIME

The percentage of outage time, due to multipath fade only, can be calculated by rearranging Equation E. The percentage of outage time P is:

Equation F

$$P = C_f \times Q_f \times 10^{(3.5 \times \log d + \log f - M_f / 10 - 5.85)}$$

where   P = probability of outage with no diversity
M_f = fade margin

SYSTEM	COUPLING LOSSES AT 2 GHz dB
Unprotected	5
Hot standby	10
Polarization protection	5
Hot standby with space diversity	6

Table 13.4   Typical coupling losses

or for a diversity system:

$P_{od} = P / I$

where $I = 0.012 \times S^2 \times (f/d) \times 10^{F/10}$
P_{od} = probability of outage with diversity
S = spacing between the antenna $5 < S < 15$ m
f = fade margin of the first antenna

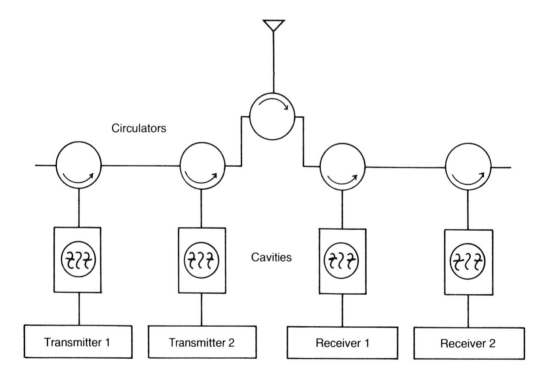

Figure 13.3   Branching and coupling of a singularly polarized protected system

SYSTEM GAINS

The smallest antennas used on microwave links are usually grid packs with about 25 dB gain. This relatively high minimum gain is used in order to ensure narrow beam width and hence good frequency reuse potential.

Table 13.5 shows the antenna gains available at 2 GHz.

At these frequencies it is possible to use grid packs that are lighter, cheaper, have lower wind-loading, and are almost as efficient as solid dishes (within 0.2 dB).

DISH DIAMETER	GAIN
1.2 m	26.4 dBi
1.8 m	30.5 dBi
2.4 m	32.8 dBi
3.0 m	34.6 dBi
3.7 m	35.5 dBi
4.6 m	37.5 dBi

Table 13.5   Antenna gain with diameter

GAIN MEASUREMENTS

In microwave radio it is normal to express gains in dBi (or gain over an isotropic antenna). Mobile radio gains are nearly always quoted as dBd (or gain over a dipole). These two measurements are directly related by:

Equation G

$$dBi = dBd + 2.15 \, dB$$

FEEDER LOSSES

Feeder losses are most significant at microwave frequencies and must be taken into account when determining system gains. Table 13.6 shows typical feeder losses at 2 GHz. You can obtain actual feeder losses from the manufacturers' catalogs. Some variations occur with the dielectric type and manufacturers. Connector losses of 0.5–1 dB can be expected at these frequencies.

At higher frequencies, waveguides are generally used, with EW-77 elliptical waveguide being most common at 7 GHz. Elliptical waveguides are semi-flexible and can support only a single polarization mode, but the ease of handling, particularly the ability to install without joins, makes the elliptical waveguide far more popular than the square or round sections.

Foam insulated 1/20″	11.3 dB/100 meters
Foam insulated 7/80″	6.4 dB/100 meters
Foam insulated 1-1/40″	4.7 dB/100 meters
Foam insulated 1-5/80″	4.1 dB/100 meters

Table 13.6   Feeder losses at 2 GHz

INTERFERENCE

Microwave systems, like cellular systems, are expected to operate in an environment of frequency reuse and have evolved a number of techniques to limit interference. The interference takes two main forms: intersystem (correlated) and intrasystem (uncorrelated). In both forms, co-channel (same channel) and adjacent-channel interference can occur. Figure 13.4 shows these interference modes.

Unlike cellular radio, microwave links have no inherent immunity to interference. Microwave links rely on careful planning and usually field survey to ensure adequate interference immunity. There are three major types of interference: carrier-beat interference, threshold degradation, and sideband-noise interference.

Carrier-beat interference is caused by small differences in frequency between the interfering carriers. This causes a beat frequency that may lie within one or more of the demodulated base band channels. This is mainly a problem with analog systems.

Threshold degradation occurs when an interfering carrier, offset in frequency from the system frequency, is of such a level that even after passing through the system filters, a significant signal level persists. The interfering carrier effectively increases the thermal noise and consequently degrades the system noise factor. This results in an effective decrease in receiver sensitivity.

Sideband noise occurs when the sideband spectral power densities and the inband power densities overlap. Upon demodulation, this interference appears as noise and degrades the received S/N ratio. This is why it is important to see that channel overmodulation does not occur.

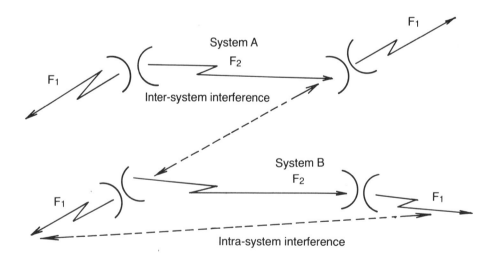

Figure 13.4   Interference modes in microwave links

MARGINS

Digital systems require a margin of 20 dB $\pm$ 5 dB for correlated interferences. To include the fade margin, 25 dB is normally allowed between the interference level and the desired-signal level. Cross-polarized signals have an inherent immunity of about 35 dB and, in well-designed systems, do not ordinarily present problems. Non-correlated interference can only be determined by measurement. A margin of 30 dB should be allowed for such interference. Often the directivity of the antenna system can be used to provide the bulk of this margin.

SYSTEM CAPACITIES

Systems are divided into three capacity groups:

*   Low capacity 2, 4, and 8 Mbit/s
*   Medium capacity 17, 34, and 70 Mbit/s
*   High capacity 102 (3 x 34) Mbit/s and above

The channel capacity for North America and Japan is 24/1.5 Mbit/s, but most of the rest of the world (Europe and Australia-Asia in particular) uses 34/2 Mbit/s.

The 32 channels are based on 64 Kbit/s per channel, with 30 channels available as speech channels. The 24-channel (or T_1) systems are also 64 Kbit/s and have all 24 channels available for speech.

For cellular links the usual practice is to use either low-capacity links, which come packaged as 2-Mbit units expandable to 8-Mbit (or DS1, DS2, US), or medium-capacity links, which are 8-Mbit units expandable to 34-Mbit (or DS3, US). The connection to the cellular base or switch is usually in 24- or 30-channel streams (depending on the system) and multiplexers are available to allow this connection.

ADVANTAGES OF DIGITAL SYSTEMS

Digital radio systems have a number of advantages over analog systems, including higher signal-to-noise and interference immunity and the ability to regenerate traffic at each repeater without adding additional noise. Higher signal-to-noise and interference immunity results in greater spectral efficiency. The ability to regenerate traffic at each repeater allows for very long hops without serious noise degradation.

The main disadvantage of digital systems is the lower immunity to frequency-selective fading, which can result in the need for an additional 18-dB flat-fade margin above that required for a comparable analog system.

MODULATION SYSTEMS

The main modulation techniques used in digital links are:

*   QAM (Quadrature Amplitude Modulation)
*   PSK (Phase Shift Keying)
*   FSK (Frequency Shift Keying)
*   CP-FSK (Continuous Phase Frequency Shift Keying)
*   FFSK (Fast Frequency Shift Keying, which is really a special case of CP-FSK)

These techniques are best illustrated by signal constellation diagrams that illustrate the phase and level relationships between the system states. Figure 13.5 illustrates PSK. Figure 13.6 illustrates QAM.

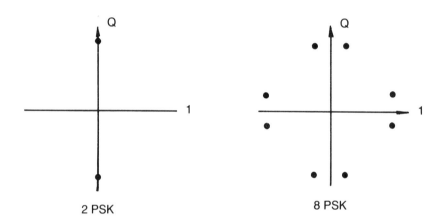

Figure 13.5   Phase shift keying (Phasor diagram)

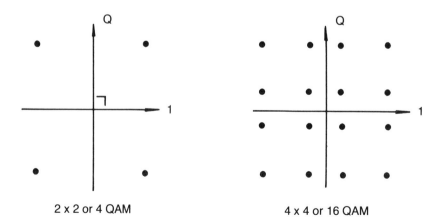

Figure 13.6   QAM modulation constellations

RACK SPACE

Microwave equipment generally requires about 2 x 600 mm rack spaces that includes space for the MUX and inverters (which are generally needed for the MUX and may be needed for the RF hardware). These racks should also have space for an IDF and alarm panels.

MICROWAVE LINKS IN CELLULAR SYSTEMS

Every base station needs to link each of its channels back to the switch. There are three ways of doing this: by conventional land line facilities, with fiber optics, or with microwave links.

Many operators prefer microwave links because they have full control over these links and often they are cheaper. Microwave links can also be readily redeployed if base stations must be moved.

The frequencies used for microwave links range from the long-haul 2- and 6-GHz bands to the relatively short-haul 11-, 18-, and 23-GHz bands. The availability of spectrum varies and some countries, notably the US, have complicated rules about which frequencies can be used under certain conditions (for example, the US has a minimum distance rule for some bands, as shown in Table 13.7). Other rules cover minimum capacity on links in certain bands.

Increasingly, the short-haul, high-frequency systems are becoming more economically attractive. The frequencies above 11 GHz are approximately three times more susceptible to weather-related outages such as rain and snow. The susceptibility decreases rapidly at frequencies below 11 GHz.

US AND JAPAN

The 24-channel units used in the US and Japan are known as T1s; each T1 bit stream occupies 1.544 Mb/s. Figure 13.7 shows a simple T1 circuit. Table 13.8 shows the hierarchy of transmission that has been established. High-capacity systems comprising multiples of 56 and 84 T1s are also available.

FREQUENCY (MHz)	MINIMUM PATH DISTANCE (km)
2110 to 2130	5
2160 to 2180	5
3700 to 4200	17
5925 to 6425	17
10700 to 11700	5

Table 13.7   Minimum path length for certain bands in the US

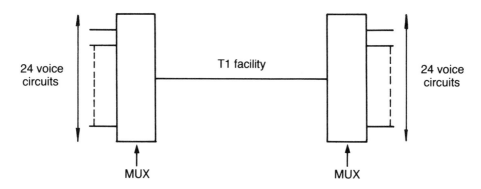

Figure 13.7   A T1 facility equipped with voice channel MUX

LEVEL	BIT RATE (Mb/s)	NUMBER OF VOICE CHANNELS	T1s
DS1	1.544	24	1
DS2	6.312	96	4
DS3	44.736	672	28

Table 13.8   Hierarchical level for data transmission

REST OF THE WORLD

Microwave elsewhere in the world is 2 Mbit based. The systems of most interest in cellular radio are the 2/8 (2-Mbit expandable to 8) and 8/34 (8-Mbit expandable to 34) Mbit systems. There is usually only a small difference in price between a 2-Mbit system expandable to 8-Mbit and a 2-Mbit system (not expandable). For this reason, if expansion is anticipated, a 2/8-Mbit system is usually chosen even though a 2 Mbit unit would suffice initially. A price break usually occurs at 8 Mbit, but an 8-Mbit unit expandable to 34-Mbit is similar in price to a non-expandable 8-Mbit system.

These microwave units may come in racks or may be intended to be wall mounted (see Figure 13.8).

Although all cellular mobile transmission paths are analog, the switches are usually digital and so it is logical for the links to the base station to also be digital. This requires an analog-to-digital converter (also known as a multiplexer or MUX) to be placed at the base station. Figure 13.9 shows multiplexer equipment at the cellular base. Some manufacturers provide MUX equipment as an integral part of their base stations, but others do not.

The switch can accept 2 Mbit data streams. Each 2-Mbit stream has 30 x 64 Kbit/s voice-channel capacity plus two control channels.

Figure 13.8   Four racks of microwave equipment. The first two racks on the left contain the MUX equipment. The next rack contains two 2-Mbit microwave transceivers, and the end rack contains a 34-Mbit transceiver. Microwave equipment is usually designed to be wall mounted. Photo courtesy Extelcom

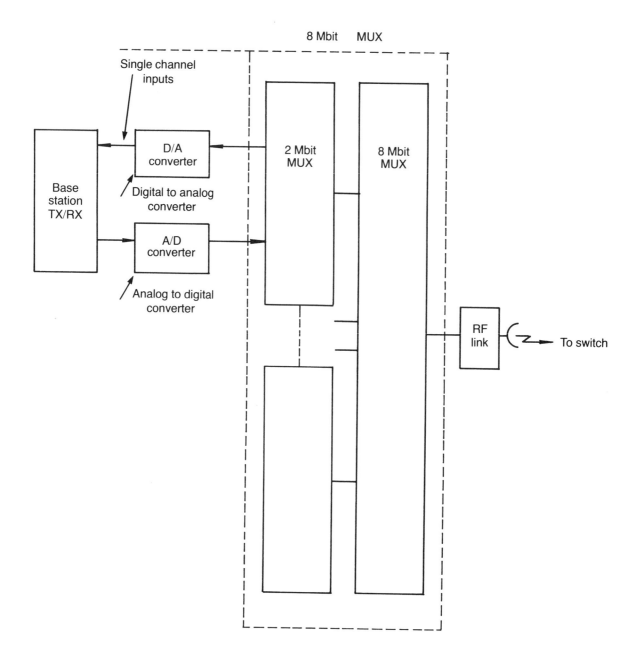

Figure 13.9   A digital link to the switch. Notice that the A/D and D/A and 2-Mbit MUX conversion may or may not be part of the base-station equipment.

DROP AND INSERT

The MUX equipment occupies a full 2-Mbit or T1 data stream each time it is accessed. Drop and insert facilities, however, can take a single timeslot and drop or insert that timeslot at any

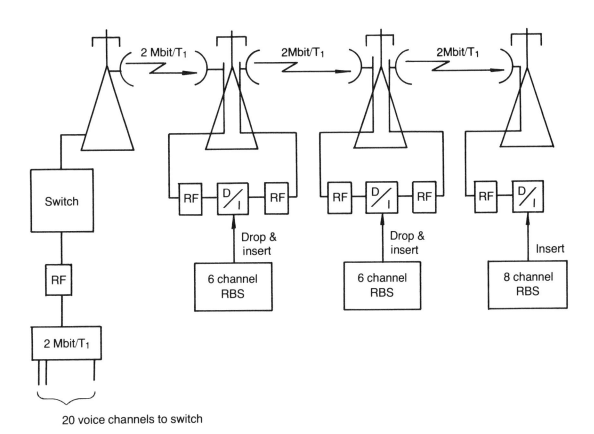

Figure 13.10 Drop and insert facilities used to fully utilize bearer capacity

location along the microwave link. These facilities might be used for a thin (low-density) route covering a highway, as shown in Figure 13.10.

The power consumption of a microwave link in the hot standby mode is from 60–200 watts, making it a relatively low power device when compared to the rest of the base station.

FACTORS IN CHOOSING MICROWAVE

The installation time for a microwave system (after obtaining approval and assuming equipment is available) can be less than one week. Approval times and equipment delivery times, though, can add up to months or, in the worst cases, years.

The reliability of modern microwave equipment is very high and the MTBF (Mean Time Between Failure) can be expected to run to many years. There are many different ways of calculating MTBF and since there is no standard, one should not place too much weight on figures supplied by manufacturers.

Although heavy rain and snow can cause outages, the probability of a complete outage on a well-designed system is very low. Weather restrictions do limit 10-GHz systems to 50 km and 18- and 23-GHz systems to 15 km (and even less in tropical regions where a few kilometers may be the usable range), but these are large distances by cellular standards.

A microwave system used for cellular should not require extensive additional sophisticated (expensive) test equipment to maintain. Consider this when ordering test equipment for the cellular network. Microwave equipment, being significantly higher in frequency than cellular bases, requires more sophisticated and hence more expensive test equipment. Where possible, test equipment for the microwave network should also be usable at cellular frequencies.

In order to have effective system management, extensive diagnostics and alarms should be built-in. Because space is almost always a problem in cellular installations, choose a system that can be wall-rack mounted without consuming too much equipment space. Often microwave racks, as sold to wireline companies, are too high for cellular base stations. The rack height should not be higher than the base-station rack height. This may require a reconfiguration of the link equipment.

SURVEY

Because most cellular base stations are either in line-of-sight to the switch or to a suitable branch point to the switch (clearly) or obstructed (clearly), the process of a survey is somewhat simplified. Because of the relatively high cost of microwave links (compared to a base station, microwave links represent about 10–30 percent of the total costs), a trade-off may occur between good cellular design and good link design when base stations are deliberately placed behind obstacles (to improve frequency reuse). In these instances, either an alternative to microwave or the use of a relay site, present themselves as options to compromising the cellular design.

Where any doubt exists, a proper detailed microwave survey should be done but, in a modern city, line-of-sight usually guarantees an adequate path over the relatively short hauls required for cellular. Long hauls (greater than 20 km) should be surveyed at RF frequencies..

TRUNKING

The call-routing and system-link configuration is known as the trunking of the network.

Cellular systems can be linked by copper pairs, fiber-optic cables, or microwave (analog or digital). Any combination of these systems can be used. A fully developed cellular network may have a complex trunking scheme involving multiple switches and route diversity. Some operators like to spread circuits to and from the PSTN over different routes and even over different systems. For example, half the circuits from the switch may go via microwave and half by fiber-optics cable. Figure 13.11 illustrates the trunking of a multiswitch system.

Today, unless the cellular operator is a wireline company with a large surplus of copper pairs, some form of digital link is likely to be the most economical way to connect the cellular switch to the base stations and the PSTN. Even where copper pairs are available, it is generally economical to use PCM to increase the trunking efficiency.

For most cellular operators, the choice is between optical fiber, digital microwave, or PCM.

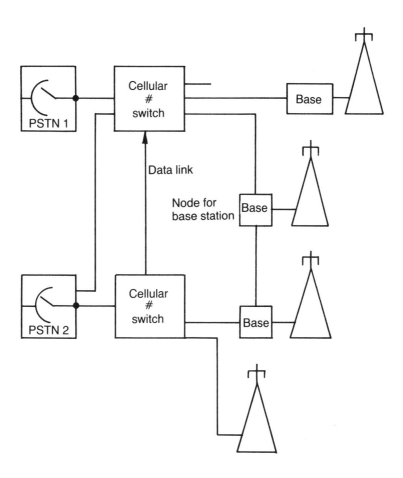

Figure 13.11 The cellular switch can be interconnected to a number of PSTN points. PSTN 2, for example, could be in a different town.

The links for cellular systems are often group connected in multiples of 24 or 30 channels. The switch is usually configured to accept incoming and outgoing traffic in these multiples, and it is usually necessary to use multiplex equipment to access traffic at the individual channel level.

In the trunking plan, some branching may be necessary to increase the total capacity of the links in the network. Branching involves using one or more base-station links as nodes (or connection points) in the linked network.

Figure 13.12 on the following page shows the connection of three radio bases to a cellular switch and also the connection of the switch to the PSTN. The figure assumes that the largest group that can be separated out from the main stream of data is 2 Mbit. In some systems, it is possible to connect at the 8-Mbit level.

Figure 13.12 also shows base B being used as a node for the 8-Mbit link from base C. One half of the 34-Mbit capacity of the link between base B and the switch is tied up and is not available for any purpose on other routes. The need and opportunity to use some base stations as nodes (or

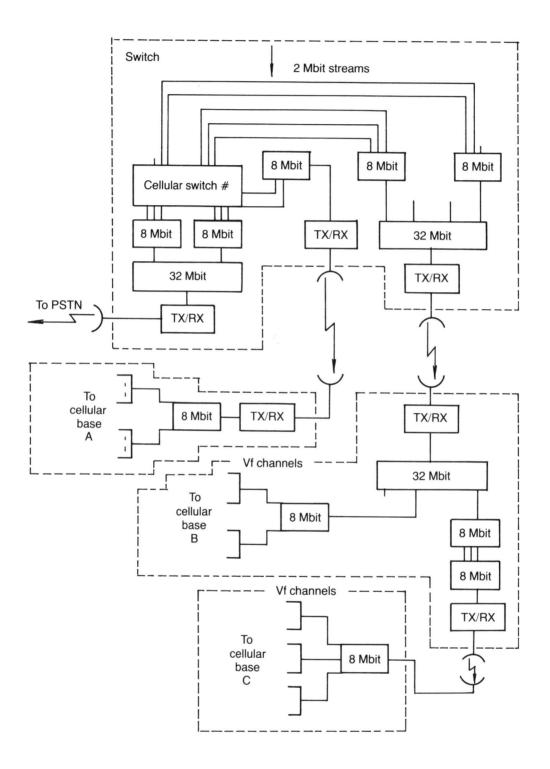

Figure 13.12 A three-base cellular network trunking scheme using 2-Mbit links. Three bases are connected to a switch via a microwave link. Base B picks up the link from base C and then through connects to the switch.

collection points) for data streams grows as the system capacity grows, regardless of whether you use microwave or fiber-optic systems.

Frequency congestion on microwave systems will often encourage the use of local nodes, as will route availability when using fiber optics. In order to attain maximum circuit efficiency, it may be necessary at some nodes to go down to VF level to avoid too much use of low-occupancy 24- or 30-channel streams.

MUX equipment frequently employs 48-volt power-positive ground; this usually requires the use of inverters at 24-volt cell sites if the base station batteries are to be used to provide backup power for the microwave.

STANDBY

Hot standby—that is, a fully active parallel system powered up and on standby—is usually used for main routes and the PSTN route. A changeover switch is provided (automatic) to allow the traffic to be switched from one link system to the other when a failure occurs. Cold standby is much the same as hot standby but the second redundant system is not powered up. Cold-standby systems usually have manual changeover. The reliability of modern microwave systems is such that it is practical in thin-route (light-traffic routes) situations to dispense with standby altogether and have only a single link.

For fiber-optic systems, the risk of loss of service is increased by the chance of having the cable cut by excavation. Fiber-optic systems, for this reason, often employ route diversity (that is, a link via two different paths).

A simple system with standby is called a 1+1 (one active plus one standby); an unprotected route is designated a 1+0.

Larger systems can provide a good degree of protection at increasingly lower costs. For example, a large system might consist of 3 x 34-Mbit bearers in service with a 1 x 34-Mbit hot standby. In this configuration, any failure of one of the main bearers can be taken up by the standby bearer. Such a system would be designated a 3+1.

SPLIT ROUTES

Sometimes, in order to further decrease the trunking costs, it is worthwhile to spread traffic over the main and protection bearer. As shown in Figure 13.12, when the PSTN route (34 Mbit) becomes full, it is possible to abandon the hot-standby configuration in favor of two parallel links with the cellular switch to PSTN traffic spread between them.

In this instance, failure of one half of the system results in a loss of half the traffic capacity of the route. This may not be acceptable on the main PSTN route, but it would generally be acceptable on a route to a node.

Where parallel systems are used, it is necessary to have two active systems employing two separate frequencies.

Chapter 14

BASE STATION MAINTENANCE

A fully developed cellular system will have a very large number of channels and radio bases. Fortunately, most cellular systems have good housekeeping software; problems such as VSWR out of range, TX failure, channel-blocking, power and entry alarms, and speech path conditions are constantly monitored and are automatically reported to the maintenance-control center.

Maintenance usually consists of board or panel replacement, and this replacement is directed by the base-station controller or by the switch (that is, apart from reporting the fault, diagnostics suggest the maintenance procedure). Faulty bus cables can cause significant problems during the base station burn-in period.

Routine maintenance is usually handled by one or two technicians who respond to a fault by replacing the suggested boards or panels. Ongoing maintenance in a growing cellular system also includes channel rearrangements to shift channels, as required, from base to base to match traffic capacity to demand. Staff should be capable of retuning the base transceivers and multicouplers.

There is a fine line between extending the capacity of a base (installation) and rearranging channels to suit traffic conditions (maintenance). In those companies where these two functions are performed by separate groups, it is necessary to clearly define the role of each group in each instance.

Routine maintenance should include approximately monthly visits to check on the condition of the batteries, air-conditioning, and site upkeep. In particular, the grass and foliage around the site should be neatly kept to avoid fire hazards. Audio levels, RF power receiver sensitivies, and the deviation on each channel should be confirmed every 6–12 months.

Because cellular bases are often on remote sites, careful attention should be paid to vandalism and attempted vandalism. Having a good security fence and some means of preventing intruders from climbing towers and structures is essential.

If this routine maintenance was all that was required, then base-station maintenance would indeed be easy. But there are other considerations. The rest of this chapter discusses these considerations in detail.

MAINTAINING QUALITY OF COVERAGE

A VSWR alarm and low TX power alarm will generally suffice to detect most RF failure or partial failure, but not all problems.

The following problems can cause reduced radiated power without detection:

- Water in the antenna
- Partial lightning damage to antenna/cable
- New (since the cellular installation) buildings, foliage growth, tower extensions, and other obstructions
- Damaged feeder
- Damaged, faulty, or waterlogged connectors

The frequency of these problems increases as the number of channels increases. If each cell has two RX and two TX antennas, then there are four antennas per cell (for up to 30 channels per cell for an omni site and 12 or more for a sectored site). The number of other passive components is proportional to the number of channels.

A modern city may easily have 50 cells (at least 200 antennas and probably many more). If the MTBF (Mean Time Between Failure) of the antennas is 20 years, then in an average year, 10 (200/20) antenna failures can be expected. Catastrophic and severe failures will, of course, be detected by the base-station controller; less severe failures may not be detected. Most of these failures will result in reduced coverage, but they may escape detection by the alarm system. This is particularly true in the case of installations with high feeder losses.

The VSWR is measured by comparing the level of the forward RF signal to the reflected signal, and if measured from the TX output, only the forward component is accurately measured. If the fault is at, or near, the antenna, the reflected component has effectively traveled the distance to the antenna and back, and has hence suffered attenuation = 2 x feeder loss. This is how serious antenna problems can escape detection.

The VSWR is most easily measured by measuring the actual reflected power through a directional coupler. Once this value has been measured, the VSWR can be calculated as follows:

$$VSWR = \frac{1 + \sqrt{\dfrac{Reflected\ Power}{Forward\ Power}}}{1 - \sqrt{\dfrac{Reflected\ Power}{Forward\ Power}}}$$

An inline RF power meter can measure both of these components. A base-station antenna should have a VSWR of no greater than 1 to 1.7.

If the reflected power is measured at the transmitter terminals, it is necessary to increase the value of the measured reflected power by the cable loss x 2 to convert the reading to the value at the antenna input. Figure 14.1 shows this VSWR measurement at the transmitter terminals.

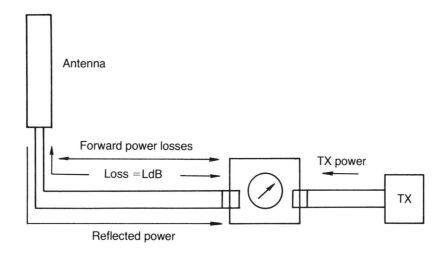

Figure 14.1   VSWR measurement at the transmitter output terminals

AIR-CONDITIONING

The integrity and proper functioning of the air-conditioning is most important to ensure reliability of the base-station equipment. It is well-established that temperature cycles contribute significantly to the fault rates of electronic equipment. It is a good idea to provide dual air conditioners so one unit is available if one should fail.

The trade-off of having the air conditioners out of service during power failures and the cost of providing emergency power should be considered. Equipment that has a low thermal operating range may not be able to function for any significant period without air-conditioning but some more robust equipment can. The fact that some equipment can operate at high temperatures, however, does not mean that it is immune to the deleterious effects of temperature cycling.

TRANSCEIVERS

Without a doubt, transceivers require the most consistent maintenance attention. Although suppliers' published figures for MTBF vary considerably, the probable MTBF in the real life of a transceiver is about one year. Published figures are invariably based on different measurement criteria of MTBF, which reflects sadly on efforts to standardize MTBF measurement.

The least reliable part of the transceiver is the RF power stage. Because the loss of the control channel puts the base off the air, it is almost always recommended that redundancy be provided for this channel. For the same reason, when a control channel has been taken out of service, it should be attended to quickly. Of course, this is not a problem with NMT900 systems, which use voice channels as control channels.

It is fortunate that transceivers are frequency-agile because they will be replaced frequently.

USING STATISTICS

The switch statistics tapes (and sometimes the billing system) produce information that can be used to diagnose base-station faults. The following sections discuss two examples.

BLOCKING

Blocking (also known as sealing) is the term used for a channel that is temporarily taken out of service by the base-station controller due to interference. This can occur while the channel is in use or when it is on standby.

The most probable cause of blocking is interference from another mobile; provided that blocking is kept relatively low (less than 2 percent), it can be considered normal. Blocking above 5 percent, however, generally indicates that there are problems, including system design errors or intermodulation problems or RF interference from other equipment. Check that new cell sites have not been added or that an existing cell on the same or adjacent channel has not been modified.

By looking at the individual distribution of channel blocking, it may be possible to identify the frequency of the interferer.

CELL OCCUPANCY

Cell occupancy statistics tell a good deal about how the system is performing. A sudden drop in cell occupancy can indicate problems such as:

- Antenna/feeder problems that cause reduced coverage (in particular, antenna downtilt or obstructions caused by new work on the tower).
- The signal-strength receiver may be out of calibration.
- The control channel may be low in RF level or off frequency.
- The receiver multicoupler and/or connecting cables may have introduced losses.
- The frequency offsets of the control may have drifted (NAMTS systems only).

CUSTOMER COMPLAINTS

Some authorities rely on customer complaints to inform them of ERP (Effective Radiated Power) problems of poor or reduced coverage. This detection system, apart from causing poor public relations, will not work if there is significant base-station overlap and other bases take over from the faulty one, or if the fault is gradual (an example, water accumulation) and subscribers come to accept the "new" coverage as normal.

To make matters worse, many customer complaints are ill-informed. "The service doesn't work at my house" may mean "It doesn't work in my underground garage" but, of course, the service is fine outside the garage.

There can be a number of reasons why a customer reports less than satisfactory service.

- The mobile or installation is faulty.
- The customer has incorrectly operated the mobile equipment.
- The customer is outside the normal coverage area.
- A base-station fault is evident.

- The customer attempted calls during the down time of the system/base.
- The customer likes to complain.

At this point there is a need for clear communication between the person recording the fault, the mobile maintenance staff, and the base-station and switching staff, to avoid duplication.

This is an easy way to proceed:

1. The fault recorder has a coverage map, with good street-level accuracy, and can ascertain whether the complaint originates from within the normal service area. The fault recorder should ask sufficient questions to ensure that the customer is using appropriate operating procedures.
2. If the complaint is outside the service area, the customer should be advised accordingly.
3. If the complaint is inside the service area, then an appointment to check the mobile should be arranged with the mobile maintenance staff. At the service center it may be possible to determine absolutely whether the fault is mobile or not, but, due to lack of adequate test equipment, it may also be that trying a replacement mobile is the only solution. It is often very difficult to entirely eliminate the possibility of a faulty antenna/cable.
4. Unless the mobile or its associated hardware has been clearly implicated, the fault report should now be passed on as a possible base-station fault. The fault could also have been a temporary switch or base-station problem (for example, outage due to maintenance). Because of this, these reports should be handled centrally and forwarded to the switching or base-station staff only when a clear pattern has emerged.
5. Analysis of consistent fault reports in areas thought to be well-covered may reveal local "dead spots." This information is most useful to system planners and designers, so it is important that reports are carefully processed and the results forwarded to the relevant personnel.

There is a tendency to "do maintenance at the lowest possible level of competence." This philosophy comes from the PSTN operators; most readers will likely have experienced the frustration that this policy causes.

In my own case, I have had to resort to threats of media complaints on two occasions to get relatively simple switching faults on my line repaired. The PSTN operator insisted on sending "the lowest-level staff," who understood only telephone apparatus faults and who therefore wanted to replace or repair the telephone. On both occasions, I clearly defined the problem (and its precise origin, which was in the switch) but it took about 6 months of insistence to have the fault cleared.

On one occasion, the fault occurred in a rented apartment; after the problem was finally rectified, it was confirmed that the fault dated back six years and three tenants. The second problem also involved a switch-related fault resulting in crossed wires. It was accurately identified, but the repair personnel spent months trying to tie it down to a phone or cable fault. This problem was fixed by a visit to the switch to point out the offending equipment.

To work effectively, maintenance staff must have a reasonable overview of the system. It is false economy to adopt the PSTN operators' approach. The maintenance staff must understand the network beyond simply replacing specific pieces of hardware.

LINE UP LEVELS

Most PSTNs have a series of test numbers that will return certain test tones. These numbers can be used to evaluate the quality of the PSTN interface and line-up levels.

The objective of the cellular operator is to make the handoff almost imperceptible. Modern cellular systems have handoff times that meet this standard, but handoffs can become conspicuous if a level change occurs during handoff. For this reason, it is necessary to accurately set all system levels. When the cellular operator uses leased lines (which may be rerouted), checks on levels should be made regularly. This is particularly true if copper cables are used for links.

After replacing or repairing any MUX equipment, it is essential to check all affected line-up levels. Line-up levels and deviation should be routinely confirmed every six to nine months.

TEST MOBILE

Nearly all system manufacturers offer a test mobile. This unit is virtually identical to a normal mobile but has incorporated an automatic answer and the ability to loop back the line to the switch.

Under the control of the switch, this mobile can automatically confirm the proper operation of each channel sequentially. It can also "check" RF propagation, because the test receiver measures the base station RF off-air. If the test mobile is deliberately given a poor antenna (for example, a dummy load) it may be possible to also use S/N measurements to detect degraded propagation. The test mobile can loop back a test tone for S/N measurements.

Because the mobile is controlled by the switch, it can be programmed for extensive after-hours testing of the channels.

Note that a single, successfully answered call to a test mobile only confirms that the base station is at least partially on air. This does not confirm satisfactory operation of the base.

To progressively test each channel, a call to a test mobile, once established, can be forced to handoff to other channels at the base station. To be effective, this should be done late at night when most channels are likely to be free.

The test mobile can be connected individually at each site so that it is local to a base.

Often, automatic test routines are available that can run either periodically (once every five minutes or so) or at certain times late at night.

Test mobiles and other "in-house" engineering services should have easily identifiable numbers so that they can be distinguished for billing purposes. Some methods reserve all numbers where the last three digits are consecutive (for example, ABCD789), or all numbers where the second-to-last last digit is 9 (for example, ABCDE9X), or a specially defined group (such as ABCD1XXABCD2XX). Similarly, "in-house" non-engineering services could also be identified like this to ultimately save the accountants many headaches.

SITE AUDIO TEST LOOPS

Some suppliers include a site audio test loop to distinguish remotely between noisy channels that are caused by the switch to base link from those that are caused by the base RF equipment.

This test tells staff located at the switch if the fault is a link or a base-station RF problem. Because each of these problems is likely to be serviced by different personnel, a good deal of diagnostic time can be saved by early identification.

INTERACTION WITH THE SWITCH

Base-station maintenance requires close cooperation with the technical people in charge of the switch, because it is at the switch that most of the alarm information is available. It is important that the lines of communication between the switch technician and the RF technician are good.

A great deal of time can be saved by having an accurate diagnosis of the problem before staff are dispatched, particularly to a remote base. There are many diagnostics available at the switch that can be used to determine the nature of a fault.

In particular, it is most important to diagnose whether the fault is from the switch, the transmission link, the site controller, or the RF. The test mobile is an important tool to differentiate between these categories.

Some manufacturers have remote-alarm monitoring equipment that can be located in the RF repair center. This equipment can reduce the dependence on switching staff to diagnose base-station faults.

System designers can assist by co-locating a switch with a base station so that the switch operators become more familiar with RF hardware. There is a tendency in cellular (as in all other enterprises) to form the "us and them syndrome" (that is, the switch operator always tends to assume that the fault is in the domain of the RF staff and vice versa). (Of course, co-siting should be done only if it does not compromise the system design, but it is generally possible to achieve such a design.)

SITE-LOG BOOKS

All sites should have locally maintained log books. In these log books are recorded all visits (entries) to the site and the purpose of each visit. Each entry should cause an alarm at the switch that must be "canceled" by the person entering the base station. The entry is recorded by both the switch attendant and the person at the base station. Table 14.1 shows a typical log book.

DATE	ARRIVAL TIME	NAME	DESIG- NATION	DEPART MENT	PURPOSE OF VISIT	DEPARTURE TIME	SIG
9.10.89	08.30	S. Davey	Technician	Maint.	Replace chn. 215	10.30	...
9.10.89	13.00	N. Kelly	Cleaner	Cleaning	Polish floors	14.00	...

Table 14.1   Typical entries in a log book

CALL-OUT PROCEDURES

After hours, it is usual to have only a skeleton staff available. The staff may amount to only one person directly on call at the switch or even one person on call-out via alarm rerouting. In either case, it is unlikely that the person on call-out will be an overall system expert, so procedures that determine the priority of call-out must be established.

In any large city there will be considerable overlap in base-station coverage, so that the outage of any one station, particularly after hours, would not be service affecting. This means it is not noticed by customers. In even larger sites, two, three, or even more non-adjacent bases may be out of service without affecting service.

Rules must be drawn up to enable the call-out staff to decide on a course of action. For example, consider a medium-sized city with 25 base stations. After discussions between switching RF and transmission staff, the following rules may be decided upon:

1. In normal working hours up to 4:30 P.M., any base-station outage is treated as urgent and must be attended to immediately.
2. Between 4:30 P.M. and 7:00 A.M., and from 4:30 P.M. Friday to 7:00 A.M. Monday, any two base stations may be out of service without call-out.
3. All urgent alarms are to be attended to by the switch operator (either an on-site operator or one with a remote terminal if provided). This operator will decide on any subsequent call-out procedures.
4. In both the switching and radio areas, at least three personnel, with a predetermined call-out priority, will be available. If called, the first party to answer will attend to the fault and determine what other call-outs are needed.
5. Each of the call-out personnel will be supplied with a pager and mobile telephone.
6. Special call-out procedures will be defined from time to time for special holiday periods.

EQUIPMENT

To service the base stations, it is necessary to have at least one dedicated vehicle and two technical staff. A station wagon or a small van is a suitable vehicle. The vehicle should be permanently outfitted with the necessary test equipment and tools.

For most installations, base-station servicing will be ongoing, on a daily basis. Spare cards and parts will normally be stored at a central area. This should probably be the same storeroom as the one in which switch spare parts are kept (because some of the parts are the same). Maintenance staff need frequent access to this storeroom, so a central location is most important.

The following is the minimum equipment necessary:

1. Two VSWR meters to 50 watts.
2. One general purpose mobile test set incorporating at least the following to 1 GHz:
    a. RF signal generator
    b. RF level measurement
    c. RF deviation measurement

    d. RF frequency measurement (accurate to $\pm$ 1 KHz)

    e. Audio modulation FM and AM

    f. Preferably a spectrum analyzer

    g. SINAD measurement

3. Three digital voltmeters.

4. One spectrum analyzer to 1-GHz RF and 10-Hz IF resolution.

5. Specific equipment for servicing the particular manufacturers' bases (for example, PCBs, PCs, special cables and plugs). Note that most special cables and plugs should be stored at each base station.

Because the time to travel to and from a base station is usually significant, the maintenance van should carry a good range of tools and miscellaneous parts (such as connectors, transition connectors, cables, and lugs). It is not usual to leave test equipment permanently on site at a base station.

The maintenance staff should also have some independent means of communicating with the switch, such as via the PSTN or a land mobile. This is important in the event of link loss or complete system failure.

QUALITY AND CALIBRATION OF TEST EQUIPMENT

Cellular radio equipment represents state-of-the-art hardware that is beginning to approach the theoretical limits of performance at room temperature. For this reason, it is essential that only high-quality and well-calibrated equipment be used when servicing or adjusting the hardware. The old service monitor of the early 1980s is probably only accurate to $\pm$ 5 or 6 dB, and even most modern ones are struggling to achieve $\pm$ 3 dB.

The equipment used to service base-station equipment should be accurate to at least $\pm$ 2 dB. This rules out most old equipment and even new equipment that has not been calibrated in the past 12 months. Errors of frequency, signal level, modulation level, or VSWR can lead to poor system performance. Additional sources of error include poor-quality connecting cables, connectors, and the use of multiple adapters.

TEST SETS

A cellular test set has many features specific to cellular radio that can contribute significantly to cellular maintenance. A good test set can completely simulate either a base station or a mobile for mobile testing. An operator with anything but the smallest of systems needs at least two test sets.

A good test set should include most of the following:

- SINAD (Signal to Noise And Distortion) measurement with the appropriate weighting
- SINAD measurement of out-of-band test tones (for example, SAT in TACS and AMPS, or phitone in NMT)
- Simulation of signaling tones

- A simple spectrum analyzer
- A frequency counter (to 1 GHz)
- A signal generator (to 1 GHz)
- Deviation measurement
- RF power measurement
- RF millivoltmeter

The ability to simulate a base station is usually an add-on feature and is necessary only if the unit is to be used to test mobiles.

Notice that mobile-specific wiring harnesses are necessary to connect the test set with each model of mobile to be tested. A typical service monitor is shown in Figure 14.2.

SINAD measurement is one of the most fundamental measurements; its method is best illustrated using a separate RF generator and SINAD meter. Figure 14.3 shows a SINAD meter. The RF generator sends a signal at a specified test frequency and deviation (usually about 1-KHz frequency and 1-KHz deviation) to the mobile. The SINAD meter has internal filters that separate the test tone (wanted signal) from the other components (assumed to be noise and distortion from the receiver).

A meter reads the SINAD directly and the RF level is adjusted until the desired SINAD is obtained. Various weighting filters are used to limit the bandwidth and pass characteristics of the noise and distortion components. This is because the mobile channel itself is filtered and the transducer (earpiece) and the ear combined do not respond to all frequencies equally. Thus, the "weighting" networks attempt to account for the subjective noise-level rather than the actual S/N levels.

Although only one service monitor is needed for in-band tests, at least two are needed for

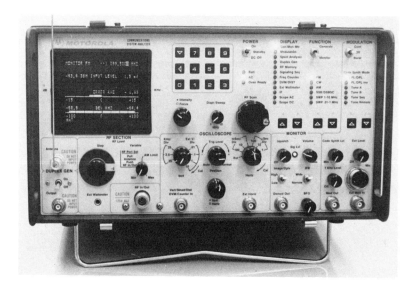

Figure 14.2   A service monitor that is suitable for base station maintenance. With optional additional equipment, this monitor can be used for mobile unit service. Photo courtesy Extelcom.

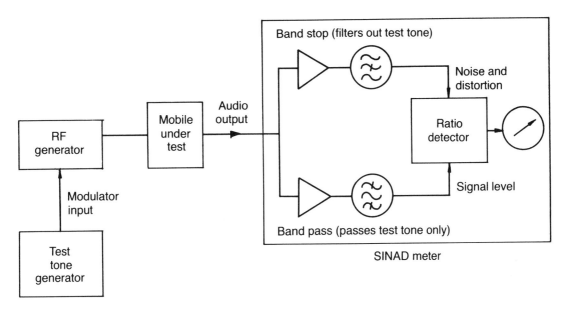

Figure 14.3   A SINAD meter

out-of-band testing; the second monitor can also be used as a reference (that is, discrepancies between them will alert the user to a calibration error). As frequency reuse increases, adjacent-channel interference increases and the need for out-of-band measurements increases.

Measuring adjacent-channel rejection requires that one signal generator be set to measure 12-dB SINAD in-band, and the second signal generated be coupled to the receiver input and set to the adjacent channel. The second generator is adjusted in level until the SINAD falls by 6 dB, as shown in Figure 14.4.

In general, it is a good policy to have at least two of each test equipment item so that the calibration of one can be checked against the other. Down time with test equipment can be excessive (repair times of 3–6 months are not unusual), and the second piece of equipment can be extremely valuable at this time.

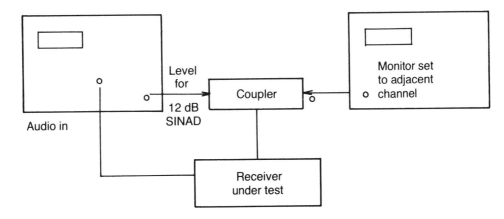

Figure 14.4   Two signal generators are needed to measure adjacent channel rejection.

A good-quality spectrum analyzer is necessary to adequately assess the transmitter performance. The analyzers normally found on service monitors do not have adequate resolution for this purpose and cannot be used. The spectrum analyzer should have a tracking capability so that it can be used to sweep filter and cavity assemblies. The analyzer needs an IF filter resolution of 10 Hz and should be able to display spurious and harmonic generation up to 4 GHz (accurate measurement above 1 GHz is not necessary).

QUANTIFYING COVERAGE PROBLEMS

Good baseline records of the original coverage must be available in order to determine if a problem has occurred since installation. There are two practical forms of this information:

- Detailed, repeatable mobile survey results
- Point-to-point readings taken from the base stations to *fixed* antennas located at strategic points

If the first method is used, a survey vehicle and equipment will be needed for maintenance (for more information, see Chapter 5, "Radio Survey").

Transmit antennas can be checked simply by measuring field strength from a TX control channel. For the receiver, a TX channel can be turned on and can be fed, in turn, to each of the base RX antennas to confirm normal functioning (note that this removes diversity and produces a temporary 6-dB loss in the receive path). This method leaves the base-station functional. If possible, it is best to return the TX channel to a frequency otherwise not in use in the network (to avoid false readings caused by distant co-channel sources).

If the second method is used, the siting of the receiving sites is critical. This method is illustrated in Figure 14.5.

If the three bases shown in Figure 14.5 had omnidirectional antennas, only one receive site would suffice; for all bases, provided a reasonable path exists for directional antennas, it is important to measure the field strength from the front of the antenna. In this instance, four

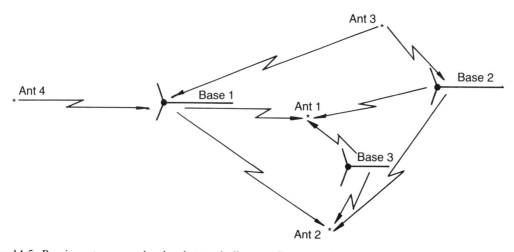

Figure 14.5   Receive antennas can be placed strategically to confirm transmit paths.

measuring antennas are necessary. These antennas must be fixed permanently, because small errors in position (as small as 10 cm) can result in large errors (of the order of 6 dB).

This system, although complex to set up, has the advantage that the test antennas can be used to transmit to the base-station receive antennas, so it is possible to confirm the path in either direction. Because the number of fixed receive antennas increases rapidly as the number of sectored cells increase, this method is not practical for big networks. It is ideally suited, however, to small- to medium-sized omni systems. Naturally, the field-strength measuring receiver used in these measurements must be accurate to a few dB and preferably easily portable so that it can be moved from site to site.

This method becomes increasingly less attractive as the number of cells grows. It is very effective, however, for small systems and in cases where omnidirectional antennas are used.

CO-CHANNEL AND ADJACENT-CHANNEL INTERFERENCE

As stated earlier, interference occurs mainly from the mobile to the base-station receiver. Sometimes, however interference occurs between two base stations.

All systems are prone to data corruption from co-channel interference. (A co-channel is a channel that has the same frequency but operates from a different site.) AMPS and TACS are prone to adjacent-channel interference. (An adjacent channel is one channel width away.) To some extent, these kinds of interference are potential problems in all cellular systems (both digital and analog).

A mobile can fail to receive an instruction if a data clash occurs from two different bases. For problems to arise, the two sources must be within 3-dB signal level of each other. If the difference is greater than this, the dominant signal prevails.

Frequency plans usually will take adjacent-channel interference into account by arranging the cells for maximum distance between adjacent channels. Figure 14.6 shows a typical frequency plan. Channels A and B, B and C, and so on, are adjacent blocks of frequency. This plan separates

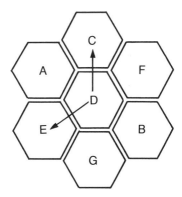

Figure 14.6   Co-channel adjacency in a 7-cell plan. Cells D and E and cells D and C are adjacent.

all cells that are co-channel by at least one cell (except D and E, and D and C), so the problem has been reduced, not eliminated.

Co-channel interference can be more severe and can cause whole base stations to fail if it occurs on the control RX channels; however, it is usually limited to a few high areas that have equal access to the two sites. Sectoring reduces this problem. A sure sign of co-channel interference is the identification of certain local areas where initiating and receiving calls is not reliable, but where reliable service is normal outside those areas.

In these instances, a frequency change is usually the best short-term solution, but the problem should be addressed as a design problem. It may be necessary to sector the site, reduce power, lower antennas, or even relocate.

THIRD-PARTY INTERFERENCE

When cellular radio was first switched on in most countries, it was common to find existing point-to-point links in the band. When the cellular radio operator was also the wireline operator, it was not unusual to find that some of these offending links were the operator's. Most of these links are easily tracked down, and after negotiation, the fixed link can be moved. In some instances, retuning to a frequency outside the cellular band is all that is necessary.

When interference persists, it manifests itself in a number of ways. The most easily recognized is interference that brings the base-station receivers out of mute (blocks channels and reduces base capacity). Another sign is calls lost during handoff. In this instance, the RBS (Radio Base Station) requests a handoff attempt and all other neighboring bases are asked to measure the field strength of the mobile and determine a base suitable for handoff.

If there is a foreign carrier with the same TX frequency as the mobile (at least in the AMPS system), it will be measured and assumed to be the mobile. If the foreign carrier records the highest field strength, the mobile is instructed to go to the base that measured the foreign carrier. Often, the mobile cannot access that base and so the handoff fails.

Although it is relatively easy to find point-to-point links that are causing interference, tracking UHF mobiles is almost impossible. A number of mobile-locating systems exist, and although they all work fairly well when a fixed mobile repeater is the source, none is really satisfactory.

The problem with mobile interferers is the intermittent operation and multipath that causes false bearings to be read; this makes it very difficult to get accurate fixes.

Because of all these difficulties, it is perhaps inevitable that some of the more persistent non-cellular users of the band will stay on, undetected, until such time as the interference to them, from the cellular system, becomes unacceptable.

SPARE PARTS

Initially, suppliers should recommend which spare parts to store. Be a little cautious, however, that the supplier will sometimes reduce the number of recommended spare parts in order to produce a lower bid price. Of course, once the system has been purchased, it is easy to recommend a few more spare parts (the same thing sometimes happens with training and test equipment).

For cellular systems, by far the most unreliable component is the RF channel equipment; adequate spare parts should be kept to allow for a one-year requirement.

Be sure to test all spare boards soon after delivery by placing them into service. A non-functional spare is of no use and a significant number of boards are delivered from the manufacturer as DOAs (Dead on Arrivals).

At least five spare antennas of each type should be on hand. These often have fairly long-lead times on delivery and are liable to fail in considerable number during the electrical storm season.

Disposable items, such as fuses and light bulbs, should be purchased in quantities adequate for two years. The frequency of failure, coupled with the cost of processing orders, makes small holdings economically unattractive. Good quantities of spare connectors should also be stored. It is false economy to save on spares, and adequate quantities should be held to ensure that they will be available when needed.

Chapter 15

BASIC SWITCHING AND TRUNKING

Until the advent of cellular radio, it was relatively easy for a switching engineer to have little or no knowledge of radio systems and for a radio engineer to know nothing of switching. In cellular radio, however, the two technologies interact inseparably. The basic concept of a switch is to connect one line (usually a subscriber) to another in such a way that any subscriber (or line) can eventually connect to any other. When the number of connections is small, this can easily be done manually. Consider the situation of the four subscribers shown in Figure 15.1. By placing the link between any two subscribers (as shown between subscribers 1 and 2 in Figure 15.1) the operator can connect them in any order.

As the number of subscribers grows, the operator's task becomes increasingly difficult; automatic telephone exchanges are needed to cope with the number of potential links.

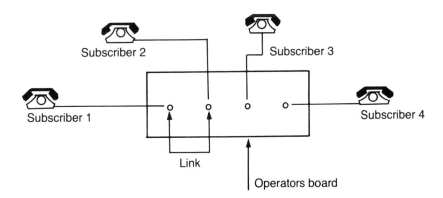

Figure 15.1   With this simple operator-controlled switch, a manual connection between subscriber 1 and subscriber 2 is made by the link.

The first automatic telephone exchange was produced in 1892 at La Porte, Indiana, US. It was electromechanical—electric switches driven by electromagnets which by mechanical movement performed the switching function. Switching was accomplished by sending pulses (dialing) to indicate the number required. In this sense, these early switches were digital.

The early switches were actuated by dialing the pulses directly, which caused them to step on to the line required (hence the name step-by-step). Later systems used various forms of memory so that the exchange switching could be done asynchronously (at a different speed) to the dialing pulses. With refinements, this type of switching and dialing remained the standard until very recently.

SPACE SWITCHES

Telephone switches were originally all "space" switches; that is, the switches physically connected one circuit to another with a connection in space. In order to make a call between two telephones, it was necessary first to physically connect the two telephones with a wire. Space switches were used to connect subscribers for the duration of the call only, so that the links could be used by other subscribers after the call was completed. Today, although a physical connection usually does not occur (because the digital switches are time-multiplex-devices), two telephones are connected by a dedicated route to each other for the duration of a call.

Figure 15.2 shows a switch in which each inlet can connect to each outlet, as well as being able

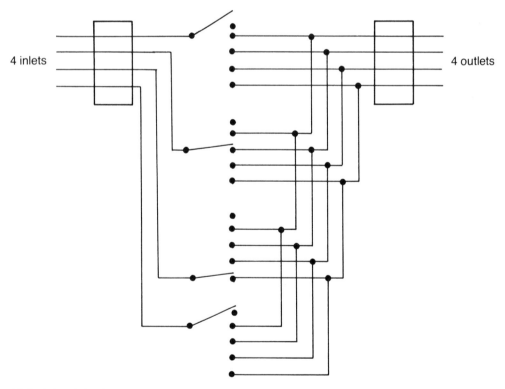

4 inlets   4 outlets

Figure 15.2   A four-inlet, four-outlet switch. Inlets 1 and 4 are in the parked or neutral position, inlet 2 is connected to outlet 2, and inlet 3 is connected to inlet 4.

to park in a neutral position. This is known as a full-availability switch, since each inlet has a path to each outlet. The number of possible paths is 16 (4 x 4 = 16).

As the switches get larger, the total number of possible paths rapidly increases. Consider a 400-inlet switch with 400 outlets; the total number of paths is 160,000 (400 x 400). Because of these huge numbers (and hence the massive amount of hardware), early switches were limited-availability, which means that each inlet could access only a limited number of outlets. For example, if the outlets per inlet are limited to 20, the total number of possible paths is 8000 (400 x 20 = 8000), which is considerably more manageable.

TIME SWITCHES

Time switches became available with digital techniques. These switches work on the principle of switching a particular inlet to a particular outlet at a certain point in time. Figure 15.3 shows how inlets are assigned their respective timeslots. The input data is then rearranged (switched) under the direction of the control store so that each incoming timeslot is connected to the desired outgoing timeslot.

In Figure 15.3, each telephone line is sampled in its respective timeslot. The telephone in

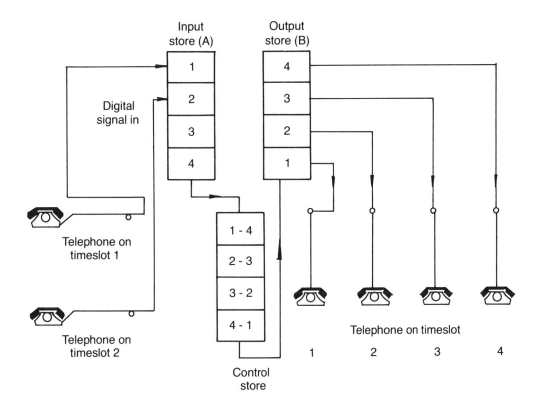

Figure 15.3 A full availability time switch uses timeslots to connect any inlet to any outlet. The timeslot translation is shown in the control store. Inlet 1 is connected to outlet 4, inlet 2 to outlet 3, and so on.

timeslot 1 on the A side is connected to the telephone in timeslot 4 on the B side. Notice that the switching is done by rearranging the timeslots, not by physical wires, so that the information can be carried by a single path between switch A and switch B.

Modern telephone exchanges (and cellular radio switches) generally use a combination of time and space switches to minimize the total hardware needed.

SPC SWITCHES

SPC (Stored Program Control) switches ordinarily use both types of switching in the one switch. As Figure 15.4 shows, modern SPC switches come in a variety of sizes and are usually designed to be housed in air-conditioned rooms. The neat suites of equipment house the processor and the switch.

Figure 15.4   A modern AXE10 from Ericsson—an SPC switch that can be used for conventional telephone or mobile telephone switching. Photo courtesy L.M. Ericsson.

SWITCH CONCENTRATORS

Figure 15.5 shows a simple concentrator switch. The switch is non-blocking, because every inlet has potential access to every outlet. Call blocking can still occur, however, because the number of simultaneous calls permitted is limited by the number of outgoing routes. In the switch shown in Figure 15.5, three simultaneous calls are allowed.

The eight inlets A–H are concentrated into the three outgoing routes (1–3) by activation of the cross-points; only one switch in any row or column can be closed at one time. This principle, called line concentration, is used extensively in all telephone switching. A radio base station acts as a line concentrator because it connects the mobile subscribers to the cellular switch in such a way that approximately 20 mobile subscribers can be connected to the switch by one channel.

The switches discussed previously have one very significant practical disadvantage: The connection between any inlet and any outlet occurs by only one path. The failure of any switch cross-point means that certain paths are no longer available. This limitation can be overcome, however, by introducing a second row of switches, as shown in Figure 15.6.

In Figure 15.6, you can see that the path between two ports (for example, B and D) can be connected by engaging the B and corresponding D row bass on any of the bars 1–3. For example B can be connected by engaging the cross connection B-1 and then 1-D or alternatively B-2 and then 2-D, and so on. This results in three paths connecting B to D. Although this configuration doubles the number of switches, it provides a very valuable redundancy in internal paths.

DTMF DIALING (TONE DIALING)

Most telephones today employ DTMF (Dual-Tone Multi Frequency) dialing. In this system, two tones are sent simultaneously to a line to indicate the desired number. Figure 15.7 shows the tone pairs and their associated numbers. The A, B, C, and D keys are not usually provided for POTS applications; they are reserved for special purposes.

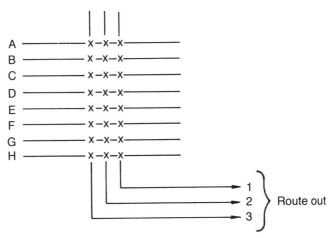

Figure 15.5   It is easiest to think of a switch as a crossbar where connections are made by activating (connecting) the cross points. This limited-availability concentrator has eight inlets and three outlets.

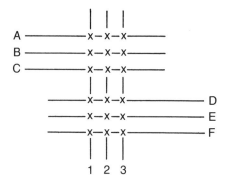

Figure 15.6 This switch has multiple internal paths between inlets and outlets.

DTMF dialing has been available for a number of decades. It was designed to take advantage of the potentially higher dialing speeds that were obtained by using code receivers with memory to store the digits and forward them as required by the switches.

In step-by-step (SXS) systems, each number dialed represents a train of pulses that cause mechanical switching in real time, making them necessarily slow. These systems are sometimes referred to as "stagger by stagger" systems by those who have seen the switches in operation.

With crossbar systems (and with modification, some SXS systems), code receivers were provided as exchange-common equipment. They were switched across the subscriber's line for the duration of dialing and could decode and ultimately store the DTMF pulses, which could then be sent further on in the network at any desired speed. Although DTMF dialing is a feature of almost every cellular telephone, it is by no means a new idea.

Each pair of tones in the DTMF scheme consists of one high-band tone and one low-band tone. This increases immunity to false decoding from voice or noise, as do other requirements such as a minimum signal-to-noise ratio and a correspondence in-level (known as twist) between tone

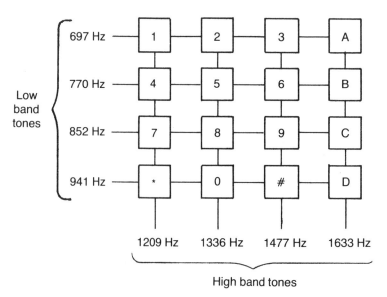

Figure 15.7 DTMF, the standard for tone dialing, consists of two tones—one "low-band" and one "high-band"—generated from the matrix shown.

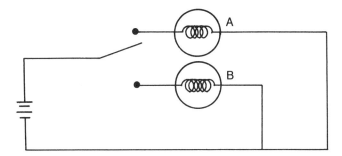

Figure 15.8   Blocking occurs in limited-availability switches, and congestion or information loss can occur in the switch. The simplest blocking switch, shown here, enables one light or the other, but not both, to be turned on at the same time.

levels for a successful decode. The tones are structured so that false decoding due to voice or noise is unlikely.

LIMITED-AVAILABILITY (BLOCKING) SWITCHES

Figure 15.8 shows the simplest limited-availability, or blocking, switch. In this example either light A or light B can be on, but not both. Because the switch can transmit information to only one of the outputs, it can lose information (for example, if a condition indicates that both A and B should be turned on simultaneously, some information is "lost" in the switch because the switch can only indicate the first state. A switch that can lose information is called a limited-availability or blocking switch.

The simplest telephone switch is an extension of the limited-availability switch, as shown in Figure 15.9. This simple uniselector switch allows a number of telephones to share a common outgoing line, but it has the disadvantage that only one telephone can use the line at any one time.

Subscribers' telephone switching stages will always be limited-availability switches. This is because one of the main functions of the subscriber's switch is to concentrate a large number of individual low-traffic telephone lines into a smaller number of high-usage lines that can be used to distribute the traffic efficiently. However, once the traffic is concentrated into parcels of about

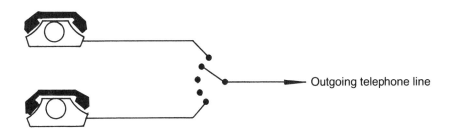

Figure 15.9   A simple uniselector line switch (which is a line concentrator with limited-availability) can be used to concentrate telephones into a limited number of lines.

0.5 Erlangs per circuit, it is efficient to use full-availability switches for onward trunking. Trunk switches are usually full-availability.

FULL-AVAILABILITY (NON-BLOCKING) SWITCHES

A full-availability or non-blocking switch is one through which it is possible to connect any idle outlet to any idle inlet, regardless of how many other connections have been made. Figure 15.10 shows the simplest full-availability switch. In this example, you can see that the switching of states A and B are independent and that information will not be lost through any limitation in the switch.

The switch concentrator shown earlier in Figure 15.5 is an example of a limited-availability or blocking switch. For example, if three of the inlets, A, B, and C, are connected to three outlets, 1, 2, and 3, respectively, then no other inlet can be connected until one of the established connections is dropped.

The switch shown earlier in Figure 15.6 is an example of a full-availability or non-blocking switch. The number of inlets and outlets must be equal in a full-availability switch.

A simple one-stage switch can easily and economically be made non-blocking for small-sized switches. Such a switch must have links from every inlet to every outlet, so the number of links increases as the square of the number of inlets. For large switches, this soon becomes prohibitive.

In 1953, Mr. C. Clos of Bell Laboratories published an analysis of three-stage switches, showing the relationship between the center switch configuration and the links. He demonstrated that for non-blocking it is necessary that each stage be non-blocking and that the number of center stages be:

Number of center switch points = 2 n − 1 = 2 x (the number of inlets / outlets per group) − 1

Figure 15.11 shows a three-stage switch.

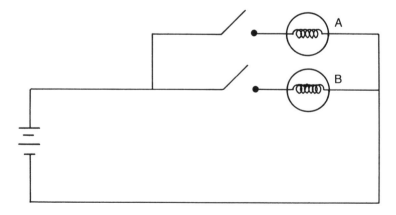

Figure 15.10 The non-blocking switch shown here allows either or both lights to be in either the "on" or "off" state. No information loss occurs due to limitations within the switch.

The path of any call may route from any inlet group to any center group by one link and from any center group to any outlet group by one link. Thus, there are K-paths through the switch from any inlet to any outlet.

It can be shown that the total number of cross-points for the switch system in Figure 15.11 is:

$$T = 2NK + K\left[\frac{N}{n}\right]^2$$

where   N = the number of inlets/outlets

n = the size of each inlet/outlet group

K = the number of center arrays

The number of center arrays of switches can be determined by imagining a switch that has all circuits busy except one inlet and one outlet. In this instance, the worst-case is an inlet group which has n – 1 active outlets and attempts to connect to an outlet group that also has only one free outlet (the one sought), but that is accessed from a different group of center switches. Figure 15.12 on the following page illustrates this. To be full-availability, the switch must still be able to switch the path between the inlet and the desired outlet, so at least one other free path must exist.

The minimum number of center switch points is:

$$(n - 1) + (n - 1) + 1 = 2(n - 1)$$

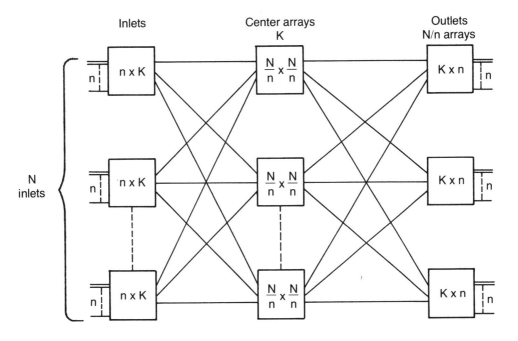

Figure 15.11 This three-stage switch has K groups of n inlets and the same number of outlets. To ensure full-availability, there must be at least two (n – 1) switch cross points at the center.

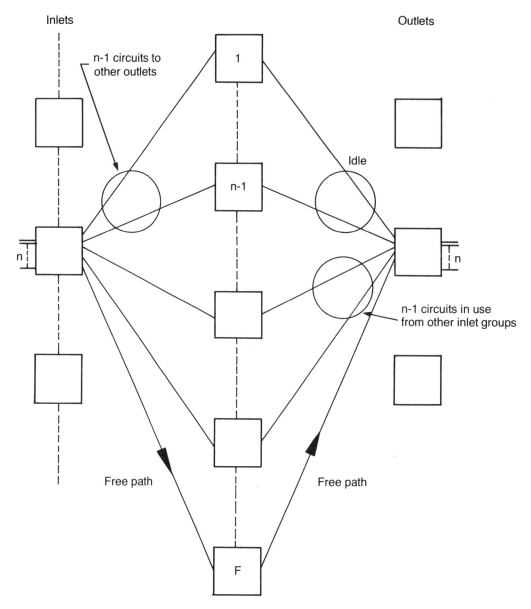

Figure 15.12 This figure shows n inlets in one group that have n − 1 circuits busy to other outlet groups. The required outlet group also has n − 1 circuits occupied. To avoid blocking, at least one free path (via center switch F) must be available.

TRANSMISSION AND TRUNKING

Telephone traffic can be carried by one, two, four, or six-wire circuits. Each of these types of circuits has its place, and it is worth considering each separately.

Figure 15.13 shows a single-wire circuit. A one-wire circuit is an Earth-return system (the Earth provides the second wire). Early telephone links were mainly single wire, or SWER (Single Wire Earth Return), because this type of circuit was cheap. The disadvantages of such circuits are many

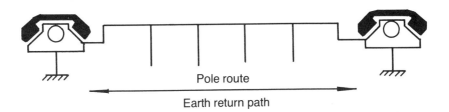

Figure 15.13 The earliest (and cheapest) transmission systems used the earth as the return circuit. This method generated considerable noise and transmission-level uncertainties.

and include variable performance due to soil resistivity, low noise immunity, and safety hazards in lightning-prone areas. Despite these problems, some single-wire circuits exist even today in rural areas.

Figure 15.14 shows a two-wire line. The two-wire line is an obvious improvement from the SWER line, and it does eliminate a lot of the SWER line problems. However, when longer routes are considered, a two-wire system has limitations when amplification is needed. Some ingenious amplifiers known as Negative-Impedance Repeaters (NIR) were developed to provide some amplification over long two-wire system routes. A conventional telephone is an example of a two-wire circuit.

The circuit shown in Figure 15.14 shows a subscriber telephone connected by a twisted pair to a telephone switch. Two-wire connections from subscribers' units to the telephone switch are usual.

This figure also shows an NIR between the two switches. The NIR provides some gain to compensate for line losses between the switches. The operation of an NIR is such that the maximum theoretical gain that can be provided on the route (end-to-end) is 0 dB (a gain of 1) before instability occurs. In practice, gains of –6 dB are more common.

On longer routes it is necessary to provide a considerable amount of gain to make up for system losses. The only practical way of doing this is to use four-wire transmission (that is, to separate

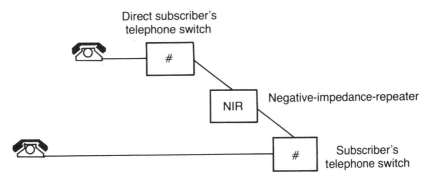

Figure 15.14 A negative-impedance repeater produces gain in a two-wire cable and provides a moderately effective means of extending transmission distances in two-wire cables.

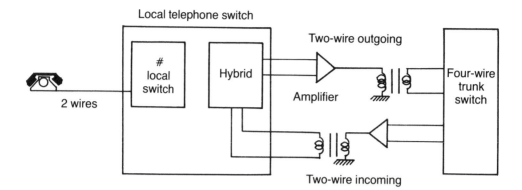

Figure 15.15 The four-wire transmission (sometimes called six-wire or four-wire E&M) enables independent amplification of the send-and-receive directions. E&M are signaling media derived over the four-wire channels.

the send-and-receive paths and amplify each). High-level (trunk-level) exchanges are usually four-wire switches, as shown in Figure 15.15. With four-wire switching, transmission without loss can be achieved in practice, with noise considerations and hybrid leakage limiting the total amount of gain achieved.

Six-wire switching can be regarded as four-wire with two "wires" used for simple signaling purposes. (Four-wire links can derive the signaling channels in a number of other ways.) The two signaling wires are known as the M lead, which transmits the outgoing signal, and the E lead, which carries the incoming signal.

SWITCH HIERARCHY

A telephone exchange consists of a switch to which 50 to 50,000 customers are typically connected. These subscribers normally have direct dialing access worldwide, which means the calls have a very wide dispersion. Consider a hypothetical town that has three subscribers' telephone switches, as shown in Figure 15.16.

If the town is relatively isolated, a fairly high percentage of the total traffic can be carried by the interexchange routes, so it would be justified to have direct trunks between those exchanges. When traffic to another area is considered (for example, to a town 200 km away), the traffic will be relatively light and the economics of providing three separate routes to the distant town may be rather poor. In this instance, a hierarchical switch, called a trunk switch, can be used effectively to concentrate the three traffic streams into one. Switch 3 in Figure 15.17 is a trunk switch.

All traffic routes from any switch in the town that are too small to justify a direct circuit can be switched through the trunk switch. In practice, the trunk switch may physically be one of the three subscribers' switches, but the trunk portion of it functions as the trunk switch shown in Figure 15.17.

This principle is also applied to international calls; a few switches collect all international traffic and disperse it to distant destinations. At the distant end, calls are routed to their destination through successively lower-ranking trunk exchanges until they finally arrive at the desired terminal exchange.

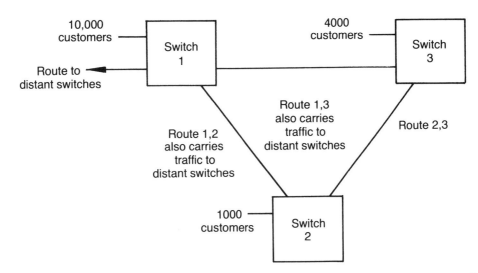

Figure 15.16 Although it may be economical to directly connect local switches in a small town, it is often more economical to route all distant traffic via one (usually the largest) local switch.

Modern digital switches do not employ a rigid hierarchical structure, but rather are able to reconfigure their routing to take advantage of the best available route for any call. Thus, a local exchange can be both a terminal and a trunk exchange and can handle transit traffic like a tandem. Exchanges that can perform this function are called nodes.

TRUNK SIGNALING

The most common form of signaling over trunks today is common-channel signaling, although voice-channel signaling is still widely used. In common-channel signaling, a different circuit is used for interswitch communications from that used for speech. Common-channel signaling enables switching to occur at very high speeds and means that voice channels are not tied up with signaling. The standard 2-Mbit, 32-channel system uses two of those channels exclusively for signaling.

In-band and out-of-band signaling use the same channels as those that are used for voice. Voice

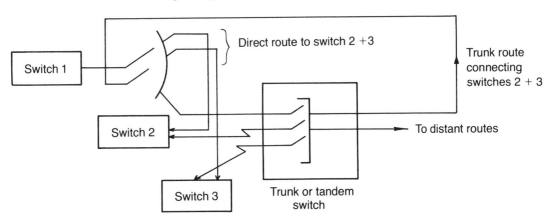

Figure 15.17 A simple trunk switch configuration

frequencies are assumed to be those in the bandwidth 300 to 3400 Hz. A commonly used in-band tone is the CCITT R1 signaling system, which uses a 2600-Hz continuous signal tone (with a notch filter to line for voice).

Cellular radio AMPS systems use 6-KHz out-of-band SAT tones to identify and control interference.

Chapter 16

TRAFFIC ENGINEERING CONCEPTS

The measurement of telephone or circuit traffic and its application to circuit dimensioning (provisioning) is fundamental to all large-scale communications systems. Traffic is measured in Erlangs (most simply, an Erlang is one circuit in use for one hour). The traffic on one telephone line can be measured with an ammeter or voltmeter, as illustrated in Figure 16.1.

When a line is looped (that is, the handset is off the hook), a DC current of about 50 mA flows. This current can be detected by an ammeter, and the DC voltage drops from the open circuit value of about 50 volts to about 5 volts. The actual loop current depends on the loop (line resistance), which can vary from 0 to 1500 Ω. When the handset is replaced, the current flow is zero and the line voltage returns to 50 volts.

A modern traffic meter uses a microprocessor connected to an interface that monitors many lines simultaneously, and it constantly scans each line to determine whether the line is in use. Figure 16.2 shows an example of a traffic measurement system.

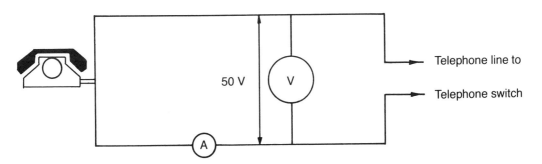

Figure 16.1   A looped (in use) telephone drops the circuit voltage from 50 volts (open circuit condition) to about 5 volts.

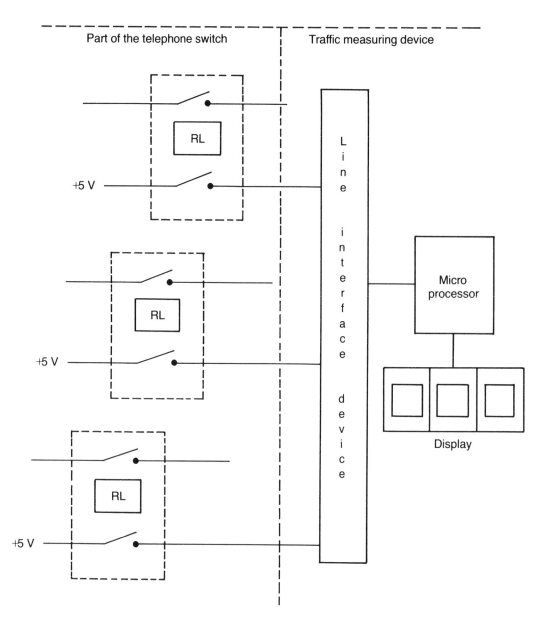

Figure 16.2 The line relay RL, which activates when the telephone circuit is seized, can be used to measure traffic. A simple integrating device is used to measure the total traffic (in Erlangs).

Traffic measurements in cellular systems should be examined on a monthly basis to ensure adequate network provisioning. Traffic can be represented by a number (of Erlangs), but remember that traffic varies with time and any practical representation is a compromise.

Fortunately, not much traffic engineering is involved in cellular radio, but you should understand some basic concepts.

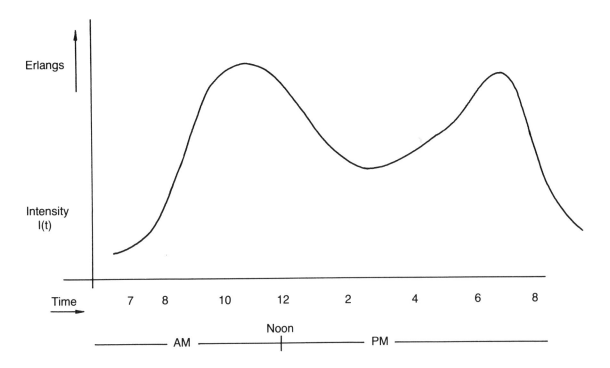

Figure 16.3   The traffic density is shown here as a function of time of day. The traffic carried by a cellular system shows two peaks characteristic of the prime "drive times." The peaks occur early in the morning and late in the afternoon.

A cellular system will have a traffic distribution somewhat like the one shown in Figure 16.3. The total volume V of traffic carried, measured in Erlang hours is

$$V = \int I(t) \cdot dt$$

This integral is a crude measure of the cellular operator's income, particularly since the call charge is largely an air-time charge. Thus, V x air-time charge = air-time revenue.

This leads to the concept of call-holding time (the time that a call is held up). From the previous equation you can see that a call lasting one hour will generate the same traffic volume as 20 three-minute calls over the same period. The call-holding time of a typical cellular subscriber is 120–180 seconds.

The average calling rate per subscriber is the total traffic at the measured time divided by the number of subscribers. Ordinarily the rate is quoted in Erlangs per subscriber per hour. This figure determines the number of subscribers who can be placed on any given system, and it varies immensely. Average calling rates from 0.01–0.045 Erlangs/subscriber have been recorded and it would appear that the calling habits of different countries are very diverse. A calling rate of 0.025–0.035 could be regarded as average.

These rates are also quoted in milli-Erlangs (0.025 Erlangs = 25 mE). To calculate the total system traffic, it is merely necessary to multiply the average calling rate by the number of subscribers.

Notice that traffic varies in many ways with time. These are the most significant variants:

- Instantaneous variation of call arrivals.
- Hourly variations that depend on demand; cellular radio usually has an early morning and late afternoon peak with a lunchtime minimum.

  The cellular peak is different from the PSTN peak, which generally occurs either side of a lunchtime minimum. The cellular peak occurs during drive times; that is, the cellular peak occurs during transit to and from work. The PSTN peak occurs after arrival at work.
- A daily pattern is usually distinguishable. There is a marked difference between weekdays and weekends, and there may be significant day-to-day variations during weekdays.
- Seasonal peaks that occur in the PSTN (for example, Christmas, Easter, Mother's Day) are likely to be reflected in the cellular network, although that reflection is sometimes negative (lower traffic as a result of holidays).
- Tariff variations can cause significant (but usually temporary) variations in the call rate. These effects, for cellular radio, usually last only three to six weeks before the original calling pattern returns.
- Long-term variations in traffic can occur over periods of months or years.

Because it is not usually economic to provide circuits (or for cellular base stations, channels) to cater to the peak demand, a compromise based on the provision of an acceptable grade of service in most instances is usually adopted.

Although traffic is measured in Erlangs (recall that an Erlang is defined most simply as one circuit in use for one hour), in practice many different units of traffic are used. The two main ones are Instantaneous Erlang and Busy-Hour Traffic (in Erlangs). Instantaneous Erlang means the number of circuits in use at the instant in question. Busy-Hour Traffic means the average number of circuits per unit time in use over the busiest hour. The traffic readings are usually taken in half-hour periods over the day, and the sum of the two adjacent half-hour blocks with the greatest total is defined to be the busy-hour traffic. This is why the busy hour usually starts either on the hour or the half hour. Table 16.1 shows the busy-hour calculation.

TIME	TRAFFIC (ERLANGS)
09.00	4.3
09.30	4.8
10.00	7.2) Largest adjacent readings – busy-hour traffic =
10.30	6.1) (7.2 + 6.1)/2 = 6.65 Erlangs
11.00	5.1
11.30	6.2

Table 16.1   Busy-hour calculation

TIME-CONSISTENT BUSY-HOUR TRAFFIC (TCBH)

The concept of time-consistent busy-hour traffic (TCBH) is the one often used for dimensioning telephone circuits. It is only a concept and has no real physical meaning. Despite that, however, circuits are provided on this basis. It attempts to obtain a weekly average traffic measurement.

This concept is best illustrated by an example. Table 16.2 shows example base-station readings taken over a week. In this example, the traffic is measured as before, for every working day. Then, for each half-hour period over the five days, the average traffic measurement is found and put in the average traffic column. From the average traffic column, the busy-hour traffic is found as before; in this case, it is (5.3 + 5.6)/2 or 5.45 Erlangs. Notice that Monday's busy hour is the highest (6.65 Erlangs). The circuits are provided only for 5.45 Erlangs, so the nominated grade of service (probability of congestion) applies only to this theoretical TCBH traffic and not to any particular day.

In congested circuits, it is likely that for significant periods of time all circuits will be busy. Because of the random nature of telephone traffic, even congested systems occasionally have free circuits and so the traffic carried is always less than the number of circuits.

Methods exist to determine the offered traffic on such circuits from the measured average traffic, but these techniques produce uncertainties that increase rapidly as the grade of service increases. Once the network becomes congested, it is difficult to measure traffic accurately enough to determine the offered traffic (and hence number of circuits needed) with any certainty. This is even more difficult in cellular radio because traffic that cannot be directed to the nearest and best cell often "overflows" to the next choice cell. Thus traffic measurements from a congested system should be used with considerable caution.

DISPERSION

Dispersion measurements are used to indicate the sources and sinks (destinations) of traffic. Although direct occupancy measurements can give the total volume of traffic, they do not give any information about the direction or origin of that traffic unless all traffic on the route has only one sink.

Dispersion measurements involve analyzing the called number to typically six digits and

TIME	TRAFFIC MON	TRAFFIC TUE	TRAFFIC WED	TRAFFIC THU	TRAFFIC FRI	AVERAGE OF 5 DAYS
09.00	4.3	2.1	4.0	3.9	2.9	3.44
09.30	4.8	4.0	5.1	3.9	4.1	4.38
10.00	7.2) 6.65	6.1	3.0	4.7	4.7	5.14
10.30	6.1)	5.4	4.5	4.8	5.7	5.3)
11.00	5.1	6.4	6.1	4.8	5.6	5.6)*
11.30	6.2	3.9	3.1	4.8	4.0	4.4
* Highest adjacent sum equals TCBH traffic						

Table 16.2   Time-consistent busy-hour averages the busy hour over a one-week period.

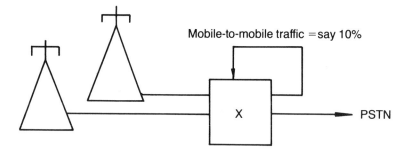

Figure 16.4   Local cellular traffic is small. 10 percent of local traffic (mobile-to-mobile) is generated by a cellular network.

recording the total holding time. This is vital in telephone switching but is not particularly important in cellular switches which generally bulk-switch all PSTN traffic to a trunk switch.

Where multiple switches are used for a cellular network, it is important to know the inter-switch traffic so that appropriately dimensioned routes can be provided. A phenomenon of local traffic applies: The percentage of local traffic (mobile-to-mobile) remains constant. As the total number of switches increases, the traffic between any two switches decreases. This is best illustrated by Figure 16.4 and 16.5.

If the local mobile-to-mobile traffic is, say 10 percent, then the dispersion to any particular switch decreases as the total number of switches increases. Figure 16.4 shows a cellular network that begins as a single switch with 10 percent local traffic. As the system grows and more switches are added, the local traffic on the original switch, and to other switches, decreases. In Figure 16.5, the three switches are assumed to be of equal size, so the traffic is distributed (approximately) evenly between switches.

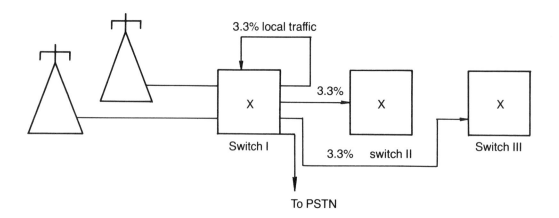

Figure 16.5   When the service expands and three local switches are networked, the local traffic remains a (nearly) constant percentage of the total traffic (all switches are assumed to be of equal size), and the gross internal traffic on any particular route decreases.

GRADE OF SERVICE (GOS)

A Grade of Service (GOS) figure is used to express the probability that a call will be lost due to switching or transmission congestion. Because it is a probability, the highest value it can have is 1.0; all calls will fail on any system that has this grade of service.

Like banks, which rely on the fact that it is unlikely that all their customers will require their money at the same time, telephone companies rely on the improbability that all their customers will attempt to place a call at one time. Banks will start a panic run if they are unable to meet the demand for funds at any time, so they must operate at a very low grade of service (that is, the probability of not having funds available to meet the demand at any time must be very low). For any one bank to do this alone would mean that it would need to keep a very large proportion of its funds available at any time in cash form. This would be very expensive. For this reason, vehicles such as the short-term money market exist to ensure that each bank has access to large sums at short notice without tying up too much of its own money.

Fortunately, telephone providers do not have quite the same problem. Usually, customers will just try again if their first call fails. But like the bank, these repeated attempts place a strain on the network, and too many repeated attempts can lead to total system collapse. The system must therefore be designed to minimize the possibility of such problems.

Telephone companies use a typical GOS range from 0.002 to 0.05. The acceptable range of call fail rates (due to equipment availability) is from 2 per 1000 to 5 per 100 attempts. For cellular purposes, the GOS varies from operator to operator, but 0.01–0.05 for base-station links and 0.002–0.001 for the switch to PSTN link are reasonable values.

DIMENSIONING BASE-STATION AND SWITCH CIRCUITS

It is normal to use a fairly high GOS for the radio path because of the high cost of the base-station channels and because an effective overflow regime exists whereby traffic offered to one base station can, if no circuits are available, be carried by a neighboring one. A GOS between 0.05 and 0.01 is ordinarily used for base stations. Table 16.3 on page 232 can be used directly to dimension bases from measured traffic. A short BASIC computer program that can be used to calculate traffic for any GOS is found at the end of this chapter. This program is not copyrighted and can be freely used.

The path from the switch to the PSTN, which carries all the mobile network traffic, is usually dimensioned at either 0.002 or 0.001 GOS because any traffic lost on this path causes calls to fail.

ALTERNATE ROUTING

Like the banker's short-term money market, there are ways of improving the grade of service without undue investment in circuits. The most common means is alternate routing, whereby traffic carried on a number of small routes can overflow to a common alternative path. This means of gaining extra circuit-efficiency is often accidently incorporated into cellular systems via the overlapping coverage provided between cells. If the most-favored (highest field-strength) base is not available, the call may be carried (overflowed) by an adjacent cell. The umbrella cell concept is an example of where this happens deliberately.

The cellular-radio provider must consider alternate routing, particularly when it is necessary to interconnect with a number of other carriers that can provide circuits and switching using a variety of different means. The simplest case of alternate routing has only one overflow route, as shown in Figure 16.6. Usually, a number of main routes (called direct routes) overflow to the same alternate route (route 2–3).

A parameter defined as cost factor is used to determine the optimum configuration where the cost of route$_n$ = C_n and the traffic carried per circuit on route$_n$ = B_n. (More strictly, B_n equals the marginal capacity, which is the marginal additional traffic carried by one additional circuit. For all but very small routes this is virtually the same as the average traffic per circuit.)

$$Cost\,factor\,H_1 = C_1 \left[\frac{C_2}{B_2} + \frac{C_3}{B_3} \right]^{-1}$$

There are various tables and computer programs that enable the cost factor to be used to determine the number of main (direct) circuits and the value of the traffic overflowed. The cost factor H_1 is known as the marginal occupancy of route 1 and is the marginal increase in carried traffic per added circuit when the offered traffic is held constant.

The number of circuits required to optimize the previous equation is usually calculated by an iterative procedure as follows:

1. Estimate B_2 and B_3 (typical values 0.5–0.8).
2. Calculate H_1 using the previous formula.
3. Find N_1 for the marginal occupancy from step 2.
4. Calculate the overflow traffic from N_1 to N_2 and determine the number of circuits required on route 2 and route 3.
5. Find B_1 and B_2 as determined from step 4 and if very different from the original values, insert the new values in step 1 and repeat.

On a computer, this is easy to do; it is quite a task if done manually.

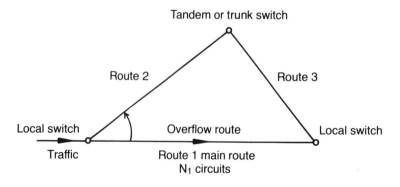

Figure 16.6   Alternate routing involves collecting small parcels of traffic into a combined trunk route.

TRAFFIC CAPACITY OF A BASE STATION

The traffic capacity of a cellular base station can be determined from the Erlang B table in Table 16.3 on the following page. (The Erlang B table is also known as Erlang's formula of the first kind, and Erlang's loss formula).

These are the assumptions of the Erlang B table:

- That the switch is a full-availability switch.
- Subscribers generate calls individually and collectively at random.
- Lost calls are cleared with zero holding time; that is, upon striking congestion, the subscriber hangs up and does not immediately attempt to redial.

The first assumption, that every channel at the base station is available for use by any incoming or outgoing caller on every attempt, is true for cellular systems.

In a traditional small telephone exchange, a good percentage of the calls will be in the local community of interest. It is more likely that the call will be local rather than distant. This means that there is a high probability that subscribers will call someone on their own switch. If the total population of local callers is small, (traditionally less than 200), then each local call reduces the probability that another local call will be made. In the extreme case of two subscribers on the switch, if a local call is made to the other subscriber, then the probability of another local call must be zero.

A small cellular base station may well have a capacity of less than 200 subscribers, but because most of these subscribers are roaming, they are individually and collectively independent. There is no reason to think that a cellular subscriber is any more likely to call another subscriber in the same cell than to call a subscriber in a different cell. Consequently, the second assumption, that subscribers generate calls individually and collectively at random, is true.

The third assumption, that repeated attempts are unlikely, also holds true, but it may be less obvious why. Because of the redial facility, a subscriber striking congestion is most likely to retry. This retry will occur on the access channel and not on a voice channel.

Congestion causes the call attempt to fail, and the caller is dropped off the system very promptly. Repeated attempts give rise to traffic on the control channel only and do not affect the voice-channel capacity.

This type of signaling is an example of common-channel signaling and is different from the co-channel signaling (that is, using a normal speech path to signal) that was used in early switching systems. In the early systems, an attempted call used speech channels for the attempt, leaving them unavailable to carry voice traffic during the call attempt.

Because all three assumptions about the Erlang B table are true, you can safely use Table 16.3 to determine traffic capacity.

NUMBER OF VOICE CHANNELS	OFFERED TRAFFIC IN ERLANG FOR THE GOS SHOWN			
	0.001	0.002	0.05	0.01
1	0.001	0.002	0.05	0.010
2	0.05	0.07	0.38	0.153
3	0.19	0.25	0.90	0.46
4	0.44	0.53	1.52	0.87
5	0.76	0.90	2.22	1.36
6	1.15	1.33	2.96	1.91
7	1.58	1.80	3.74	2.50
8	2.05	2.31	4.54	3.13
9	2.56	2.85	5.37	3.78
10	3.09	3.43	6.22	4.46
11	3.65	4.02	7.08	5.16
12	4.23	4.64	7.95	5.88
13	4.83	5.27	8.83	6.61
14	5.45	5.92	9.73	7.35
15	6.08	6.58	10.63	8.11
16	6.72	7.26	11.54	8.87
17	7.38	7.95	12.46	9.65
18	8.05	8.64	13.38	10.44
19	8.72	9.35	14.31	11.23
20	9.41	10.07	15.25	12.03
21	10.11	10.79	16.19	12.84
22	10.81	11.53	17.13	13.65
23	11.52	12.27	18.08	14.47
24	12.24	13.01	19.03	15.29
25	12.97	13.76	19.99	16.12
26	13.70	14.52	20.9	17.0
27	14.44	15.28	21.9	17.8
28	15.18	16.05	22.9	18.6
29	15.93	16.83	23.8	19.5
30	16.68	17.61	24.8	20.3
31	17.44	18.39	25.8	21.2
32	18.20	19.18	26.7	22.1
33	18.97	19.97	27.7	22.9
34	19.74	20.76	28.7	23.8
35	20.52	21.56	29.7	24.6
36	21.30	22.36	30.7	25.5
37	22.03	23.17	31.6	26.4
38	22.86	23.97	32.6	27.3
39	23.65	24.78	33.6	28.1
40	24.44	25.60	34.6	29.0
41	25.24	26.42	35.6	29.9
42	26.04	27.24	36.6	30.8
43	26.84	28.06	37.6	31.7
44	27.64	28.88	38.6	32.5
45	28.45	29.71	39.5	33.4
46	29.26	30.54	40.5	34.3
47	30.07	31.37	41.5	35.2
48	30.88	32.20	42.5	36.1

Table 16.3   Erlang B table

(continued)

Table 16.3   *(continued)*

NUMBER OF VOICE CHANNELS	OFFERED TRAFFIC IN ERLANG FOR THE GOS SHOWN			
	0.001	0.002	0.05	0.01
49	31.69	33.04	43.5	37.0
50	32.51	33.88	44.5	37.9
51	33.33	34.72	45.5	38.8
52	34.15	35.56	46.5	39.7
53	34.98	36.40	47.5	40.6
54	35.80	37.25	48.5	41.5
55	36.63	38.09	49.5	42.4
56	37.46	38.94	50.5	43.3
57	38.29	39.79	51.5	44.2
58	39.12	40.64	52.4	45.1
59	39.96	41.50	53.4	46.0
60	40.79	42.35	54.4	46.9
61	41.63	43.21	55.4	47.9
62	42.47	44.07	56.4	48.8
63	43.31	44.93	57.4	49.7
64	44.16	45.79	58.4	50.6
65	45.00	46.65	59.4	51.5
66	45.84	47.51	60.4	52.4
67	46.69	48.38	61.4	53.3
68	47.54	49.24	62.5	54.3
69	48.39	50.11	63.5	55.2
70	49.24	50.98	64.5	56.1
71	50.09	51.85	65.5	57.0
72	50.94	52.72	66.6	58.0
73	51.80	53.59	67.6	58.9
74	52.65	54.46	68.7	59.8
75	53.51	55.34	69.7	60.7
76	54.37	56.21	70.7	61.7
77	55.23	57.09	71.7	62.6
78	56.09	57.96	72.7	63.5
79	56.95	58.84	73.8	64.4
80	57.81	59.72	74.8	65.4
81	58.67	60.60	75.8	66.3
82	59.54	61.48	76.8	67.2
83	60.40	62.36	77.9	68.1
84	61.27	63.24	78.9	69.1
85	62.14	64.13	79.9	70.0
86	63.00	65.01	80.9	71.0
87	63.87	65.90	81.9	71.9
88	64.74	66.78	83.0	72.8
89	65.61	67.67	84.0	73.7
90	66.48	68.56	85.0	74.7
91	67.36	69.44	86.1	75.6
92	68.23	70.33	87.1	76.6
93	69.10	71.22	88.1	77.5
94	69.98	72.11	89.2	78.4

(continued)

Table 16.3   *(continued)*

NUMBER OF VOICE CHANNELS	OFFERED TRAFFIC IN ERLANG FOR THE GOS SHOWN			
	0.001	0.002	0.05	0.01
95	70.85	73.00	90.2	79.4
96	71.73	73.90	91.2	80.3
97	72.61	74.79	92.2	81.2
98	73.48	75.68	93.3	82.2
99	74.36	76.57	94.3	83.1
100	75.24	77.47	95.3	84.1

CIRCUIT EFFICIENCY

It is important to note that when a small number of channels are used at a base station, the circuit efficiency (which can be defined as subscribers per circuit) is very low.

Converting traffic (in Erlangs) to subscribers is relatively straightforward. First, it is necessary to know the average calling rate of a subscriber. This is typically six calls/day of 150 seconds duration, which translates to 0.03 Erlangs/subscriber (given suitable assumptions about the busy-hour pattern). Hence the number of subscribers per base is:

Traffic capacity of the base
Calling rate of subscriber

To show how the circuit efficiency varies with total circuits (channels), Figure 16.7 shows subscribers/channel versus number of channels for a 0.01 grade of service, assuming a calling rate of 0.03 Erlangs/subscriber (note that 0.03 Erlangs is often written as 30 mE).

Clearly small cells (less than 10 channels) are very inefficient in terms of subscribers/channel (and hence in dollars/channel) and should not be used unless it is unavoidable. For this reason, most operators use 7 channels (one rack of equipment) as the minimum installed capacity of a base station. At this size, the channel efficiency of 11.9 subscribers is still only about one half that of a fully equipped base.

Sectoring divides the total number of channels per cell by the number of sectors. A 21-channel, three-sector cell is, in fact, three times a 7-channel cell and therefore carries 250 subscribers (3 x 7 x 11.9 = 250). Compare this to a 21-channel omni site that has two less control channels and can carry 430 subscribers (21 x 20.5 = 430), and it is easy to see that sectoring reduces circuit efficiency.

The Erlang loss formula is based on the probability of congestion B, being:

$$B = \frac{P(N)}{P(0) + P(1) + P(2) + ... + P(N)}$$

where P(N) is the probability that the Nth circuit is busy when offered traffic A, such that:

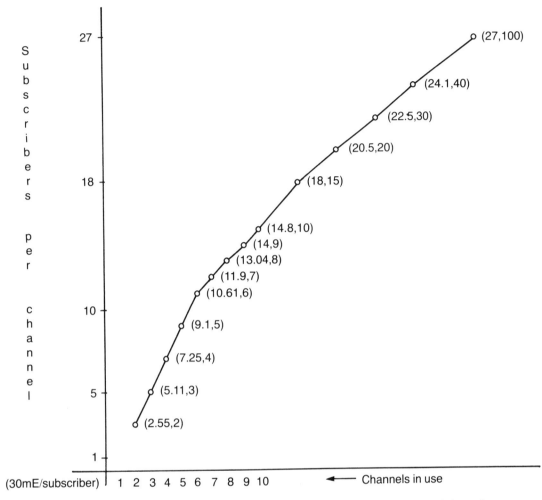

Figure 16.7   The number of subscribers per channel increases rapidly as the number of channels in use increases.

$$P(N) = \frac{A^N e^{-A}}{N_i}$$

This simplifies to:

$$B = \frac{\dfrac{A^N}{N_i}}{1 + A + \dfrac{A^2}{2} + \ldots + \dfrac{A^N}{N_i}}$$

As previously mentioned, the Erlang B table can conveniently be put into a simple computer program to allow ready calculation of any grade of service. The following BASIC program can be run on any personal computer.

```
10 REM program to produce Erlang B circuits
20 PRINT "Erlang B table"
30 INPUT "offered traffic"; A
40 INPUT "GOS"; G.
50 C = 1
60 N = 0
70 N = N + 1
80 C = 1 + N * C/A
90 B = 1/C
100 IF BG THEN GOTO 70
110 F$ = "No. of CCTS."
120 PRINT USING; F$; N
130 GOTO 30
140 END
```

This program can be used to calculate base-station link and switch-junction circuits from known or estimated traffic and nominated GOS.

Chapter 17

MORE ABOUT SWITCHING AND TRUNKING

There are many possible trunking schemes that can be used for cellular radio; the most common schemes view the cellular switch as either a terminal or a trunk switch. The type of trunking scheme and the type of access code are related in cellular systems. Those that use access codes similar to long distance codes (that is, they begin with a 0 or a 1) will switch as though they were in fact distant switches. Systems that use local telephone numbers will be connected to the PSTN in the same manner as a local telephone switch. The advantages of these schemes are discussed in Chapter 18, "Numbering Schemes."

Because of the wide community of interest in both schemes, the cellular switch is best connected at a high level in the switch hierarchy. Regardless of which trunking scheme is used, there are some common considerations of network security.

Figure 17.1 shows a typical trunking scheme of a mobile switch, with dual routes to each base and to each trunk switch. Because the mobile switch is a high revenue-earning switch (and is often classified as a premium service), it is a good idea to use route diversity with half the circuits on each route and with automatic changeover in the event of failure of one route. This can be achieved economically using modern digital links. Similarly, when an expansion to more than one switch is contemplated, it is worth considering placing the second switch in a different physical location.

SWITCH HIERARCHY

It is recommended that the mobile switch be placed high in the network hierarchy. The advantage of high-level trunking is that, in general, fewer switches are required for the average call, so trunking costs and transmission losses are minimized.

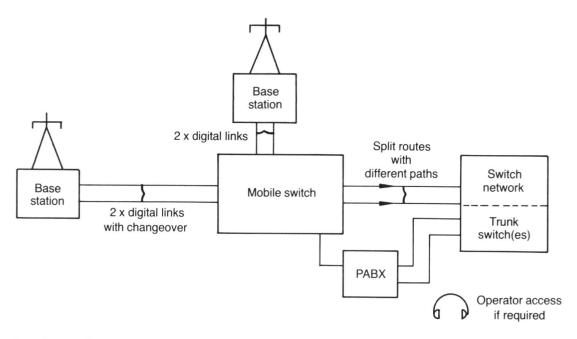

Figure 17.1 A simplified cellular network consisting of switches, base stations, links, and PSTN

To visualize the relative position of a mobile switch, consider a typical city network, as depicted in Figure 17.2. The mobile switch ordinarily draws its customers from all over the city and therefore has no geographical center of interest (except perhaps the CBD). If the mobile switch is connected low in the hierarchy—for example, at a primary center—then most calls must be routed through a number of switches before reaching their destinations. This switching delay is true for both mobile-originated and mobile-terminated calls.

The worst possible connection level from the viewpoint of network efficiency is at the local switch level, where, the two-wire transmission is the first problem. All mobile circuits are four-wire and so a two- to four-wire hybrid must be used on all routes resulting in unnecessary losses. Further, as Figure 17.2 shows, the community of interest of a mobile customer is the whole city, and most calls must be routed to a different area using the network hierarchy. This involves more switching paths and hence more loss than would be the case for higher-level switching. The mobile switch ordinarily signals other exchanges in the standard format of the country (for example, CCITT's signaling system number 7 or R2).

SWITCHING

The mobile switch is essentially an SPC (Stored Program Control) switch that enables connections to be made between the mobile bases and the rest of the telephone network.

Smaller switches may consist of a simple single-stage time switch with full redundancy (or N + 1 redundancy) of the switching stage for reliability. Bigger multistage switches use a combination of space and time switching. Figure 17.3 shows the structure of a mobile switch.

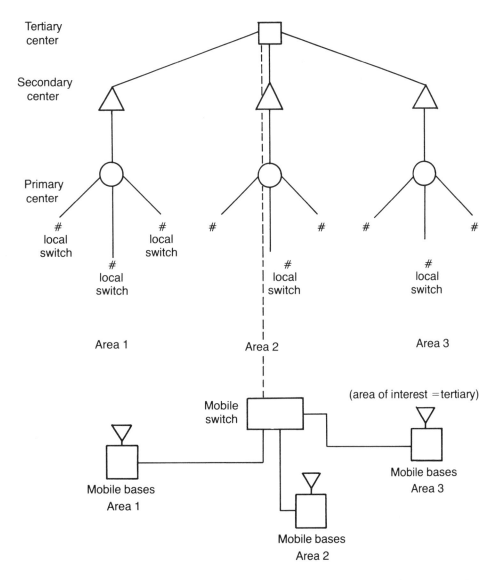

Figure 17.2   The mobile switch has a community of interest that includes the whole of the local service area of the PSTN. The switch should therefore connect at a high level in the trunk network.

The switch is composed of four main parts: the terminating circuitry, the switch, incoming and outgoing interfaces, and controllers. These elements are discussed in the following sections.

TERMINATING CIRCUITRY

This circuitry connects the switch with the outside world. It transmits instructions (data) and voice (usually coded as data). The terminating circuits, like all other parts of the switch, are controlled by the processors and the stored programs.

Alternate routing of trunk groups is often provided, and four alternate routes are typically available. Trunk routes from the switch can be either unidirectional or bidirectional; the numbers

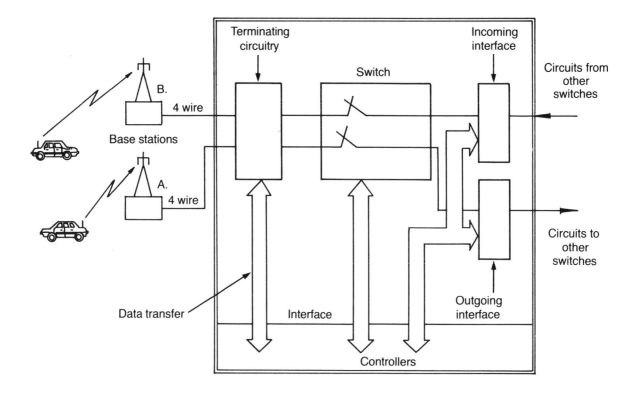

Figure 17.3 Simplified block diagram of a mobile switch

and size of each type of route is usually limited. At least 50 such routes are ordinarily available from a medium-size switch. This allows traffic to be distributed in a cost-efficient way and can be used to avoid wireline toll charges when the system operates over a number of charge zones.

THE SWITCH

In cellular radio systems, the switch is normally a non-blocking (full-availability) switch. (Non-blocking means that every inlet has a path to every outlet.) Earlier switches and some simpler switches are limited-availability types, which means that each inlet has only a fixed number of paths (often in multiples of 10) to any outgoing route. In practice, this arrangement means that the switch can block a call, even though there are free outgoing circuits, because all internal paths in the switch are in use.

INCOMING AND OUTGOING INTERFACES

Incoming and outgoing interfaces connect the switch to the outside world and provide the necessary signaling and signaling translation so the switch can communicate with other switches.

The signaling between the mobile switch and the PSTN can take many forms. These are the forms most often encountered:

- Dial pulses
- MF
- MFC R2
- DTMF
- CCITT system number 7

To make matters even more complicated, the "standard" signaling systems can have many variations, and it is usual that the signaling at the cellular switch must be tailored to the local PSTN version of the standard signaling format. This usually involves considerable software costs.

The cellular switch will have limits on the usage of the outlets/inlets for their various functions. These functions provide for limited (but efficient) numbers of:

- Both-way junctions
- Unidirectional junctions
- Total number of separate trunk groups
- Maximum trunks in a group
- RF cell groups
- Channels in each RF cell group
- Intersystem data links
- Teleprinters
- Tape decks
- Voice recordings
- Three-party conference lines
- Tone receivers
- Alternate routing patterns

Any or all of these functions may have inherent limitations on the number of outlets available for each function; these limitations do not usually present problems, but they do restrict some fully loaded switch configurations.

CONTROLLERS

The controllers consist of processors, memory, software, and hardware that enable the switch to perform the functions required of it. The processors in a cellular switch normally control many non-switching functions such as system monitoring, diagnostics, bill record-keeping, and alarm monitoring.

Notice that although billing records are kept, most cellular switches cannot process these records. Billing records are usually transferred to tapes that are then processed using external equipment. Housekeeping routines ensure that the use and availability of resources (including trunks, receiver channels, and scratch pad memory) are continually checked.

Figure 17.4   A complete Motorola EMX500 with a subscriber capacity of 15,000 consists of only eight racks of equipment. The racks are usually wired as a single suite. Tape drives and digital voice announcement equipment are included. Photo courtesy Motorola Communications and Electronics, Inc.

CELLULAR SWITCHES

A cellular switch can be thought of as a trunk switch that performs a large number of non-switching functions. For this reason, a cellular switch is usually considerably larger (physically) than a trunk switch of the same capacity. Figure 17.4 shows a cellular switch, and Figure 17.5 shows the wiring of a cellular switch.

The cellular switch is characterized by these basic parameters: total number of ports, total PCM ports, and Busy Hour Call Attempts (BHCA). These parameters are discussed in the following sections.

Figure 17.5   A cellular switch is quite complex internally, as can be seen from the mass of interconnecting cables behind the neat cover panels of a Motorola EMX500. Photo courtesy Extelcom

TOTAL NUMBER OF PORTS

The total number of ports is the sum of the inlets and outlets (whether to the switched network, the base stations, or peripheral equipment such as recorded announcement machines). These inlets/outlets are in multiples of 24 or 30 channels.

Note that small switches using digital links to the bases consume port capacity as a function of link size rather than channels in use (that is, a 2-Mbit, 32-channel link uses 30 ports regardless of the number of active channels). Figure 17.6 illustrates using MUX equipment to save inlet ports. The link between the MUX and the cellular switch should be less than 250 meters unless repeaters are used. Back-to-back MUXs can be used as repeaters.

TOTAL PCM PORTS

The total number of PCM ports equals the total number of PSTN and radio-voice-channels. There may, however, be other ports in addition to the PCM ports.

BUSY HOUR CALL ATTEMPTS (BHCA)

Varying from about 1000 to 1,000,000, BHCA measures the maximum number of call attempts that can be handled during a busy hour. This maximum, reflecting the processing power of the switch, will limit the number of customers that can be connected to the switch.

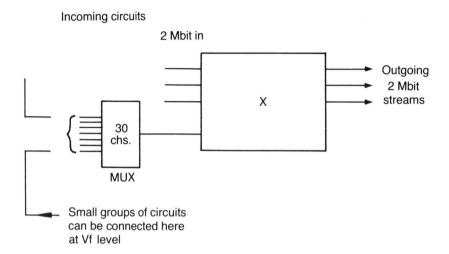

Figure 17.6   Small groups of circuits can be VF-connected to save inlets. Note that every 2-Mbit stream occupies 30 inlets regardless of how many inlets are actually active.

ROAMING

For automatic roaming, it is necessary that the switches communicate with each other and exchange information. This exchange is usually done with a dedicated digital link that ties the switches together. At the time of writing, such a setup was practical only between switches from the same manufacturer (although some bilateral arrangements have allowed some manufacturers to interconnect with one other).

NMT switches have an interconnect specification and can all be connected together regardless of the source of manufacture.

SWITCH PERIPHERALS

Switch peripherals are pieces of equipment associated with the cellular switch and under the control of the switch CPU. This equipment includes system statistics and billing tapes, PSTN links, PABX (optional), and other items detailed in Figure 17.7 and described in the following sections.

SYSTEM TAPE

The system tape loads the system software and system parameters that enable the cellular network to operate in its current configuration. Additional network equipment will be associated with a system update to inform the switch of the new configuration.

The system tape contains details of both the subscriber's file and the system files. The subscriber's file capacity can vary from a few hundred to 200,000. The subscriber's file contains

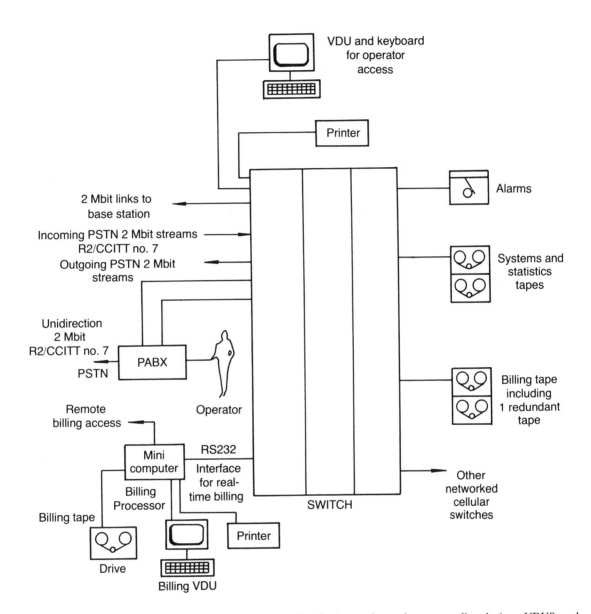

Figure 17.7 The switch CPU has to control peripheral equipment such as printers, recording devices, VDUS, and perhaps a PABX interface.

operational data about the subscriber in the processor memory. The stored subscriber data is as follows:

- Home area
- Service status (for example, non-payment)
- Serial number
- Class of service

- Last known location
- Call forwarding and no-answer transfer information

When subscriber information is updated, it is added to the processor memory immediately unless a call is in progress, in which case it is temporarily delayed. The updated subscriber data is also written onto the system tape. The updated information is usually entered using a local or remote keyboard.

The system file contains the following information:

- Routing information on all trunks
- Cell parameters (most of which are downloaded to the base station in an operational system)
- Cell numbers, channels, and channel addresses
- Handoff information (neighboring and directed retry calls)
- Hardware configuration, including peripherals and internal equipment
- System parameters

This information can be modified in real time from a keyboard.

STATISTICS TAPE

The statistics tape is manufacturer specific and stores details about the system performance including system statistics, traffic and equipment statistics, and subscriber statistics. The statistics tape records such things as system outages, channel outages, channel blocking, congestion, channel usage, handoffs, call attempts, call completions (cellular and land line), traffic, and call-holding times. Often, this tape can be read only by the supplier and may therefore be of limited use to many operators. This is particularly true if the statistics tape is used as a management information system. For this reason, elaborate management software is often incorporated in the billing systems.

PSTN LINKS

PSTN links are usually digital streams of 24- or 30-channel unidirectional traffic (that is, the streams are either incoming or outgoing rather than two-way, although two-way steams may also be permitted).

Typically, the traffic will be about 70 percent outgoing and 30 percent incoming, depending on local calling habits. The cellular system must be designed to be compatible with the PSTN and so must use the same signaling format. The two most common standards are R2 (which has almost as many variant forms as there are countries using it) and CCITT signaling system number 7. The equipment supplier must have a detailed description of the actual signaling system used in the local PSTN so that the hardware and software can be tailored accordingly.

PABX

A PABX is generally not required but can be provided as an additional subscriber service to allow a personalized answering service. In some locations (for political or technical reasons), having a PABX as the PSTN interface may be necessary. Note that unless the PABX uses the same signaling system as the PSTN, additional and probably expensive decoding hardware and software will be needed at the cellular switch.

ROAMERS FILES

Roamers files contain block-validation information (for example, all subscribers in a certain foreign group are permitted access) and negative-file information (for example, stolen units and bad debts).

BILLING SYSTEM

The billing system is a separate entity from the switch and may have no physical connection because the billing can be done by processing the raw data on the billing tape at a remote location. The billing computer for an average-sized cellular operator is a minicomputer. If the operator is a wireline provider, however, the billing may well be integral with the wireline billing and use a mainframe computer. For real-time billing (that is, billing available instantly when requested), a data interface between the computer and the switch is necessary. Most billing systems will have limited real-time capacity so real-time billing is usually reserved for a subgroup of flagged customers. These customers are usually short-term renters.

The link may be RS232; like other such world "standards," RS232 has many non-standard forms, so the type of RS232 should be ascertained from the switch provider. The "standard RS232" is EIA-232D, published in January 1987, which conforms with CCITT V.24 and V.28 and ISO (International Organization for Standardization) IS 2110. These three references give functional, electrical, and physical standards, respectively.

The billing system may or may not also have an integral MIS (Management Information System) that also logs and analyzes system functions, such as channel usage, outages, traffic, and other housekeeping. The MIS may be interactive (that is, system commands such as subscriber validation can be input from the billing/MIS computer). Most billing systems expect a number of remote terminals to be operating simultaneously off the host computer.

ALARMS

All alarms, both local and from the base stations, are reported to the cellular switch. Often the cellular switch features remote access to the alarm status to allow remote monitoring. Some systems also have remote access to each base station.

Alarms are usually divided into two or more categories, including major alarms (which affect service to a degree likely to be noticed by the user) and minor alarms (which may lead to partial reduction in capacity). The alarms are classified by the operation depending on the severity of the disruption to service of a particular fault. In a large city, the complete loss of one base station could be a minor problem, whereas in a small city that has only one base station the same loss would be a major problem. Lower levels of severity exist when, for example, a redundant unit is faulty and is switched out of service, or a simple channel is blocked in a large base station.

ROUTING

Dialed digits can be deleted or prefixed as required to ensure correct onward-routing. Alternate routing may be specified for various trunk groups, enabling the most economical route to be selected. This is particularly important when the switch operates over a number of PSTN charge zones.

CALLS TO/FROM MOBILES TO PSTN

For incoming calls (from the switched network to a mobile subscriber), the significant digits of the called number must be communicated to the control circuits. For example, if the mobile subscriber's number is OAB-CDEFGH and OAB signifies a mobile number, only the last six digits are required. The mobile switch is thus structured as a group-switching stage.

HANDOFFS

Handoffs involve switching the call from a channel on one cell to a channel on another. The procedure is internally quite complex and sometimes results in a small number of calls lost within the switch. Typically, the switch handoff success rates are about 98 percent and losses attributable to other factors are dominant.

A handoff is initiated by the base station when the base scanning receiver detects low signal-to-noise ratio, low signal power, or a foreign SAT tone. The base station controller then requests a handoff from the switch.

CALL SUCCESS RATES

As in all telephone switching systems, many call attempts do not result in completed calls—the called party is busy or not in attendance (the main causes of incomplete calls) or because the call failed for system reasons. In a cellular switch, completed calls can be expected to be distributed as follows:

- 30 percent land to mobile
- 65 percent mobile to land
- 5 percent mobile to mobile

These have success rates of:
- 50 percent for land-to-mobile calls
- 85 percent for mobile-to-land calls
- 50 percent for mobile-to-mobile calls

The low success rate of attempted calls to mobile units largely accounts for the smaller traffic in that direction. This is due mainly to the mobile being unattended, out of range, or switched off at the time of the call attempt.

DISCONNECTION

Disconnection ordinarily occurs when either party hangs up, but the release can be missed if the subscriber drives out of range (for example, into a garage) before concluding the call. Because it is possible to use VOX (voice-operated switching) as a power-saving feature, loss of carrier alone does not necessarily mean that the mobile is out of range. For this reason, it is necessary to have an audit instruction that instructs the mobile transmitter to key up. If it fails to do so, the call is then disconnected.

UNCHARGED LOCAL CALLS

Many networks worldwide have uncharged local-call access. This approach presents no problems except in those countries that have taken it further and offer local-access (uncharged) calls only, as an option, and have subsequently installed telephones without charging or metering equipment.

These numbers are normally barred access to any trunk numbers and thus cannot reach mobile telephone numbers unless the mobile telephone system has a local telephone number. In either case, they cannot be charged for local calls originating from local-call-only numbers. This problem is usually overcome by adding a PABX with access permitted to "local" subscribers, who then book a manual call to the operator, as illustrated in Figure 17.8. Because this method is clumsy, it is a good idea in this situation to establish a new category of mobile subscribers who can (optionally) elect to receive local calls from non-metered numbers at their own expense.

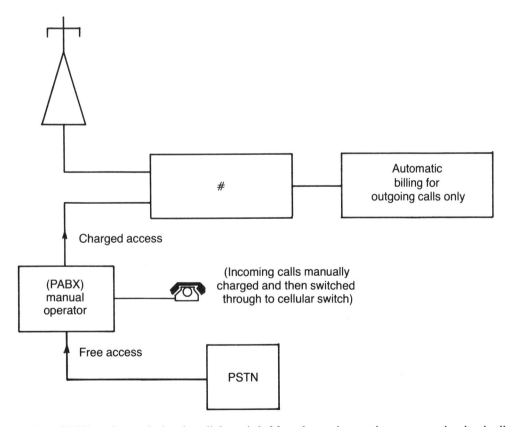

Figure 17.8   A PABX can be attached to the cellular switch. Manual operation may be necessary when local calls are (generally) uncharged and automatic line identification is not available.

In order to allow access by uncharged subscribers, a special access code is needed to trunk the barred subscriber directly to the mobile switch. The call would be marked from the barred terminal exchange as a non-metering call. It would be diverted automatically to the mobile switch, where it would be connected normally (except that a tone would be sent to indicate to the mobile subscriber that the call is a reverse-charge call).

The worst aspect of using operators to manually connect calls is the same problem that led to a strong drive from the 1950s through the 1970s for automation—the cost escalates linearly as the system grows. Because cellular switches can be as large as 50,000 lines, the number of operators needed can be staggering.

During the normal day shift (and dependent somewhat on caller patterns and habits), operators are needed at the rate of 1 per 100 customers manually connected. The operators, of course, handle only incoming calls from the PSTN to the cellular switch. Outgoing calls from a mobile are fully automatic.

DTMF overdialing could be offered as a solution to this situation and would require the land-line user to dial the PABX (which automatically answers) and then dial further digits to indicate the mobile number sought. This solution has two drawbacks: The first is that a network without IDD access is unlikely to have DTMF telephones; the second is that, without automatic-line identification (again unlikely to be available), only reverse-charge (to the mobile) calls can be made.

SWITCH LOCATION

Wireline operators have traditionally placed their main switches at what is known as the "copper center." The cheapest location for a land-line switch is the one that minimizes the total length of cable to the subscribers (and other exchanges). The location that minimizes the amount of copper (or total cable length) is known as the copper center.

Using a strictly mathematically accurate configuration to minimize the total length of copper is complex. However, the copper-center concept is relatively simple, as you can see in Figure 17.9.

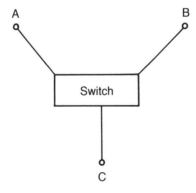

Figure 17.9   The copper-center concept. A PSTN subscriber switch is placed at the copper center to minimize the total length of cable used.

Here, the switch is located at the "center of gravity" of the copper mass feeding A, B, and C. For this reason, conventional wireline switches are usually found at the population centers.

It should be noted that cellular switches do not have subscriber links. Instead they have "trunk" links to the bases and the PSTN. Because only a relatively few links are involved—and with modern technology the cost of these links is not strongly distance-dependent—the concept of "copper center" is no longer dominant in choosing a switch site.

A wireline operator almost invariably chooses to place the cellular switch in an existing high-security building, which for the wireline operator is the cheapest location. Such a site will be even more economical if the chosen building has a trunk switch (that is, the trunk link costs will be low).

Also consider that the links to the base stations and the switch site should be well-located with respect to the bases. Base-station links are generally microwave or optical fiber. When microwave links are used, ease of link access to the switch site is a primary consideration. Whether microwave or fiber-optic links are used, the link cost is not strongly dependent on the link length.

NON-WIRELINE SWITCH LOCATIONS

When non-wireline switches are used, there is a wide choice of switch locations. Rental properties can be used for switch rooms but are not recommended for the following reasons:

- Insecurity of tenure
- Lack of security where other tenants are involved
- Cost of moving the switch should it ever be necessary
- Unscrupulous landlords, who know the cost of relocation, could demand unreasonable rents
- Lack of control over the building, its expansion, and use

Purchased properties are much more practical for switch rooms. Because there is no real need for a central location, costs can be controlled by locating the switch outside the expensive inner city area. A site that minimizes the total number of links will probably also minimize the cost.

Chapter 18

NUMBERING SCHEMES

There are two basic types of numbering schemes used in cellular radio: "local" and "access-code." In a local numbering scheme, the mobile has a number similar to a PSTN subscriber. In an access-code scheme, an access-code (similar to a long-distance call access code) is used. In either case, the numbering scheme should be consistent with the scheme used in the fixed network; that is, the mobile subscriber is regarded as either a local or a long-distance subscriber to and from the fixed and mobile networks. (Otherwise, extensive customer re-education is necessary.)

For example, if the subscriber has a number code of CDEFGH and an access code of OAB, then a fixed-telephone subscriber must always dial OAB CDEFGH. This requirement allows, in most countries, full automatic roaming (provided that the number and access-code are unique). The exception to this requirement is that the subscriber who has an access code can, beneficially, be regarded as a local subscriber of the nearest major city for mobile-originated calls. The subscriber can originate calls in this city without inserting the city's access code. The following example shows how these calls are made.

- To a mobile (via either the fixed or mobile network) dial:

    OAB CDEFGH

- From the mobile (to the fixed network) dial:

    (access code) + xxxxxx (subscriber's digits)

    or

    xxxxxx (subscriber's digits only)

Note that the nearest main town (which is usually where the switch is located) does not need to be physically very near at all; it can be a few hundred kilometers away.

Some continuous-service areas are very large and the temptation (or need) may arise to delete the local access code to more than one city within a continuous-service area. Deleting the code can be achieved by categorizing mobiles as local to a particular area.

To have more than one "local-call" area, the bases must be designated (by allocation of SID codes) as belonging to a particular local area so that all mobile units originating calls from that area are classified as either local to it or not. The mobile units indicate the home area by the first digit(s) of the MIN2 code.

An alternative approach requires the access code to always be dialed. Although this means dialing redundant digits, it makes the numbering scheme completely compatible with the fixed network scheme (that is, the mobile is always regarded as "long distance").

Where service areas overlap, the best course is to delete the option to omit the local access code. Otherwise, a good deal of confusion will arise over which local service area the mobile subscriber is attempting to access. Always requiring mobile telephones to dial the full access code solves this problem.

Figure 18.1 shows the result of overlapping coverage in areas with different dialing schemes.

ACCESS NUMBER VERSUS LOCAL NUMBER

A local numbering scheme has the advantage that, when used in the home city (where most of the traffic will be), there is no need to dial extra digits for an access code. The disadvantage of a local numbering scheme is that, in the roaming environment, it is not possible to distinguish a called number as a mobile unit except by adding an additional and usually complex pattern of digits.

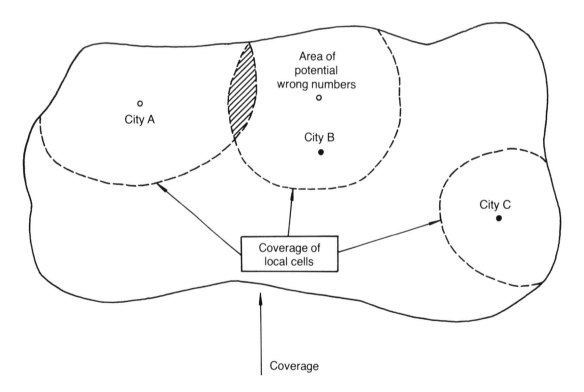

Figure 18.1 Wrong numbers can occur if coverage areas with different dialing plans overlap.

For example, consider the situation shown in Figure 18.2. A roaming mobile is in an automatic roaming system, where a call is sent from the PSTN to a mobile with no access code. If a PSTN subscriber in city C calls the mobile, which is ordinarily in city B but is roaming in city A, then the call is recognized as a call to a local subscriber in city B and switches through accordingly, until it reaches the mobile telephone switch. In an automatic roaming system, the mobile in the foreign area is either logged on in city A automatically or logged on when the subscriber makes the first call. In either case, the home switch in city B is informed that the customer is now operating in city A. The mobile switch holds the call and ultimately switches it through to the mobile in city A. Thus, links from C to B to A are tied up on this call.

When an access number is used, more economical trunking occurs. Figure 18.3 shows roaming with an access code. When an access code is used (for example, OAB), a call originating in city C to any number beginning with OAB is directly routed to the nearest local mobile switch (in this case, in city C). This switch then analyzes the called number and determines the location of its home switch by examining the digits following the access code. The switch then calls via the common-channel link to the home switch. If the called subscriber is available, then the call switches through. If, however, the subscriber has roamed, this information can be transferred back to mobile switch C so that it can establish the call directly with city A (not via city B).

Of course, if all local PSTN switches thoroughly analyzed the called number, it would be possible to identify mobile phones whether or not an access code was used. Then the procedure described earlier for a mobile with an access number could be applied to a local-number-only scheme. This procedure is not practical with electromechanical switches, however, and it would be costly even in electronic exchanges.

The advantages of common-channel signaling should now be apparent: Valuable speech links are not needed for signaling, and switching can be done after the final route is determined.

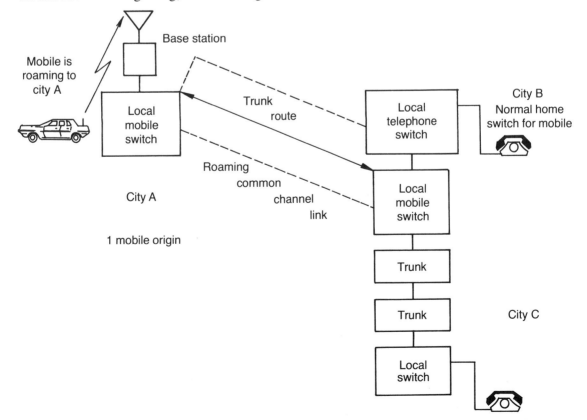

Figure 18.2   Roaming can unnecessarily tie up the trunk network if a cellular roaming access code is not used.

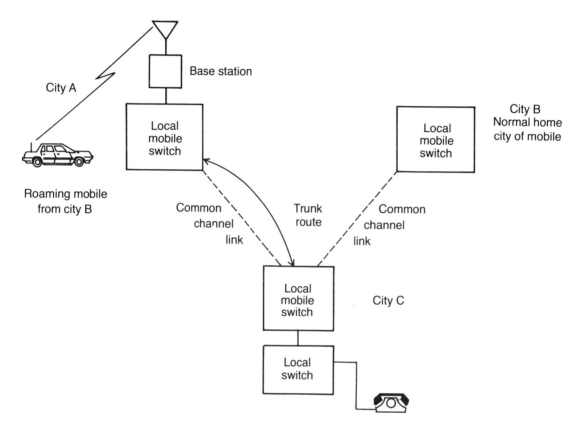

Figure 18.3 The mobile telephone switches can efficiently perform the trunk switching when an access code is used.

Complicated numbering schemes that are not user-friendly should be avoided at all costs. Many subscribers have considerable difficulty mastering the techniques required to use a mobile telephone, and this difficulty would greatly increase if a new numbering scheme had to be mastered as well. The disadvantages of the simplest numbering scheme (a local number) have been discussed. It appears that the common (country-wide) access code (that otherwise is consistent with the conventional network numbering scheme) is the most favorable approach. In addition to the advantages already mentioned, the access code also alerts anyone calling a mobile that there is a toll charge for the call. This information is valuable where calling-party (A party) charging is applied.

MIN NUMBERS

The mobile has a 10-digit identification code made up of two parts: MIN1 and MIN2.

MIN1

MIN1 is the first three digits of the number code. These digits should be allocated according to the country of origin, following CCITT Recommendation E.212, which provides world allocation of three-digit mobile country codes (MCCs) and sets up a unique worldwide identity for all mobiles. International roaming is practical only if all countries follow this coding system. (Note that these numbers are different from the international dialing codes.) Table 18.1 lists the CCIR mobile codes for various countries.

ZONE 2	
CODE	COUNTRY OR GEOGRAPHICAL AREA
202	Greece
204	Netherlands
206	Belgium
208	France
212	Monaco
214	Spain
216	Hungary
218	German Democratic Republic
220	Yugoslavia
222	Italy
226	Romania
228	Switzerland
230	Czechoslovakia
232	Austria
234	United Kingdom
238	Denmark
240	Sweden
242	Norway
244	Finland
250	Soviet Union
260	Poland
262	Federal Republic of Germany
266	Gibraltar
268	Portugal
270	Luxembourg
272	Ireland
276	
278	Malta
280	Cyprus
284	Bulgaria
286	Turkey

ZONE 3	
CODE	COUNTRY OR GEOGRAPHICAL AREA
302	Canada
308	St. Pierre and Miquelon
310	United States of America
330	Puerto Rico
332	Virgin Islands (USA)
334	Mexico
338	Jamaica
340	French Antilles
342	Barbados
344	Antigua
346	Cayman Islands
348	British Virgin Islands
350	Bermuda
352	Grenada
354	Montserrat
356	St. Kitts
358	St. Lucia
360	St.Vincent & the Grenadines
362	Netherlands Antilles
364	Bahamas
366	Dominica
368	Cuba
370	Dominican Republic
372	Haiti
374	Trinidad and Tobago
376	Turks and Caicos Islands

Table 18.1.   MIN1 numbers recommended by CCIR

(continued)

Table 18.1   *(continued)*

ZONE 4	
CODE	**COUNTRY OR GEOGRAPHICAL AREA**
404	India
410	Pakistan
412	Afghanistan
413	Sri Lanka
414	Myanmar
415	Lebanon
416	Jordan
417	Syria
418	Iraq
419	Kuwait
420	Saudi Arabia
421	Yemen Arab Republic
422	Oman
423	Yemen Democratic Republic
424	United Arab Emirates
425	Israel
426	Bahrain
427	Qatar
428	Mongolia
429	Nepal
430	United Arab Emirates (Abu Dhabi)
431	United Arab Emirates (Dubai)
432	Iran
440	Japan
450	Korea
452	Viet Nam
454	Hong Kong
455	Macao
456	Kampuchea
457	Laos
467	Democratic People's Republic of Korea
470	Bangladesh
472	Maldives

ZONE 5	
CODE	**COUNTRY OR GEOGRAPHICAL AREA**
502	Malaysia
505	Australia
510	Indonesia
515	Philippines
520	Thailand
525	Singapore
528	Brunei Darussalam
530	New Zealand
535	Guam
536	Nauru
537	Papua New Guinea
539	Tonga
540	Solomon Islands
541	Vanuatu
542	Fiji
543	Wallis and Futuna Islands
544	American Samoa
545	Gilbert and Ellice Islands
546	New Caledonia and Dependencies
547	French Polynesia
548	Cook Islands
549	Western Samoa

(continued)

Table 18.1   *(continued)*

ZONE 6	
CODE	**COUNTRY OR GEOGRAPHICAL AREA**
602	Egypt
603	Algeria
604	Morocco
605	Tunisia
606	Libya
607	Gambia
608	Senegal
609	Mauritania
610	Mali
611	Guinea
612	Ivory Coast
613	Upper Volta
614	Niger
615	Togolese Republic
616	Benin
617	Mauritius
618	Liberia
619	Sierra Leone
620	Ghana
621	Nigeria
622	Chad
623	Central African Republic
624	Cameroon
625	Cape Verde
626	Sao Tome and Principe
627	Equatorial Guinea
628	Gabon Republic
629	Congo
630	Zaire
631	Angola
632	Guinea-Bissau
633	Seychelles
634	Sudan
635	Rwanda
636	Ethiopia
637	Somali
638	Djibouti
639	Kenya
640	Tanzania
641	Uganda
642	Burundi
643	Mozambique
645	Zambia
646	Madagascar
647	Reunion
648	Zimbabwe

ZONE 6	
CODE	**COUNTRY OR GEOGRAPHICAL AREA**
649	Namibia
650	Malawi
651	Lesotho
652	Botswana
653	Swaziland
654	Comoros
655	South Africa

ZONE 7	
CODE	**COUNTRY OR GEOGRAPHICAL AREA**
702	Belize
704	Guatemala
706	El Salvador
708	Honduras
710	Nicaragua
712	Costa Rica
714	Panama
716	Peru
722	Argentina
724	Brazil
730	Chile
732	Colombia
734	Venezuela
736	Bolivia
738	Guyana
740	Ecuador
742	Guiana
744	Paraguay
746	Suriname
748	Uruguay

CCITT Recommendation E.213 states the following:

- The numbering plan should allow standard telephone charging and accounting principles to apply.
- Each administration should be able to develop its own numbering plan.
- It should be possible to change the international roaming identity without changing the telephone number allocated to the mobile.
- Roaming without constraints should be possible.
- This numbering plan refers only to interconnection with the PSTN (that is, it does not apply to mobiles that are not interconnected).

Some countries, notably the US, use a local 10-digit access number for cellular phones. This system can cause a good deal of confusion when international roaming is contemplated. But because the US is virtually the only country doing this, much of the problem can be overcome by noting that the 10-digit codes used in the US never have 0 for the seventh digit.

MIN2

MIN2 consists of seven digits and usually corresponds to the mobile units telephone number.

A unique and recommended way to recognize a roamer from the US is to ensure that countries outside the US that use 6- or 7-digit local numbers always make the seventh digit zero. This restricts users to 6-digit mobile numbers although the PSTN number can still be 7 digits (the seventh digit would be redundant to the mobile switch). For example, Figure 18.4 shows a number scheme for the Philippines that follows this scheme.

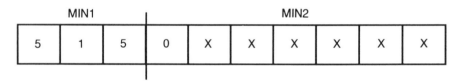

Figure 18.4   A number scheme for the Philippines

Chapter 19

MOBILE INSTALLATION

A badly installed mobile phone often leads to unreliable service and a dissatisfied customer. Therefore, a very important part of a cellular telephone system is the mobile telephone installation center. The main contact to customers is through the installation center, so care should be taken in designing the center's layout and operation. The center provides installation, maintenance (sometimes repair), and customer training.

In order to provide good service, the installation center must be properly equipped, have adequately trained staff, and sufficient floor space.

BUILDING AND EQUIPMENT

Give careful thought to the size, location, and layout of the building. The building location is most important. Customers want a convenient location and usually one not too far from the city center. It is likely that arrangements will need to be made to transport some customers to and from their places of work, which becomes more difficult, but even more necessary, when the center is remotely located. If possible, locate the center close to public transportation. In large cities, the only way to ensure reasonable access for the customer is to have several centers. The individual centers can be virtually autonomous or there can be a main installation center that supports the others with hardware and other services.

The installation center should project an image that is consistent with the corporate image of the cellular company. The center's image is the one most closely identified by the customer with the cellular supplier. Appearances are important. A modern neat building with plenty of parking is ideal for a mobile installation center.

The size and location of the building will be determined by the expected traffic, but, regardless of size, there are a number of features that a good installation center should have. The following sections detail these features.

SERVICE BAYS

Each service bay should be 9 meters x 4 meters and have overhead clearance of 3.5 meters. The service bays should be located as far away as possible from the customer-reception area and should have independent access.

Customers should not be permitted in this area because they could

- Steal equipment and tools
- Be injured and sue the owner
- Cause damage to (or even steal from) other vehicles

A wide entrance and a grill roller-door can be used to prevent unauthorized access without cutting off daylight or air flow into the building. The grill can be electrically operated by a remote-control unit, and an external solid roller-door can be used at night. The door should be high enough to permit access by large vehicles.

It is advisable to ask customers to leave their cars at the reception parking area and have the installation staff drive the cars into the service bay. Be sure to use protective covers and mats to protect the customer's car from getting dirty during the installation.

Service bays should have the following facilities and equipment:

- Bench space about 1 meter wide in front of each car
- Adequate conventional phones for testing
- Battery trolley and charger for jump-starting cars with dead batteries
- Area for metalwork
- Adequate power outlets
- Adequate lighting
- Washbasins, detergents, degreasers

As an adjunct to the service bays, it is useful to provide a special heavy-duty work area where welding, grinding, and metalwork can be done.

TOOLS

An installation center should have the following tools. (These are the minimum tools needed.)

- Electric drill and drill bits with hole saws up to 25 mm (or to suit antennas used)
- Center punch (should be automatic)
- Crimping tools (to suit cables)
- Power extension leads
- Pliers, sidecutters, and long-nose pliers (a variety of these)
- Two soldering irons: a heavy-duty one and a light one for printed circuit boards
- Flashlight and inspection lamps
- Allen keys (hexagon wrenches)

- Screwdriver set (normal slot blade and Phillips types)
- Electric screwdrivers (normal slot blade and Phillips types)
- Ball-peen hammer
- Files, including rat-tail for smoothing holes
- Tin snips
- Power extension modules with safety switch
- Wire brush
- Socket set and wrenches
- Hacksaw
- Vise with 70 mm grips
- Vise with 30 mm grips

POWER OUTLETS

To ensure an adequate number of power outlets, place sockets every two meters inside the reception area, and all service and workshop areas. Many operators prefer overhead power feeds above the service bays. Chains suspended from the ceilings support the extension cables. This setup is convenient for powering tools used during mobile installation. Power outlets should be mounted about 150 mm above the floor level (or higher if required by local wiring codes).

TEST EQUIPMENT

The amount of equipment needed depends on the size of the operation and the level of repair. In the simplest case of a very small operation, only a digital voltmeter and a VSWR meter, a Polaroid or similar camera, and installation tools might suffice. A very large center that has repair facilities needs elaborate test equipment.

A large center needs the following test equipment:

- 10 AMP regulated power supplies (12-volt or 0–24 volt regulated)
- 1 (preferably 2) cellular-radio test sets
- 1 set of harnesses for the cellular-radio test set for each mobile model serviced
- Digital voltmeters
- Battery-power load testers
- VSWR meters (usually incorporated in an RF power meter)
- Fixed test jig for mobile units
- Soldering equipment and tools
- Oscilloscope, twin beam 50 MHz
- Distortion and noise measuring set
- Frequency counter to 1 GHz
- Frequency generator to 1 GHz
- RF wattmeter to 50 watts and 1 GHz
- Directional coupler
- RF attenuators 6 dB, 10 dB, 20 dB
- Dummy load 50 ohms, 20 watts continuous

If the repair goes beyond board level, a good deal more equipment would be needed.

The Helper Instruments Company has just released an interesting piece of test equipment for installations. It makes SWR measurements a breeze. The equipment consists of a small signal generator (800–1000 MHz adjustable) and a sensitive SWR meter. This device can be connected to an antenna and an SWR reading can be obtained immediately. This device not only saves time, but could become an essential instrument in a small installation center.

STOREROOM

Adequate storage is essential and careful attention should be paid to the security of this area. The actual space demand is a function of customer turnover and the number of models kept. Figure 19.1 shows a typical transportable mobile telephone. Mobile telephones are fairly bulky when delivered in their original protective packing. The amount of storage space needed grows as the diversity of models increases. Correspondingly, storage for spare parts increases.

Finally, because mobile telephones and their spare parts are expensive, the storeroom should be secure and should be located in the least-accessible area to the customer.

RECEPTION/TRAINING AREA

A pleasant reception and training area with comfortable seating should be provided for customers. Usually, this area should have a VCR (for training) and a demonstration unit.

Every effort should be made to ensure that the customer does not wander away from this area. Providing tea- and coffee-making facilities is a good idea. Because unforeseen delays do occur, a good supply of up-to-date, general-interest magazines should be made available. The reception area should be the best room in the building, with good-quality furniture and fittings. If space permits, a separate interview area is a good idea.

OFFICE

Because cellular radio installation involves a good deal of paperwork, a separate office is needed. The office should have at least one private area for personal interviews with staff and clients. This office should also be of a high standard.

PARKING

Adequate parking for customers as well as for staff should be provided. Because some customers will want to leave their vehicles all day, sufficient parking space will be needed.

STAFF FACILITIES

Staff lunchrooms and washrooms should be located for convenient access by staff. These should be pleasant and of a high standard. Locate public washrooms near the reception area, where they will be convenient to customers, not adjacent to the service bays or storeroom.

Figure19.1   An NMT450 transportable telephone. Photo courtesy L.M. Ericsson.

EXAMPLE CENTERS

The size of the installation center can be determined once the installation rate is known. The relative importance of each of the areas discussed earlier depends on the size of the operation. For more information, see Chapter 23, "Budgets."

SMALL INSTALLATION CENTER

Figure 19.2 shows a floor plan for a small mobile installation center. The office must serve as the customer reception area as well as an office and should be equipped accordingly. The access door to the workshop should always be closed, and there should be signs posted to keep customers out of the area. Large one-meter wide workbenches are an asset. Skylights above the service bays can also be useful.

MEDIUM-SIZE INSTALLATION CENTER

Figure 19.3 shows the floor plan of a medium-sized installation center that can handle 12–20 installations per day. Notice that the reception area is deliberately located away from the service bays. Adequate parking outside the building equal to at least twice the number of service bays should be provided. An area for heavy work (metalwork, welding, grinding) is provided separately.

CUSTOMER EDUCATION AND TRAINING

Teaching customers how to use a mobile unit is essential. Instruction is best done on a demonstration unit. Except in very small centers, the trainer need not be—and preferably should not be—the installer.

Customers should not only be trained in the use of the system, but they should also be given an idea of the radio service area and its limitations, along with some understanding that mobile units will not always work in all locations. The trainer may find it helpful to indicate that the service variability will be much like that of FM reception. It is a good idea to give the customer a coverage map. The trainer should be patient (many customers are slow learners).

CLEANLINESS

Because the staff will work mainly on the more expensive car models, it is very important that they observe cleanliness precautions at all times. Before beginning work, they should cover the seats with a lint-free cloth or plastic and cover the floor mats or carpeting with paper or plastic.

Even the most expensive cars can leak oil. Therefore, precautions, including surfacing the floor of the work area with a grease-resistant material, should be taken. Adequate facilities for degreasing should also be provided.

Smoking should not be permitted in the workshop, storeroom, and service-bay areas.

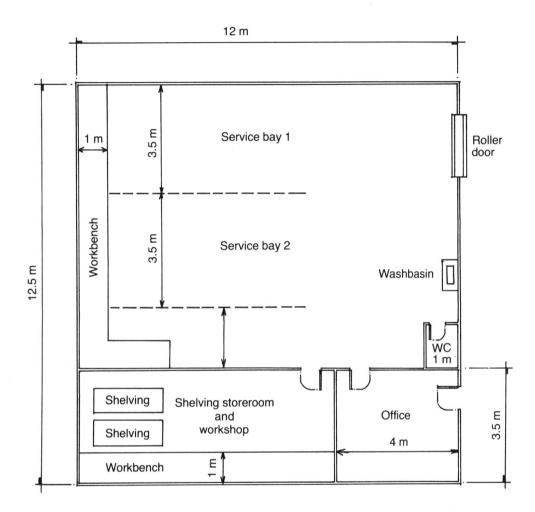

Figure 19.2   A small service center designed as a one- to three-person operation needs only two service bays.

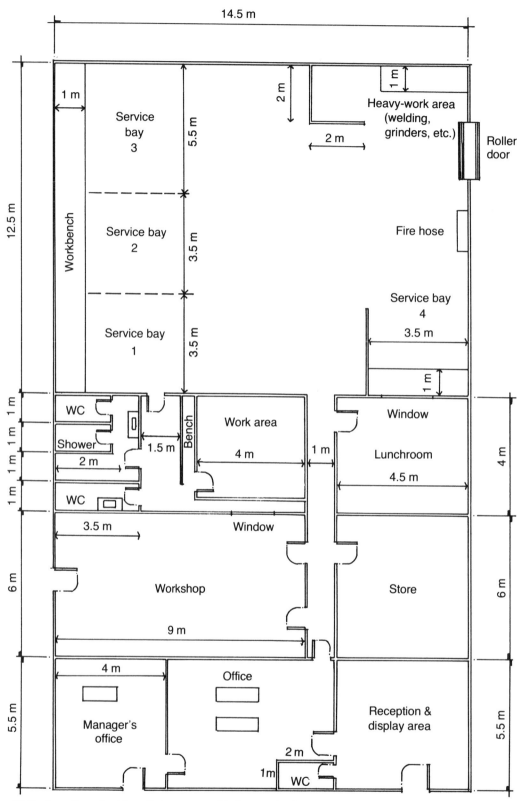

Figure 19.3 A medium-sized installation center with four service bays and adequate room for 812 staff may require around 400 square meters of floor space.

MAINTENANCE POLICY

Same-day repair is usually provided, and some operators offer repair services on demand. Additional staff are required to provide service on demand; be sure to plan for this contingency when staffing the center.

Large centers work better if customers leave their cars all day. Otherwise, scheduling can be a problem because many customers are not particularly punctual.

Consider an after-hours service policy. If a customer requests service after-hours, a suitable "call-out" fee should be charged. This fee is designed to cover costs (at commercial rates) of the staff on call for a period of two hours (minimum), plus other charges at normal rates (including the actual hours used). The call-out fee equals a minimum of two hours at overtime rates.

Another possible service is a 24-hour service hotline. Providing this service is usually easier for wireline operators than non-wireline because they can use existing manual switchboard staff. Wireline operators are well advised not to use their general telephone number as a service hotline because the operator usually will not understand the complaint. A dedicated line or number to the desk of a suitably trained operator is therefore normally necessary.

STATISTICS

A good deal of valuable management information is available at the installation center. The information should be recorded and formally forwarded to management on a monthly basis. Such information includes:

- Product demand, which should be divided into handhelds, transportables, and fixed installations; and into systems sold outright and by lease, reinstallations, rented systems, and in-house systems.

   The monthly summary should total all installations and any unsatisfied demand. Additionally, option changes and conversions should be noted.
- Performance, measured as the average installation time (for each handheld, transportable, and fixed in-car unit), repair time, and number of faults not cleared the same day.
- The cumulative services installed (those sold, leased, rented and in-house).
- Fault analysis, under the headings transceiver unit, handset, antenna, harness, battery feeds, antenna feeder, damage, and "no fault found." This analysis helps identify fault trends.

CUSTOMER DATABASE

In all but the smallest business, it is necessary to maintain a customer database system. A personal computer is suitable for the job. The database holds details of the customer equipment purchased and prepares an invoice.

The database should include the following information for each customer:

- Name
- Address
- Telephone number
- Equipment purchased
- Cost of equipment
- Equipment Serial Number (ESN)
- Mobile telephone number
- Installation data

The database software should be able to retrieve information about a customer based on any of the above parameters. Thus, for example if a stolen unit is recovered, the owner can be located from the ESN.

Most database software also generates information on commissions and profit of sales from the basic customer data. Good software customized for this purpose is available commercially at reasonable prices.

INVENTORY CONTROL

Because of the high cost of holding inventory, a stock-ordering policy based on the "just in time" principle is very cost effective. This approach relies on holding minimum inventory and ordering "just in time" to avoid shortfalls. The complexity of this approach rises directly as the diversity of the stock increases. To maintain control, use a software program that monitors the inventory and generates orders automatically. This job should not be done manually.

LIABILITY

Signs indicating that "all care but no responsibility will be taken" should be prominently displayed. Remember, however, that this disclaimer is not legally valid in most countries. Consequently, the installation center operator should conduct the operation to minimize the possibility of future litigation.

Take particular care when a car is delivered with obvious recent damage. Do not begin mobile installation until the damage has been brought to the attention of the owner and acknowledged as pre-existing. If the owner is not available, take photographs that clearly indicate the extent of the damage before any installation work is evident. If work is to begin without contacting the owner or the owner's agents, then ask a number of witnesses (not installers) to confirm the condition of the vehicle before any work begins.

STAFFING

Staffing requirements vary greatly depending on the number of mobile units to be installed per day. Depending on its size, an installation center can be a one-person operation or have up to about 50 staff. Operations with more than 50 staff are not recommended; customer convenience dictates that opening a new center at a different location is preferable to further expansion. Regardless of the size of the operation, the duties required are much the same. Figure 19.4 shows the typical structure of an operation designed to install 10 vehicle units/day and service an existing customer base of 8000.

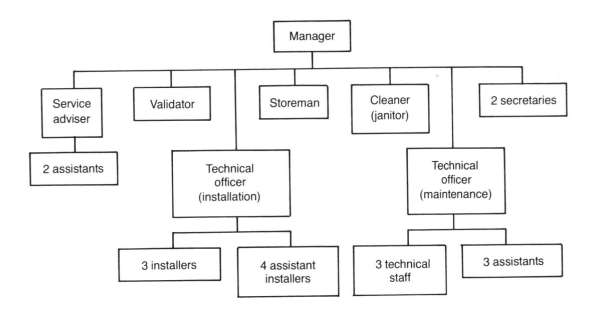

Figure 19.4   Typical structure of a large mobile installation center

THE SERVICE ADVISOR

The service adviser teaches customers how to use the mobile units and records details of the customers and their equipment. The service adviser handles complaints (which should come directly to this area) and should be able to distinguish equipment malfunction from operating errors. The service adviser arranges appointments for installation and repair.

THE TECHNICAL OFFICER (INSTALLATION)

The technical officer (installation) is in charge of installation, scheduling, equipment orders, and installation standards.

THE TECHNICAL OFFICER (MAINTENANCE)

The technical officer (maintenance) supervises all repairs, ensures there is adequate stock on hand, and oversees the quality of maintenance.

VALIDATOR

The validator position is required only if the installation center belongs to the network provider or a reseller. The validator is responsible for the validation (entry of customer data into the main switch) of new connections.

BATTERIES AND TALK TIME

A major limitation of handheld and transportable equipment is the battery "talk time" (transmission time). The first handhelds had a talk time of just over half an hour and standby times of about 8 hours. Figure 19.5 shows a typical AMPS transportable.

Improvements in battery technology have brought about both a reduction in battery size and an increase in effective talk time. Current (1989) handheld models achieve up to one and one-half hours of talk time or 14 hours standby. Some retailers offer a spare battery and advertise two hours of talk time, not one.

The batteries generally used today are wound electrode nickel-cadmium (NICAD) cells. The earliest versions of these were able to deliver about 700 mAH and improvements have led to units capable of 1100 mAH for handhelds. Although there has been steady improvement in these batteries, a real breakthrough will probably involve different technologies.

A new battery for handhelds was recently announced; it can substantially prolong the talk time by combining nickel-cadmium (NICAD) cells and user replaceable zinc/air batteries into a single unit. The zinc/air battery has a significantly higher mAH capacity than an equivalent NICAD, but of course it is not rechargeable. The zinc/air battery provides some load sharing, but its main function is to provide the standby load and to charge the NICAD battery during periods of non-use.

Figure 19.5   A Motorola AMPS transportable. Photo courtesy Motorola Communications and Electronics, Inc.

The NICAD cell is used during the talk cycle. This combination can increase talk time by a factor of 10.

Batteries for transportable units can currently deliver about 3.6 hours of talk time and 20 hours of standby and are thus not as limited as those for handhelds. Nickel-cadmium batteries are commonly used with portables and some transportables, and proper cycling is the only way to obtain efficient use of these batteries.

Unlike lead-acid batteries (for example, car batteries), which have a prolonged life if kept fully charged, nickel-cadmium batteries exhibit a memory phenomenon. That is, if a nickel-cadmium battery is lightly used and recharged frequently, it tends to remember this usage pattern and will later be unable to deliver its full ampere-hour capacity. Therefore, these batteries should be fully charged and then used until fully discharged whenever possible, particularly when the battery is new. A few deep recharging cycles will contribute to a long battery life.

Batteries that show evidence of "memory" discharge very quickly but can often be restored by two or three repeated deep discharge cycles.

It should be noted that in the race to achieve improved talk time, some manufacturers have reduced the power output of their handhelds below the standard 0.6 watts. This results in additional problems with "talk out," which is the main range limitation on a mobile. Some sets have output powers as low as 0.35–0.4 watts; this noticeably reduces performance in fringe areas.

THE HANDS-FREE OPTION

The hands-free option allows the telephone user to talk without picking up a handset. It was developed quite early in cellular mobile history but was perfected only around 1987. Earlier versions used VOX operations, which were often less than satisfactory, particularly in the often noisy vehicle environment. Modern hands-free options are full-duplex (that is, the send-and-receive functions work simultaneously) and are "compulsory" in some countries where using a handset while driving is not permitted.

Studies have shown that drivers using a conventional mobile handset are under demonstrable induced stress that is not present when they use a hands-free option. Whether for reasons of image or perhaps because of possible legal liability, most manufacturers are moving toward including the hands-free option as a standard feature.

SENSITIVITY AND PERFORMANCE

There is very little difference between the mobile telephone models in terms of specified sensitivity or on-air performance. This uniformity exists because the technology used in virtually all mobile phones approaches state-of-the-art. Typically, a mobile telephone has 3 watts RF output (vehicle-mounted and transportable) and –116 dBm to –118 dBm (max) sensitivity. TACS mobile units are rated at –113 dBm but, as they are measured differently, the two figures are in fact equivalent.

What does vary in mobile telephones is quality and consistency; some manufacturers meet their specifications more often than others and so their telephones may perform better on average.

Some reputedly authoritative sources suggest otherwise and even suggest comparing different phones on a demonstration basis if the phones will be used in a fringe area. This test is only fair if done with several phones of each type on the same day. On average there will be little difference, but a few individual units will noticeably out-perform the rest. Remember that in fringe areas the propagation is time- and space-dependent, so a particular phone that performs well on a particular day in such an area may not necessarily perform as well on the following day.

Even the most reputable manufacturers will admit that not all their mobile phones are created equal; unscrupulous suppliers can sometimes even use one of their better mobile phones against a bad example of the competitor's to prove superiority. Rented mobile phones will frequently have short talk times (battery life) due to indiscriminate battery charging; this should not be used to evaluate the talk time of the model rented.

A–B BAND AREA

Mobile telephones are usually programmed as A or B band units, meaning that they will search for only the A or B control channels (as programmed). This limitation presents no problems in the home service area, but when roaming it may be necessary to switch from A to B (or vice versa) to enable the mobile to lock onto the control channels of the host roaming operator. (For more information, see Chapter 28, "Roaming.") Locking on by using the mobile phone keyboard is relatively easy. Note that problems will occur if a mobile on the A-band system is programmed as A preferred. Although this setting will work in most situations when operating in a fringe area for the A-band system, the mobile may move to a B system control channel and stay locked onto that system.

MOBILE ANTENNA INSTALLATION

The mobile antenna is a most important link between the mobile unit and the rest of the telephone network. Pay special care to the installation of the antenna. Approximately 50 percent of faults are antenna related. Antennas are subjected to extremes of temperature, physical stress, vandalism, corrosion, and sometimes poor installation.

Increasingly, the on-glass, or through-the-glass, antenna is becoming the customer's first choice. Cellular mobile antennas normally come pretuned and should not be trimmed unless instructions are provided otherwise. This precaution holds true as well for some on-glass types, even though the match provided will be a function of the glass thickness.

Base-feed antennas are commonly used in cellular radio. Most roof-mount antennas are of this type. End-feed antennas are the basis of "through-the-glass" antennas, where capacitance is formed by two plates on either side of the window. Figure 19.6 shows typical configurations for both base-feed and end-feed antennas.

The base-feed antenna has an input impedance of approximately 50 ohms and is the most common antenna used in cellular radio. The matching coil is sometimes formed by coiling the antenna wire around a former about 30 mm in diameter. The base-feed antenna is durable and inexpensive. This type of antenna is best mounted either on the center of the vehicle roof or centrally on the trunk.

The end-feed antenna is the basis of the "through-the-glass" antenna; the capacitance is provided in part by two metal plates mounted either side of a glass window. Additional series capacitance is usually provided to enable fine-tuning of the input impedance (to 50 ohms or VSWR = 1.0). The biggest problem with on-glass antennas is that they sometimes simply fall off due to the effects of heat, moisture, pollution, temperature, and (not infrequently) the automatic car wash.

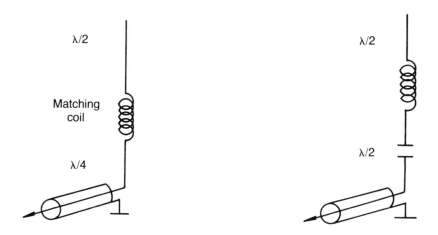

Figure 19.6   Base-feed antenna and end-feed antenna

Antennas are fixed in place either with double-sided tape or with acrylic bonding material. Double-sided tape can be most effective, but remember that it can be used only once (so care must be taken in positioning). To ensure a good bond, the area around the antenna fixing point should first be carefully cleaned with a commercial window cleaning product or other cleaning materials recommended by the antenna supplier. It is a good idea to seal around the edges of the double-sided tape with a suitable sealant. This type of fixture is removed using a razor blade or similar cutting tool.

Acrylic bonding is a very powerful bond whose application should be strictly to the manufacturer's specifications or it may not be possible to remove the antenna later. It is definitely not advisable to substitute other bonding materials because of incompatibility of their coefficients of expansion. Carefully follow the curing instructions, especially the amount of time needed before the joint becomes load bearing.

Glass-mount antennas must avoid the defroster wires and can be rendered useless when the vehicle has metallic window tints. It is possible to buy test kits for use with standard digital voltmeters that detect metallic components by their effective capacitance.

The antenna cable will frequently be run via the ceiling or floor of the vehicle. It is important to ensure that the cable is well hidden. If this is not possible, ensure that the cables are well restrained and will not catch on luggage or be easily tampered with.

When running the cable along the floor, be careful of the screws which are sometimes used to hold down the trim. These can puncture a carelessly laid cable. If the cable runs via the ceiling, firmly attach it using a suitable tape to ensure rigidity so that the cable will not move because of the vibrations. When the cable is fully run, cut off any excess cable because it is quite lossy at 800 MHz.

Attach the coaxial connector using a suitable crimping tool and stripping tool—it is not acceptable to improvise with pliers.

The position of the center conductor is important if a good connection is to be made; take care to ensure it is fitted properly and check again after installation.

TRANSCEIVER MOUNTING

The transceiver should be located to avoid low spots (where water may accumulate), and it should be firmly restrained in an appropriate bracket. Never allow the unit to float.

Provide good access to the RF, control unit, and battery connectors. Mount the transceiver so that access to these connectors for purposes of maintenance is easy.

DRILLING HOLES

Pay special attention when drilling holes in vehicles. Always ensure that the location is a safe one and that the drill will not puncture the gas tank or other functional parts. When the location has been determined, measure the placement of the antenna hole—don't guess at it. Then, use a spring-loaded center punch to start the hole. Drill a pilot hole and then use a hole punch to punch out the antenna hole. Remove burrs from the hole with a paint brush or vacuum cleaner.

When fitting the antenna, make sure that a good water-seal is achieved and that there is a good electrical contact between the antenna base and the vehicle body.

ANTENNA MOUNTING

Cellular antennas suffer considerably from vandalism; it has been found that the less conspicuous the antenna, the less likely it is to be vandalized. Moreover, smaller antennas also seem to survive a car wash a little better.

The mounting and antenna type is largely determined by customer preference, but the installer should be aware that a poor selection can seriously degrade the performance.

Take particular care with fiberglass bodies or panels where no ground plane is readily available. In this instance, use ground-plane-independent antennas, such as a coaxial dipole, or use a standard antenna with a ground plane formed by a layer of conductive tape.

The best place to mount a mobile antenna (for reception) is in the center of the vehicle roof. Most customers object to this, however, and so less optimum locations need to be used. Almost as good as the roof-mount are the on-glass, rear-window, elevated-feed center-trunk, and fender mounts.

Rooftop antennas are often very difficult to mount, and so it is probably fortunate that most subscribers do not want this mounting. Some vehicles have a double skin on the roofing that can buckle under the tension of an antenna and result in water leakage.

If rooftop mounting is requested, however, it is essential to make sure that a center-mounted antenna is actually centered. Take care to double-check the measurement of the proposed hole. Patching a hole drilled in the wrong area is very expensive.

Drilling a hole through the roof requires some care not to catch the headliner underneath. It is advisable to use two people for this operation—one to drill and one to ensure the integrity of the headliner. Using an area of the roof close to the interior light (if it is in the center) can simplify installation by making internal access easier.

An elevated-feed antenna raises the height of the antenna to a level comparable with that of a roof-mounted antenna. Because these antennas are, in effect, dipole-fed antennas, they are ground-plane independent and so can be mounted almost anywhere on the vehicle. Figures 19.7 and 19.8 show elevated-feed antennas.

ANTENNA GAIN

Antenna gain cannot be adequately tested except by using digital sampling techniques that collect large amounts of data on antenna performance over closed loops. Comparing two antennas in the same general location using just listening or S-meter tests is not sufficient. Because of the standing-wave pattern, the same antenna will give very different results when moved as little as a few centimeters. Some manufacturers have exploited this situation to claim that they have produced "high-gain antennas" or "fringe-area antennas," knowing that their claims are difficult to check.

Any claims of more than 4.5 dB gain for cellular antennas should be treated with a good deal of suspicion, because such antennas are most difficult to manufacture. True 4.5+ dB antennas will normally have at least three segments and are somewhat unwieldy.

"True" 4.5+ dB gain means only gain measured normal to the antenna and along its most sensitive axis. Due to multipath, much of the RF energy will impinge on the antenna off-axis, and hence the net gain will be less than the "nominal" gain.

Figure 19.9 illustrates this effect. The higher the gain of the receive antenna the more narrow will be the lobes of the field pattern and so the more multipath will reduce the effective gain.

To successfully test the gain of a mobile antenna, it is necessary to sample the field strength along a closed loop of roads at least 6 km in diameter. It is preferable to average the results in both the clockwise and counterclockwise directions because, apart from a center-roof mounted antenna, almost all other mounting positions are highly asymmetric. The gain will be very much dependent on the relative positions of the transmit antenna and the mobile receive antenna in relation to the body of the car.

Sampling must be done at a rate at least equal to the standing-wave pattern wavelength (approximately one sample every 0.2 meters). Such sampling requires a good deal of specialized equipment that is ordinarily available to cell-site designers only.

Based on a study done by the author in Australia in 1987, the relative performance of antennas

Figure 19.7   An elevated-feed antenna

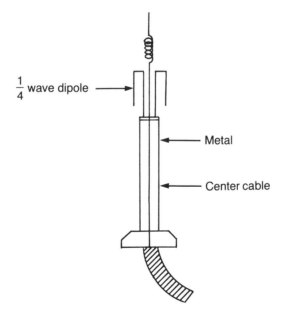

Figure 19.8   An elevated-feed antenna showing that the elevation rod is an extension of the coaxial feeder. It is ground-plane independent.

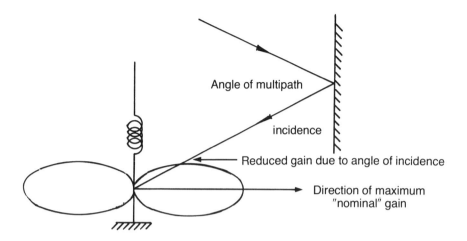

Figure 19.9   "High-gain" mobile antennas are limited in effectiveness. The "nominal" antenna gain is the maximum gain along the major axis. Other angles of incidence reduce this gain.

mounted elsewhere than the center of the roof have been determined. Table 19.1 shows these results.

All of the measurements in Table 19.1 involved sample sizes of about 60,000 obtained by driving around closed loops of about 6 km.

Table 19.1, shows that gutter-grip mounts (and for the same reason other similar mounts that use brackets that lift the antenna away from the ground plane) and magnetic mounts vary widely in efficiency and should be thoroughly tested before being recommended for customer use.

For the fender mount, the gain is very directional. To ascertain the degree of directionality, measurements were made in the two extreme directions of directly toward and directly away from the transmitter. Figure 19.10 indicates how these measurements were taken.

The results of the measurements were as follows:

$$G_{FORWARD} = -0.5 \ dB \ (forward \ gain)$$

$$G_{BACKWARD} = -5.5 \ dB \ (backward \ gain)$$

Although losses due to fender mount are not high, in the particular case where the body of the vehicle is between the base and the mobile antenna, significant losses occur. Despite this potential disadvantage, however, the fender mount is still a good and very popular option.

Through-the-glass antennas are now popular and have a performance somewhere between the fender-mount and gutter-grip-mount options. Through-the-glass antennas do tend to be susceptible to car wash and heat stress. Take care to avoid mounting them near rear-window-demister elements.

GAIN OF ANTENNA RELATIVE TO CENTER ROOF MOUNT	MOUNTING POSITION
- 6 dB to 0 dB	Magnetic base (various bases) Roof mount
- 10 dB to - 4 dB	On fender with magnetic bases (various)
- 5 dB	Gutter-grip mount above driver's door (not ground plane independent type)
- 1.5 dB	Fender mount*
- 1.5 dB	Through-the-glass mount**
* Although the average loss was low, the losses recorded traveling directly toward and away from the base are interesting (see Figure 19.14). ** Results from similar tests in US.	

Table 19.1   Effects of mounting position on antenna gain

The gutter-grip, a relatively common mounting method, has a loss of 5 dB and is not recommended unless used with ground-plane-independent antennas. It is also advisable to avoid the "easy" option of running the antenna feeder between the body and the car door because feeder damage usually happens eventually.

The more conventional method of measuring field strength of an antenna in polar form is also possible. This method is the most frequently used, probably because it is the simplest. In its basic form it involves connecting the transmitter to a vehicle-mounted antenna and placing the vehicle on a turntable. The vehicle is then rotated and the far-field strength is recorded at various angles of turntable rotation. A polar plot as shown in Figure 19.11 is then made.

· Although this method is adequate for base-station antennas, it is nearly meaningless in the mobile environment. The very complex radiation pattern caused by the interaction of the vehicle and the antenna is not only difficult to interpret but also varies considerably with the angle of incidence of the received signal.

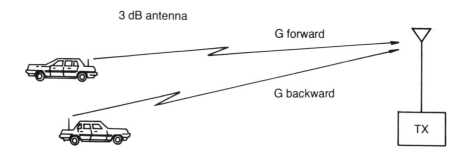

Figure 19.10   Fender-mount antennas are least effective when the vehicle body is between the antenna and the base station.

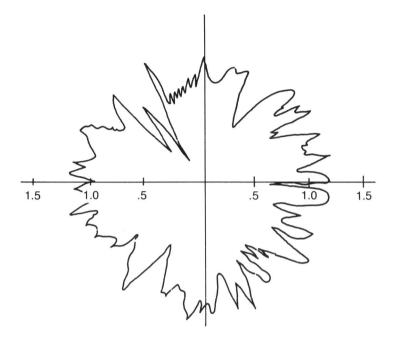

Figure 19.11 A polar plot of a typical mobile antenna

An approximation that the signal always arrives in the plane perpendicular to the receive antenna is usually made when the polar diagrams are drawn. Of course, in the real world of multipath propagation, this assumption is not necessarily valid. Assumptions made as a result of these diagrams, particularly about gain and directivity, are highly suspect.

ANTENNA RAKE

Many customers are very concerned with appearances and like to use adjustable-slope antennas that conform with the lines of the vehicle. The general feeling in the industry is that this will result in significant gain losses compared to a conventional vertical mount. To determine the magnitude of this gain loss, a study of raked 3-dB antennas was undertaken by the author in 1988. Some measurements were taken that indicate that small angles from the vertical can be tolerated without significant loss.

Table 19.2 shows the results of a number of measurements on sloping antennas. The method used in Table 19.2 gave readings to an accuracy of ± 0.5 dB, which accounts for the 0 dB reading for 10-degree angle of rake. These tests were conducted at the same time as the antenna location tests. The tests were conducted over the same closed routes, and the sample size was comparable.

ANGLE OFF VERTICAL	LOSS COMPARED TO VERTICAL ANTENNA (dB)
10	0
20	2
30	2.5
40	3.5

Table 19.2   3 dB roof-mounted antenna gain degradation when mounted off-vertical

It is clear from these results that the small angles of rake, (less than 10 degrees) sought by most customers do not seriously reduce or impair gain.

Figure 19.12 shows a vehicle with antenna rake.

NUMBER ASSIGNMENT MODULE (NAM)

The Number Assignment Module (NAM) can be a PROM (Programmable Read-Only Memory), an EPROM (Erasable Programmable Read-Only Memory), or an E^2PROM (Electrically Erasable Programmable Read-Only Memory) that contains details about the customer, the system, and the options chosen.

The PROM is a 32 x 8-bit memory which is non-volatile and cannot be erased. The NAM programmer is simply a PROM burner. It is called a burner because the links inside the PROM are programmed by applying a current of sufficiently high level to burn out link structures internally, to encode the required information. Some mobiles can have their NAM programmed from the telephone keyboard (when an E^2PROM is used), but most mobiles require a NAM programmer. Consequently, the NAM programmer is a normal accessory in a cellular installation center.

Programming from a NAM programmer is made easier by having a master NAM (one that contains all the standard default values, which can be copied so that new NAMs need only to have their specific parameters added).

Figure 19.12 Some owners like the trendy look of a raked antenna. Small angles of rake result in only small losses.

The code is stored in hexadecimal (a number system based on 16) and the information is usually input in hexadecimal. The following list shows typical NAM data.

- SIDH. System Identification of Home Mobile Service Area (–5 digits HEX); defines the cellular operator's system and is used by the mobile to distinguish roamer areas
- MIN1. Mobile Identification Number (three-decimal-digit country or system code)
- MIN2. Telephone number of the mobile unit
- SCM. Station Class Mark (two digits); indicates if the mobile has VOX and its power output
- IPCH. Initial Paging Channels in Home Switch (four digits) usually 0333 or 0334; defines the starting point of the mobile unit's control-channel search
- ACCOLC. Access Overload Class (two digits); divides the subscribers into 16 subgroups
- PS. Preferred System Mark (1 digit); used when this function is not keyboard-accessible
- GIM. Group Identification Mark (two digits)

(Additional information on manufacturers' options can also be placed in the NAM.) Most mobiles today have E^2PROM and NAMS, and are programmed from the keyboard.

Chapter 20

TOWERS AND MASTS

Towers (self-supporting structures) and masts (guyed structures) cost about the same if they are both short (that is, at heights ordinarily encountered in cellular radio). As the structures get higher, the costs of guyed masts tend to increase linearly (for the same cross-section); the cost of self-supporting towers increase exponentially. Because guyed masts require a good deal of land, they are used mainly in rural areas.

Towers and masts have an important advantage over poles and buildings in cellular applications: They can be used initially with omnidirectional antennas mounted on top for maximum coverage, and then later, when cell sectoring is used, the sectors can be mounted at lower levels on the towers or masts according to the coverage sought (that is, each sector can be mounted at a different height to give independent control over the coverage of each sector).

In cellular installations, tall towers are not usually needed and, except for rural areas, poles should be adequate for heights up to 30 meters. A number of imaginative designs have evolved; most use a triangle on the top for mounting.

If a triangle with 3.5-meter sides is mounted on the top of the pole or tower, it is possible to attach up to 3 TX/6 RX antennas to it. To improve isolation, transmit-and-receive antennas are often mounted so that the transmit antenna is vertical and the receive antenna is upside down. You must always ensure that water drainage is adequate on the inverted antenna because antennas usually have drain holes only at the bottom. Sometimes you must drill additional holes and plug the original holes.

The triangle mounting bars should be about 1.5-meters high to vertically separate two antennas. The triangle configuration is often chosen because of its simplicity for construction purposes.

When a square-section tower is used, the platform can also be square. Figure 20.1 shows a 40-meter tower, specified by the author for Extelcom in the Philippines. This tower has a 3.5 x 3.5 meter square platform on top.

Figure 20.2 shows an antenna installation in Indonesia that has a triangular mast section with three triangular platforms. Notice the directional, radome-enclosed Yagi antennas. This is part of the Jakarta NMT450 network.

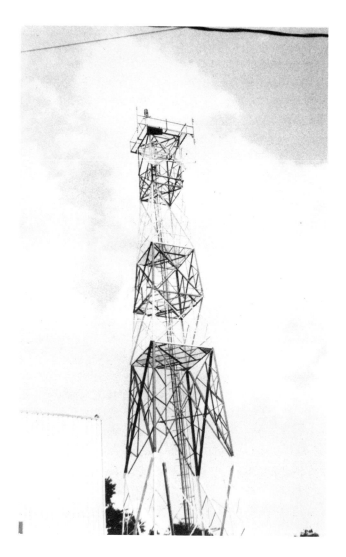

Figure 20.1   A cellular/microwave tower with a 3.5 x 3.5 meter square platform for cellular antennas.
Photo courtesy Extelcom

As mentioned earlier, towers and masts require different amounts of land. Figures 20.3 and 20.4 show the area needed for a mast.

Figure 20.5 shows the amount of land needed for towers of different sizes. Table 20.1 shows the amount of land needed specifically for three-leg and four-leg towers. In this table, T and W are the land dimensions used in Figure 20.5.

No structures of any kind should be built closer to the tower than the edge of the boundaries defined in Table 20.1 because the support of the surrounding soil against turning moments may be diminished. These dimensions are a guide only; the design of a tower or mast depends on such factors as wind loading, local building codes, and local planning-authority regulations.

The choice of monopole, mast, or tower for cellular radio is often made for the operator by the

Figure 20.2   These are the triangular mounting platforms used for the NMT450 system in Jakarta. This system uses the "Stockholm Ring" six-cell configuration. The radomes contain Yagi's for the NMT450 system. Photo courtesy L.M. Ericsson.

local-government rules or environmental considerations. Sometimes, however, there is a choice, so it is worthwhile exploring the alternatives.

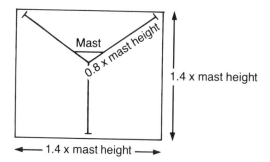

Figure 20.3   Optimal land area for mast

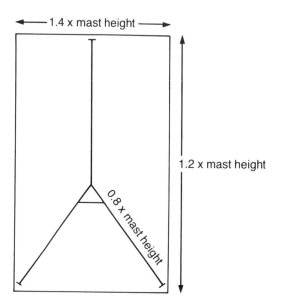

Figure 20.4   Minimal land area for mast

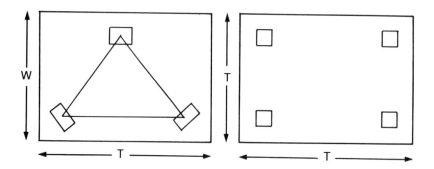

Figure 20.5   Dimensions of land for towers of different sizes

3 LEGS				4 LEGS		
TOWER HEIGHT	T	W	APPROX WEIGHT (tons)	TOWER HEIGHT	T	APPROX WEIGHT (tons)
10 M	7	7	0.7	20 M	7	1
20 M	8	7	1.7	30 M	9	2.2
30 M	10.2	9	3	40 M	10	4
40 M	11.5	10	6	50 M	12	8
50 M	13.8	12	10	60 M	13	12
60 M	15.5	14	14	70 M	14.4	16

Table 20.1   Land usage and weight for three- and four-leg towers

MONOPOLES

In general, a monopole is more aesthetically pleasing, although very few neighbors are likely to welcome any support structure. The monopole, like a building, has a fixed platform height and usually comes in a very limited range of sizes (typically, 15–50 meters). It also has an internal ladder and cable tray. Its main structural advantage is the small land area required, typically 9–16 square meters (3–4 meters square).

Monopoles can be erected in about one day provided the base has to be poured and cured. Because they are available in a limited range of sizes, they often can be ordered almost off the shelf and have much shorter delivery times than towers.

Monopoles are usually fabricated in tapered sections of about 10 meters apiece and fit together simply by stacking the sections (see Figure 20.6).

The support is simply a cage, designed to withstand large turning-moments, embedded in concrete. Shafts typically 8 meters deep and tapering from 2 meters at the bottom to 3 meters at the top form the foundation. Other monopoles can have much wider bases with correspondingly smaller shafts. Bolts 2 meters long and 57 millimeters in diameter, embedded in the concrete, attach to a flange at the bottom of the shaft. Up to 50 bolts can be used to hold the flange.

These structures can be designed to give the torsional stability required for low-frequency microwave bearers (maximum 1/2-degree twist).

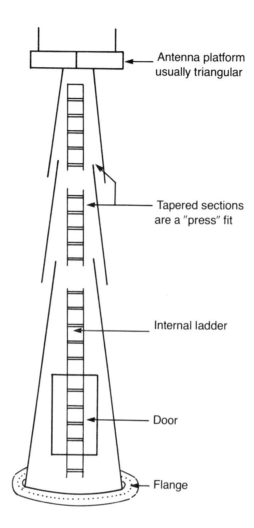

Antenna platform
usually triangular

Tapered sections
are a "press" fit

Internal ladder

Door

Flange

Figure 20.6   The construction of a monopole

GUYED MASTS

Guyed masts are practical only where land is inexpensive. They often prove to be the cheapest solution in rural environments.

Guyed masts are usually constructed of sections of triangular cross-section about 6 meters long. The sections are typically 0.5 to 1 meter wide per side and are designed to be bolted together. The strength of a mast is essentially in the guying, so proper tensioning of the support cables is vital. Concrete anchors hold the guy wires. For cellular applications, the standard cross-sections should prove adequate to accommodate the cellular and microwave antennas. Figure 20.7 shows a 50-meter mast in Asia.

Figure 20.7   A 50-meter mast in Asia. Notice the long grass (a fire hazard) in the foreground and recent evidence of fire in the very near foreground.

HYBRID STRUCTURES

Sometimes it is not easy to decide if a structure is a mast or a tower. Figure 20.8 shows a structure that started as a tower and later sprouted a mast on top. This structure is on a rooftop in Manila, the Philippines. The mast may have been built to reduce the total loading on the roof.

FABRICATION

The cost of a mast is related more closely to the weight of steel than to the height. Table 20.2 shows that the weight of a mast is almost linearly dependent on its height. It can be seen by comparing Table 20.2 with Table 20.1 that self-supporting masts increase more rapidly in weight than guyed masts, particularly for structures higher than 30 meters.

Figure 20.8   A hybrid mast/tower constructed on a rooftop

MAST HEIGHT (meters)	APPROX. WEIGHT (tons)
30	1.5 to 3
50	3 to 5
70	4 to 7
100	10 to 12

Table 20.2   Mast height and weight

TOWERS

Towers are self-supporting structures that are most practical when land is expensive. Figure 20.9 shows a tower that supports a number of microwave dishes (the solid dishes) and gridpacks (the wire-formed dishes).

Towers require less land than guyed masts and are capable of supporting a large number of antennas, a factor where you plan to rent tower space to other users. When other microwave facilities are planned (for example, for a wireline carrier), a self-supporting tower is probably the best choice.

A three-sided tower is usually the best value (load-carrying ability per dollar). For the same strength, however, it has a wider base and requires more land than a four-leg tower. A four-sided tower has an extra face and can carry more antennas. For a given strength, it is also smaller.

Towers members can be of various types, including solid, tubular, and channel sections. Tubing is the cheapest material for tower construction; it is available in a large number of sizes and needs little work to make it suitable for towers. Tubing does, however, have a long-term maintenance liability. Moisture can build up inside the tubing and cause corrosion, and in extreme environments, can freeze, thereby splitting the tubing. In coastal areas or in areas near heavy industry, this type of construction could prove to be a liability. Such towers need to be designed with weep holes, and the holes need periodic cleaning and unblocking.

Round-bar members can be made of round solid sections; they do not have the corrosion problem of tubular sections. They do, however, require substantially more steel for the same strength and thus weigh and cost more.

The most common material used in tower structures is the channel section, which can be made from formed-plate or angle sections. Formed plate is cut from rolled plate; it is cut to length with the bolt holes punched while still in the plate form. It is then cold-formed into 60-degree or 90-degree channels along the center axis.

Deformed plate is made from milled 90-degree sections. For 60-degree sections, the plate is bent another 15 degrees on each flange. This plate is cheaper than formed plate, but often it is not precision-formed, which can lead to problems with bowing.

Figure 20.9   A tower in the tropical Philippines. This one carries a number of microwave dishes and gridpacks.

SOIL TESTS

Before a tower, mast, or pole can be erected, it is necessary to conduct a soil test. This involves taking core samples of the ground on which the structure is to be built and then having the samples analyzed. Using the results, the design engineer can determine the load-bearing capacity of the soil and its ability to resist the turning moments of the footings. Only after this test is complete (from one to four weeks) can design of the structure foundations begin.

OTHER USERS

If the structure is in a particularly prominent position, you should consider, before the tower is designed, the prospect of obtaining additional revenue from leasing tower space to other users. A modest increase in cost at the design stage can significantly improve the load-carrying ability of the structure.

When planning for other users, include them in the overall design by assigning their number, antenna type, and positions on the tower at the design stage. The structural design should also include detailed drawings of the proposed positions of other users so that they can be allocated at a future date without the need for new load calculations.

In general, cellular operators need not fear that including other users will cause interference, provided that they operate outside the cellular band and do not transmit very high power, as is the case with UHF TV, for example.

Other users' services sometimes sprout up almost spontaneously in certain prominent areas and are known in the trade as "antenna farms." These "farms" can appear almost anywhere; Figure 20.10 shows one such farm thriving on a rock face in Baguio, in the Philippines.

LIGHTNING PROTECTION

Lightning protection works by providing a path of low resistance for a lightning strike. For this reason, the lightning conductor must be the highest point on the tower and have a good path to the ground. That path is best provided by copper straps.

Since the days of Benjamin Franklin (who invented the lightning rod), some people have argued that the protection of a lightning rod comes from "discharging" the atmosphere around the tower and preventing strikes by preventing static build-up. This belief has been soundly disproved by every generation since Franklin, but it still persists. Even today some manufacturers claim to produce such devices. There is no evidence that pointed lightning rods work better than rounded ones.

Figure 20.10   An "antenna farm" on a rocky outcrop in Baguio, in the Philippines. These antennas are mainly TV antennas mixed with a few links. Notice that the high-gain antennas are often mounted off-vertical, where they won't work well.

The "zone of protection" can be defined as the 90-degree cone around the antenna, as shown in Figure 20.11. The protected antennas should be inside the "cone" of protection described by the lightning rod.

With the number of antennas on even a medium-sized cellular installation, it is often difficult to find a place to put a lightning rod where it will not cause significant pattern distortion. For this reason, many operators dispense with the lightning rods and rely on DC ground-potential antennas (antennas designed to withstand lightning discharges). This approach seems to work reasonably well.

As a minimum requirement, each leg of the tower should be connected to a grounding rod, and the rods should be connected with a buried copper bus-bar in order to be well grounded. Grounding the tower effectively and tying it to the main building ground is important. Figure 20.12 shows such an arrangement.

Because the antenna feeders have large-diameter copper shields, they can make very attractive lightning paths. To reduce possible equipment damage, it is good practice to ground the feeders at the top and bottom of the tower, as well as at the entry point to the building structure.

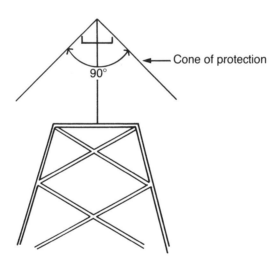

Figure 20.11   Lightening rods can be envisioned as providing a "cone" of protection to an area beneath them.

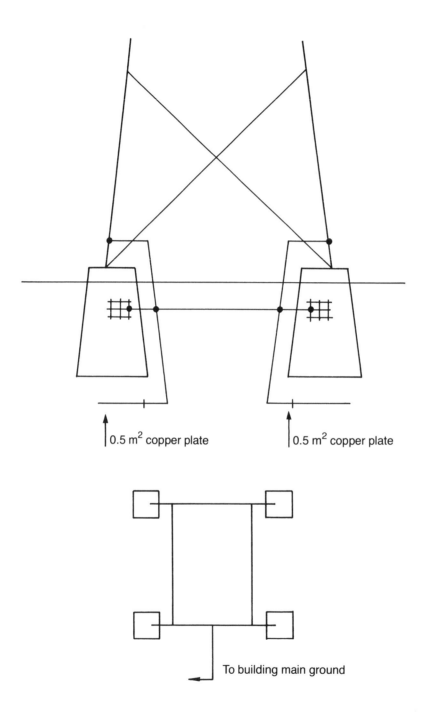

0.5 m² copper plate   0.5 m² copper plate

To building main ground

Figure 20.12   A tower should be grounded at each leg and the grounding rods should be connected together.

ANTENNA PLATFORMS

Often in cellular applications and particularly where sectored cells are employed, it is convenient to provide a platform for the antennas. This platform provides a safe working place and should have hand-railings, grated floors, and kickplates (to prevent tools from falling over the edge). The platform will normally be the same shape as the tower cross-section (three or four sides) and should have sides of 3.5 meters (whether triangular or rectangular).

Alternatively, extension arms can be used. Extension arms are suitable for small base stations, particularly those with few antennas (for example, omni sites with less than 16 channels).

TOWER DESIGN

The antenna structure must be designed by a structural engineer, but it is worthwhile to consider the design parameters. The structure must account for gravity loads (dead loads) that include structure weight, antennas, and ice, as well as live loads, such as those caused by wind and seismic activity. Invariably, wind, ice, and tower fittings will provide the dominant loads on the tower.

WIND LOADS

Until very recently, the dynamic load caused by wind was not fully understood, and towers were designed to withstand a known static load, which was increased by a safety factor (often doubled) to account for dynamic effects. In the light of recent studies, it is clear that early designs tended to overdesign the bases and underdesign the top portions of the structures. Particularly in typhoon and hurricane areas, the top portions of old designs are now being strengthened.

In general, as more collapsed structures are studied and more detailed information of long-term wind peaks becomes available, the minimum requirements of codes for determining wind loads have consistently increased. Old structures should therefore be used only after a thorough survey and inspection.

Wind speeds, recorded by national authorities, are of interest to a tower designer. The designer should know the peak gusts (instantaneous readings) and fastest-minute-wind (the highest velocity sustained for one minute). These two figures are connected by a ratio of approximately 1.3:1.

Fifty-year peak wind velocities are sometimes interpreted as ones that are expected to occur 50 years apart. This is not an accurate interpretation. A better interpretation is that 50-year peaks are ones that occur with a probability of 2 percent each year. Therefore, the fact that an old tower is still standing may simply be good luck!

TYPICAL SPECIFICATIONS FOR A 40-METER TOWER

The tower designer must know a number of things about a tower before beginning the design process. The following lists contain the considerations required for a typical 40-meter tower:

- Four-sided (or three-sided).
- 40 meters.
- Designed to EIA RS222C, Australian Design Standards, or other preferred standard.
- Stress factor (that is, suburban or rural safety factor).

For example, in the Australian design code for suburban areas this stress factor is:

1.7 x factor on steel.

1.75 x factor on foundations (this factor can be found from the relevant design code).

- Zone specifications and wind loading, depending on location.
- Maximum allowable twist (0.25 degree for 7-GHz microwave or 0.15 degree for 10 GHz).
- Maximum allowable tilt (1 percent for 7-GHz microwave or 0.5 percent for 10 GHz).
- Platforms and walkways at the levels where access to microwave dishes will be required.
- A platform of about 3.5 x 3.5 meters at the top, with guard rails 1.5 meter high and suitable for mounting cellular antennas at the edges. The mounts will be used to attach antennas with tubular supports up to 70 mm in diameter using three heavy-duty clamps.

  Up to 10 antennas can be mounted at the top with an equivalent flat-plate area of 0.23 m^2, weighing 50 lbs for 60-degree, 17-dB sector antennas. (See manufacturers' catalogs for particular antenna types.)
- Cable tray will be accessible from the ladder and will be 0.6 meter wide.
- Safety guard around the ladder, which will be internal with respect to the tower.
- IAO standard paintwork and an aircraft warning beacon at the top.
- Tower orientation.
- Specify the microwave dishes, type (solid or grid pack) and mounting level. Allow for future expansion (even if expansion is not planned, it will probably be required; a good rule is to estimate the future requirement and then double it).
- Tower footings should be confined within a square plot of land (as specified earlier in this chapter).

HOW STRUCTURES FAIL

A free-standing structure such as a tower is most vulnerable in the compression leg (the side away from the direction of the wind). A mast is similarly subject to compression failure, but because of the multiple guying points, has a more complex failure mode. Failure in both instances will probably be due to buckling.

The stress is very sensitive to wind velocity. It varies as the square of the velocity for static loads and as the velocity to the power of approximately 2.5 for dynamic loads. Wind speed varies more or less regularly with height and has an approximately parabolic gradient from ground level to 400 meters.

A less predictable factor is turbulence, although this is probably the major factor in structural failure. Turbulence is poorly correlated along the length of the structure (it is randomly distributed) and varies rapidly with time. In modern studies, the very unpredictable nature of turbulence is taken into account, and it has been found that some turbulence patterns are significantly worse than others.

Topology plays a part and many large towers will be situated on hilltops to gain additional elevation. Hilltops unfortunately produce increased airspeeds over their crests and a 10-percent hill slope can produce a 20-percent increase in airspeed or a 40-percent increase in wind loading. This is the reason that windmills and wind generators are usually placed on hilltops.

Stiffness (the ability to resist deflection) is a sought-after characteristic in structures and an important factor for reliable microwave operation. Stiffness is often obtained, however, only by using more metal, which increases the cost and weight. For economic reasons, modern structures are designed to minimize the amount of materials used, so a trade-off occurs. Adding extra dead loads (for example, equipment and antennas) reduces stiffness.

TOWER, MAST, AND MONOPOLE MAINTENANCE

The unscheduled replacement of the antenna support structure can be both costly and disruptive to the installed service and should be avoided if at all possible. The collapse of a tower or mast, particularly in a populated area, can be at best embarrassing and at worst a catastrophe.

Antenna support structures require regular routine maintenance, which is often neglected on the grounds that the structure has been up for years and has not shown signs of fatigue to date.

To appreciate the need for competent inspections, it is necessary to first understand how and why structures fail. These are the major causes of failure:

- Poor design, which inadequately allowed for static or, more frequently, dynamic wind loads
- Overloading of the structures with too many antennas and feeders
- Corrosion, particularly where hollow structural members are used
- Insufficient attention to guy-wire tensions and conditions (corrosion)
- Inattention to the indicators of stress
- Guys corroded or improperly tensioned

As an interesting example of the problems facing masts, the mast shown earlier in Figure 20.7 and again in Figure 20.14 is worthy of a closer look. The long grass in the foreground of Figure 20.7 represents a fire hazard, and evidence in the extreme foreground indicates a recent fire.

This mast uses passive reflectors (the large plates at the top) to deflect a microwave link to the ground-mounted receiving dishes illustrated in Figure 20.13. These dishes are protected by conical radomes. The cracking and spalling concrete seen in Figure 20.13 at the base of the mast is a sign of excessive stress.

Masts are held up by guy wires that are anchored into concrete blocks. Signs of stress were evident at all of the anchor points at the site in Figure 20.13. Figure 20.14 illustrates cracking and spalling at these points at this site. All of the anchor points inspected on this structure showed signs of spalling. This mast was well painted and relatively rust-free, but as Figure 20.15 illustrates, little attention was given to mechanical details. The buckle linking, the guy wire to the anchor point in Figure 20.15, shows that the bolts were not tightened and washers were not used. The large central bolt is about 40 mm in diameter.

Routine inspections should be carried out about once a year for structures located near the coast and every two to three years at sites more than 100 km from the sea, as well as after severe storms or periods of prolonged heavy icing.

Figure 20.13   The dishes with conical radomes (to help water run off) mounted at the base of the tower shown in Figure 20.7. Notice the spalling concrete base.

INSPECTION

Very few cellular companies are large enough to employ a full-time, qualified structures inspector. Those that can will invariably be wireline operators.

Because of the special nature of support-structure maintenance, the cellular operator will generally find that there are few companies with the necessary expertise and that the availability of those companies is limited. Having found a competent operator, it is therefore a good idea to arrange the maintenance on a contract basis. The company should have a good structural engineer and experienced inspectors who can climb and inspect every portion of the structure. The inspection process should begin with a review of the existing documentation about the structure and its fixtures. It should then proceed step be step, using a checklist like the one provided at the end of this chapter.

Figure 20.14   Cracking and spalling of the guy-wire anchor blocks

If only cellular or mobile two-way (PMR) antennas and microwave links are mounted on the tower, the inspection can be carried out without disturbing the operation. The inspector should avoid prolonged periods of exposure (more than 10 minutes) within 1 meter of the antenna. The relevant local RF radiation limits should be observed.

STIFFNESS

A structure that is too flexible is subject to excessive stress and is liable to failure. All structures have resonant modes about which they vibrate. The primary mode for a free-standing tower involves its whole length and results in maximum movement at the top. The tower will sway under wind loads and the period of this sway is a measure of its stiffness. This period is the time to complete one full cycle (that is, from the vertical position through to the maximum deflection and back to the vertical is one half a period). This period can be measured by observation (difficult and inaccurate), by a video camera (better), and by an accelerometer (best).

Accelerometers are usually located at three or more positions along the length of the structure; the results are relayed to the ground for later analysis. Equipment records motion in two directions, as well as torsion. The optimum period is a function of the structure height, strength, design, and mass. For a 180-meter tower, a two-second period is good; a four-second period would indicate excessive flexibility.

Because early design codes did not fully appreciate the effects of dynamic wind loading,

Figure 20.15   The buckle connecting the guy wire to the anchor block in Figure 20.14. Notice that the bolts were not tightened and that no washers were fitted. The structure was, however, relatively rust-free.

underdesign of the top portions of the structures was common (together with overdesign of the lower portion). As a result, the flexibility of the top portions often, over time, causes high levels of stress. Strengthening the top portions is thus often required at a cost of approximately 10 percent of the structure cost. Cellular operators will probably encounter this problem only if they use an old, existing tower; design techniques today properly account for the distribution of stress. A good indicator of stress is localized flaking paint and, in some instances, corrosion. Flaking paint is best detected soon after a storm when the recent stress highlights the problem.

REPAIR

Any repair and maintenance indicated by inspection should be undertaken as soon as practical. Finding suitable contractors to do the work may be difficult.

Towers should be painted once every five to seven years, depending on the environment. Painting and touch up for corrosion can be done by many contractors, particularly by those who specialize in heavy industry or bridges.

Replacing bolts, adjusting antennas, and low-stress members can be done by a suitably qualified rigger.

Stress problems are more serious, however, and require the intervention of a structural engineer. The stress problems could be due to weakened members but are more likely design-related. After analysis, the structural engineer can recommend the necessary modifications. The replacement of high-stress members requires the services of a specialist structures contractor.

Stress can be reduced by lowering the wind loading of attachments, but it more often involves adding structural members. The structural engineer usually considers various alternatives to reduce stress and recommends the most cost-effective one.

Welding of strengthening members often destroys galvanizing and other protective coatings, so protective coats will be needed.

TOWER INSPECTION CHECKLIST

A tower inspection should include the following steps:

TOWER

1. Check foundations, ground points, and straps.
2. Check for corrosion and condition of painting.
3. Check welds for cracks, using ultrasonic equipment where necessary.
4. Check for signs of stress, particularly flaking of paint or bowing of members.
5. Check all bolts for proper tension and corrosion. (Some may actually be missing).
6. Check guys for proper tension and possible corrosion. In some areas, anticorrosive agents must be applied.
7. Note the position of all fixtures, and when these positions differ from the records, note the details (including photographs).
8. Check for bent or fractured members.
9. Check for tower twist or distortion (sometimes twist can be detected by checking that the tower lines are true).
10. Check the condition of the galvanizing.
11. Check for corrosion in hollow members; this can sometimes be detected by hitting members with a hammer and listening for falling rust. (In some instances and particularly in corrosive environments such as along the coast or in heavy-industrial areas, a low-stress member can be removed and replaced by a new one. This member can then be examined in a laboratory for strength and corrosion.)
12. Keep a permanent log of the inspection.

GROUNDING

13. Check that all clamps and ground straps are secure and in good condition.
14. Check that bolts are covered by an anticorrosive material.
15. Check that the lightning rod is secure and in an effective position relative to all antennas (higher than any antenna and at least three wavelengths away). (Some cellular installations dispense with lightning rods and use DC ground antennas instead. This is sometimes essential where space on the tower top will not allow for a reasonable separation between the antennas and the rod.)

ANTENNAS

16. Check that all antennas are vertical or at the correct angle of downtilt.
17. Check the physical condition of the antenna; it should be free of cracks, dents, and burns.
18. Check that all bolts and clamps are secure.
19. Check that the antenna grounding is secure.
20. Check that the feeder grounding is secure.
21. Check that the feeder support is adequate and not causing wear or fatigue.

22. Check for any slippage of the feeder.
23. Listen for any audible signs of gas leakage in pressurized systems.
24. Check that the antenna "tail" connector is properly sealed.

ANCHORAGE AND FOUNDATIONS

25. Check that concrete anchors are free of spalling (flaking) or cracks.
26. Check that anchor bolts are tight.
27. Check that grounding is secure.
28. Check that anchor rods are not rusted or corroded.
29. Check for any signs of anchor slippage or creep.

GUY WIRES

30. Check for any signs of rust or broken strands.
31. Check that the connectors to guy wires are in good condition.
32. Check that the turnbuckles are in good condition.

TOWER LIGHTING

33. Check that all beacons are in working order.
34. Check that all beacons are in good condition.
35. Check that beacon drain holes are clear.
36. Check that beacon reflectors are in good condition.
37. Check that beacons are free of signs of moisture.
38. Check that beacon lenses are clean.
39. Check that beacon wiring is in good condition.

ICE SHIELD

40. Check that the ice shield is secure and undamaged.

Chapter 21

INSTALLATIONS

Most operators opt for turnkey installation, at least for their first system. A turnkey installation is one where the installer (usually the supplier) contracts to provide, design, and install a complete system and hand it over to the operator when it is fully functional (that is, ready to "turn the key" and start).

With turnkey installations, the operator must rely heavily on the supplier; although this may be necessary for new operators, the operator should take steps to avoid prolonged dependence. The disadvantage of being dependent on a supplier becomes obvious when a contract is prepared. An inexperienced operator must either write a very open-ended contract or employ a consultant. Both approaches have disadvantages, but the first is the most fraught with problems.

Suppliers usually dread an open-ended contract because they must specify their offers without any firm knowledge of what other suppliers might be specifying. A supplier who specifies a well-designed and complete network may lose the contract simply because another supplier cut corners to achieve the lowest bid price. A minimum design may seem more attractive to an inexperienced operator because omitting one base station can reduce the cost presented in a proposal by about $500,000 (including links).

The disadvantages of open-ended contracts become apparent when the operator begins evaluation. Because the suppliers are not contracting for exactly the same thing it becomes very difficult to compare the offers. It takes a very experienced cellular engineer several weeks to effectively compare two dissimilar proposals.

Relying on the supplier often runs smoothly through the design and installation phases but again becomes awkward during the commissioning and acceptance phases. At these stages it is necessary to certify that the work to date has been done adequately and in accordance with good practices and the terms of the contract.

Unless an independent appraisal of the work is available, the operator will find it very difficult to have faith in the acceptance. Even when a supplier goes to some length to ensure a fair acceptance test procedure is followed, the operator can never be completely sure of the value of that acceptance.

Depending on a supplier for ongoing expertise to keep the system operational also presents problems. When such expertise comes from the supplier, it is very expensive, and the operator does not have complete control over the availability and selection of expert staff. The alternative, using consultants, also has difficulties. With consultants, the operator also has little control over the availability and selection of expert staff. If the peak demands of the operator and the consultant coincide, the operator may not receive the highest-priority service. Therefore, any arrangement with consultants should include a retainer and a guaranteed response time.

Large consulting firms often assign different experts to different phases of the project. This division of labor causes a discontinuity in direction, and it may be that none of the consultants will have an adequate overview of the operator's system to be fully effective. It is wise to require that at least one specific consultant be available for the duration of the project.

Finally, the operator must assume that the consultants are competent because, almost by definition, the operator is not in a position to determine if a particular consultant is competent or not. For this reason the operator should acquire the expertise needed to design and run a network as early as possible.

TRAINING

Most systems offer a training program. These programs typically cost $500 per day per participant; the total cost for a start-up system would be about $100,000. At this price, make sure to get good value for the money spent.

Begin by ascertaining that personnel have the educational background appropriate to the courses offered. The training courses usually assume a good technical knowledge of radio transmission for the RF courses, and good knowledge of switching equipment for the links and switch-training courses. The courses are usually detailed and complex and will soon leave behind those who do not have adequate technical background. Personnel must also have a good command of the instructional language (usually English). Insist too that training instructors be experts with a good command of the instructional language.

Courses should be provided in a timely manner; remember that a course undertaken and then not applied for six or more months will largely be forgotten. Course participants should be able to apply their new knowledge within two months of the course. The very fact that the knowledge will soon be needed increases retention.

By participating in the installation phase, employees can gain valuable insights into the functioning of the system. It is not unusual to specify in a turnkey contract that the operator's staff provide some of the labor for installation, thus ensuring active participation from the start. The areas of particular interest to an operator are the design phase (site selection), survey technique (RF path survey), site preparation, and, finally, installation and testing.

Because it will likely be necessary to constantly expand the network, some knowledge of what is involved in such expansions is also invaluable. As discussed earlier, it is nearly impossible to forecast demand (particularly on a new system) and even more difficult to forecast the load on a particular base station. It will be necessary to move channels from one base to another after the system has been switched on so that the channels are placed where the traffic is directed. The operator should be able to relocate channels as required. This means that suitable test equipment must be provided.

THE OPERATOR'S RESPONSIBILITY

When the operator accepts a turnkey system, payment becomes due and it is usually difficult to get the contractor to return for more than minor adjustments. Because the contractor's view of the project is somewhat different from the operator's, it is a good idea for the operator to be particularly alert during acceptance.

The operator's priorities are as follows:

- A good, efficient system with good coverage
- Ability to meet operational targets
- Ability to meet the market requirements at the least cost
- A competitive system
- Low maintenance costs
- Ability to expand efficiently and at minimum cost

A contractor's priorities, however, are somewhat different, namely, to:

- Install the system on budget
- Meet the contract specification
- Perform as a credible contractor and win subsequent expansion contracts. (Except in NMT systems, once a switch has been selected it is not usually economical to change suppliers).
- Meet time constraints

The order of these priorities is not always necessarily the same and some operations may have additional priorities, but it is easy to see that the objectives of the contractor and the operator are somewhat different.

For example, if a radio survey indicates that a particular area has marginal coverage and there is a possible (but not definite) need for an additional base, the operator, considering mainly the cost, may decide to take the risk and save money. The contractor, on the other hand, seeing the potential damage to the firm's reputation should poor coverage result, may decide that the doubtful base should be put in as a precaution. Alternatively, the contractor may ignore the problem in order to produce a lower quote. Whatever the contractor's decision, it will be based on different considerations than the operator's.

Contractors are often stressed by the demands of operators and the inability of manufacturers to supply on time. It is very easy, under such duress, to see the fine details such as labeling and documentation as relatively unimportant. The unwary operator who does not carefully check both the overall performance and the detail ultimately pays the price.

ACCEPTANCE TESTING

The operator should be responsible for acceptance testing because this is the one opportunity to ensure that all is well before paying for the system. At the end of this chapter are checklists that detail the items that must be checked before a base station is accepted into service. This checklist should take about one hour to complete.

Acceptance can be either absolute or conditional. If the base is inspected and found to have only a few minor shortcomings (for example, missing labels on equipment racks, some handbooks missing, and some spare parts not available), then it may be appropriate to issue a conditional acceptance (that is, the work is accepted subject to the shortcomings being cleared up within an agreed period—say one month).

But it is not appropriate to issue any kind of acceptance if major shortcomings are found. Examples would include:

1. The radio link to the switch is not functional.
2. There are no commissioning test sheets.
3. The installation is untidy.
4. Grounding straps are not provided.
5. The battery electrolyte levels are incorrect.

Because a poor standard of installation results in high maintenance costs over the life of the system, it is necessary to be very firm about acceptance procedures.

COMMISSIONING

As part of the acceptance, the accepting officer should be involved in the commissioning phase, usually in the last two weeks for a base station and in the last four to six weeks for a switch of the installation. This phase involves testing and aligning the system to ensure that everything operates within specification. The accepting officer should verify that all tests were properly done and recorded. The best way to do this is to be directly involved in the commissioning. Also, this phase is the most instructive part of the installation, and it is an opportunity for the operator's staff to become familiar with the equipment.

A physical check (using a vehicle and a mobile) that coverage is adequate and that handoff occurs correctly should be done for each base station site. The coverage of the site should be confirmed manually (using the mobile to see the limits of its range) or as a measured field strength. Serious discrepancies between actual coverage and predicted coverage may well point to some problems with the antenna, feeders, or system parameters. For this reason the operator should have access to a field-strength meter (using high-speed sampling). This is a lot cheaper than having the installer do those measurements, and it reduces the operator's dependence on the supplier.

MOVING AWAY FROM TURNKEY INSTALLATION

As the system evolves, the operator will likely gain confidence and be able to undertake a good deal of the work involved. Moving away from turnkey installation often results in large reductions in installation costs, and on that basis it should be at least considered by all operators.

Providing towers, huts (shelters), power, and clearing and preparing sites is work that can most readily be done first. The staff involved in this work should use the expertise available from the original turnkey project to become familiar with the requirements. The following sections discuss preparing the site for installation and establishing a staging area.

SITE PREPARATION FOR INSTALLATION

It is essential to have a properly prepared site before installation can begin. A well-prepared site will be cleared and sealed in such a way as to ensure that a path to and from the site is free of dirt, dust, or loose particles. Other work that is likely to produce dust (for example, building extensions, preparing the site, and landscaping) should not take place simultaneously with installation. Power (preferably three-phase) should be available on the site. Note that the power requirements are significantly larger than normal domestic requirements. The electrical grounding should be in place, connected, and tested. There should be a communications link back to the switch and, preferably, to other places as well. This link could be a telephone, a two-way radio, or even the engineering order wire on the microwave. The tower (support structure), antennas, and feeders should all be in place. Air-conditioning should be installed and operational, and all doors should be fitted and functional. The doors should have adequate security locks.

STAGING AREA

It is also necessary to provide a staging area where equipment can be unloaded, stored safely, checked, and sorted for dispatch to particular sites. The area needed for storage is quite large. Figure 21.1 on the next page shows two trucks loaded from a staging area ready for dispatch to a base station. In all, the equipment for a switch and three base stations required nine medium-sized flat-bed trucks.

The equipment comes in cartons that weigh from 200 to 500 kg. A normal forklift is required. The cable used for the RF feeders is usually LDF50 or similar. This cable comes in large drums, and it is necessary to provide some means of attaching a spindle through the drum to access the cable. Figure 21.2, also on the next page, shows such a frame and a smaller drum of rope used on site to measure the length of the cable run. The required length is marked on the rope and then an equal length of cable is run off back at the staging point.

Figure 21.1   The equipment used for cellular installations is bulky and heavy and needs a staging area for sorting.

Figure 21.2   A frame is needed to hold the RF feeder cable drum so that the required lengths can be conveniently cut.
Photo courtesy Extelcom

ACCEPTANCE-TEST SHEETS

The following acceptance-test checklists can be used by acceptance-test personnel for cellular radio base sites.

Site (Name)   _____

Location   _____

Switch (Location)   _____Tel._____

Installed by   _____

Installation supervisor   _____Tel._____

Inspected by   _____

☐ on completion   ☐ work still in progress

Acceptance date   _____

Signed   _____

Conditional acceptance date   _____

Signed   _____

(Subject to rectification of items indicated on attached sheets)

Date in service   _____

POWER RECTIFIERS

		OK/NOT OK	COMMENTS
1.	Mounting and layout	_____	_____
2.	Cabling and terminations	_____	_____
3.	Alarm extension	_____	_____
4.	Designations	_____	_____
5.	Commissioning test results	_____	_____
6.	Handbooks	_____	_____
7.	Load sharing	_____	_____
8.	Safety signs	_____	_____
9.	DC distribution	_____	_____

BATTERIES AND DISTRIBUTIONS

		OK/NOT OK	COMMENTS
1.	Fusing	_____	_____
2.	Battery spacing	_____	_____
3.	Electrolyte level	_____	_____
4.	Battery vents	_____	_____
5.	Battery lead burning/connections	_____	_____
6.	Battery cabling	_____	_____
7.	Hydrometer and thermometer	_____	_____
8.	Cell voltmeter and millivoltmeter	_____	_____
9.	Designations	_____	_____
10.	Drip trays	_____	_____
11.	Safety equipment	_____	_____
12.	Safety signs	_____	_____
13.	Water supply	_____	_____
14.	Fuse alarm extension	_____	_____
15.	Accessibility for testing	_____	_____
16.	Battery function and continuity (test)	_____	_____
17.	Floor loading of batteries within limits	_____	_____

EXTERNAL PLANT

		OK/NOT OK	COMMENTS
1	Tower, mast, or pole	_____	_____
2	Gantry	_____	_____
3.	Guys and anchors	_____	_____
4.	Lightning protection	_____	_____
5.	Tower, mast, or pole grounding	_____	_____
6.	Mains surge protection	_____	_____
7.	Tower lighting	_____	_____
8.	Corrosion protection	_____	_____
9.	Safety signs	_____	_____
10.	Site RF radiation records	_____	_____
11.	Equipment shelter	_____	_____
12.	Base security, fences, locks, gates, etc.	_____	_____
13.	Cable window and seals	_____	_____
14.	Grounding of equipment rooms	_____	_____
15.	Access to tower restricted	_____	_____
16.	Antennas correctly mounted	_____	_____

INTERNAL PLANT

		OK/NOT OK	COMMENTS
1.	Mounting and layout	————	————
2.	Cabling and terminations	————	————
3.	Alarms and telecontrol	————	————
4.	IDF cabling	————	————
5.	IDF labeling	————	————
6.	System performance	————	————
7.	Commissioning test results	————	————
8.	Station logs	————	————
9.	Designations	————	————
10.	Handbooks	————	————
11.	Drawings	————	————
12.	Test cables and extender cards	————	————
13.	Spares	————	————
14.	TRX plug crimps	————	————
15.	RF N-connectors	————	————
16.	Independent link to switch (control center) functional (telephone line or radio link)	————	————
17.	Lightning arrestors on all incoming cables	————	————
18.	Door alarms functional	————	————
19.	Suitable fire extinguishers	————	————
20.	Redundant control channel functional	————	————

ITEMS TO BE CHECKED

POWER RECTIFIERS

Mounting and layout

☐ Correct positioning of cabinets.
☐ All mains terminals covered.
☐ All cabinet components supplied, including tops and coverplates for unused positions.
☐ No cracked or non-working meters.

Cabling and terminations

☐ Cabling runs are satisfactory, cable ties used where necessary.
☐ Correct size of cable used for current to be carried for maximum size of installation.
☐ Cable crimps not loose.
☐ No undue mechanical stress by heavy cables on circuit breakers.

Alarm extension

☐ Correct settings and operation of all alarms provided on power cabinets (mains fail, float low, high volts, etc.)

Designations

☐ All circuit breakers labeled.
☐ Cabinets (if more than one) are numbered.
☐ Switch plates clearly marked.
☐ Modules are all numbered, if modular-type rectifiers.

Commissioning test results

☐ Should be provided by the installers and be on site.

Handbooks

☐ Should be left on site by the installers (and usually contain the test results).

Load sharing

☐ If power supply is modular , all modules should supply approximately the same current (but not necessarily equal). Turn the rectifiers on one at a time and note reconfigured load sharing and current limiting functionality.

Safety signs

☐ Mains hazard.
☐ −Ve 24 V ground (where applicable).
☐ Any others that are required by local regulations.

DC distribution

☐ Cabling from power cabinets to radio cabinets for correct current rating.
☐ Trays used where necessary to support the cable correctly between cabinets.

☐ Bus bars and battery feeders insulated and suitably protected against accidental short circuits.

BATTERIES AND DISTRIBUTIONS

Fusing

☐ Current rating of fuses supplied for battery capacity and current drains.
☐ If indicators supplied, these should be clearly visible.
☐ Spare fuses available.

Battery spacing

☐ Clearance over and around cells is sufficient for maintenance work (SG readings, etc).

Electrolyte level

☐ Correct in all cells.

Battery vents

☐ These can be frail and can be easily broken in installation. Check none are broken.

Battery lead burning/connections

☐ Each cell is correctly connected to the next via the lead V-connection. There should be no cracks or breaks capable of producing a high-resistance across the connection.

Battery cabling

☐ Cables correctly tied and supported on cable trays.
☐ Correct size cable used.
☐ No loose crimps.

Hydrometer and thermometer

☐ Correct size hydrometer is on site.
☐ Thermometer is present in each pilot cell.

Cell voltmeter and millivoltmeter

☐ Both on site.
☐ Spiked voltmeter 0–3 V to measure cell voltage.
☐ Millivoltmeter to measure the volt drop across the inter-cell connections.

Designations

☐ Each cell and battery is correctly labeled.
☐ Pilot cell is indicated.
☐ Fuses are labeled.

Drip trays

☐ Size and capacity are adequate to catch and contain one cell full of acid if a cell container cracks.

Safety equipment

☐ Check presence of equipment as required by company regulations (can include rubber gloves, rubber apron, face shield, first-aid kit, etc.)

Safety signs

☐ Acid precautionary warning signs posted.

Water supply

☐ Fresh, clean water and a small washbasin available.

Fuse alarm extension

☐ Alarm given when fuse OC or CB operated.

Accessibility for testing

☐ Batteries should be placed in such a position to allow cell replacement without impediment.

Battery function and continuity

☐ Gradually reduce the rectifier output voltage and note changeover to battery-powered operation. A voltage drop of not more than 2 volts should occur and current should remain the same. This checks battery and battery feeder continuity as well as the changeover mechanism. In systems not yet commissioned, the rectifier power should be turned off for this test. The battery voltage will initially drop rapidly and will then rise stabilizing about one volt above the minimum.

Floor loading of batteries within limits

☐ Battery loads (kg/m^2) should be confirmed as being within floor structural limits.

EXTERNAL PLANT

Tower, mast, or pole

☐ State which, and give height. Check all ironwork is galvanized, no evidence of early rust, and all nuts and bolts are in place.

Gantry

☐ Feeders between structure and building are adequately supported by a gantry or tray.

Guys and anchors

☐ Check masts having multiple guys and concrete anchors. Guys should be examined for correct tightness and rusting, and concrete anchors for flaking and cracking.

Lightning protection

☐ Usually provided at top by antenna. Ensure lightning rods will not interfere with antenna pattern.

Tower, mast, or pole grounding

☐ Structure to be strapped to ground at ground level. Feeders to be grounded top and bottom of tower and at cable entry. Tower grounding connected to building ground.

Mains surge protection

☐ Usually provided on mains powerline into building.

Tower lighting

☐ Provided on structures near airfields, in accordance with local aviation regulations.

Corrosion protection

☐ All ground strap connections are sealed with anti-corrosion kit.

Safety signs

☐ Relevant safety signs are prominently displayed at foot of structure and on site fence.

Site RF radiation records

☐ When required, this record contains the maximum radiation from each antenna, and the safe working level of radiation on the tower top.

Equipment shelter

☐ The equipment shelter is properly and completely finished.

Base security, fences, locks, gates, etc.

☐ The site fencing and security are ensured.

Cable window and seals

☐ Cable window and seals are properly fitted to prevent water seepage.

Grounding of equipment rooms

☐ The equipment room is adequately grounded and the ground resistance less than 10 Ω. The test results should be available.

Access to tower restricted

☐ The tower is separately fenced and locked.

Antennas correctly mounted

☐ Antennas are correctly spaced and either vertical or at the correct level of downtilt.

INTERNAL PLANT

Mounting and layout

☐ Positioning of cabinets and supporting framework is as per design drawing.
☐ Blank panels, covers, etc., provided as required.
☐ Feeder supports provided above cabinets and up to the cable window.
☐ Frame racks firm and secure.

Cabling and terminations

☐ Neat distribution and positioning of all inter-cabinet cables; tied down at regular intervals.
☐ Cable plugs to be complete, no missing components.
☐ Plug labels correctly supplied and marked.
☐ Combiner-module and combiner-star connector tails are free and unstrained; N-connectors to be tight.

Alarms and telecontrol

☐ All alarms as specified are correctly returned to the mobile switch or monitoring center.

IDF cabling

☐ Neatness and tying of cables.
☐ Terminations correct for the type of termination applicable at IDF.

IDF labeling

☐ All circuits correctly labeled appropriate to the type of IDF system.
☐ Record book correctly made out.

System performance

☐ All channel modules within specification.
☐ All combiners within specification.
☐ Feeder-antenna return loss within specification.
☐ Base station controller, redundancy functional.
☐ PCM or link system.
☐ System line-up levels on each channel or port correct.
☐ All alarms functional.
☐ Final call-through test on each channel module prior to cutover.

Commissioning test results

☐ To be left on site by installers for subsequent use by maintenance staff.

Station logs

☐ To be provided at cutover by installation staff for batteries, attendance, or other as locally required.

Designations

☐ Cabinet numbering.
☐ Module numbering.
☐ −Ve ground signs.

Handbooks

☐ To be left on site by installers for maintenance use.

Drawings

☐ Copy of floor layout and cabling records to be left on site with handbooks.

Test cables and extender cards

☐ Available where needed.

Spares

☐ Check any spare parts ordered are available before cutover.

TRX plug crimps

☐ Check crimping of wire to plug connection.

RF N-connectors

☐ If available, use N-connector gauge on all feeder connectors and tails. Some center pins may be out of specification and can cause damage to the socket.

Independent link to switch (control center) functional (telephone line or radio link)

☐ Check for proper functioning of voice link to switch or control room.

Lightning arrestors on all incoming cables

☐ All cables entering and leaving the building should have suitable lightning arrestors.

Door alarms functional

☐ Test all door alarms and confirm the proper working.

Suitable fire extinguishers

☐ Suitable non-conductive fire extinguishers are in place.

Redundant control channel functional

☐ Turn off the operational control channel and confirm the proper functioning of the changeover.

Chapter 22

EQUIPMENT SHELTERS

Equipment shelters should be designed to maximize flexibility for expansion, to minimize operational costs, and to provide a clean, safe working environment for the staff. There are many different ways to achieve these objectives. This chapter presents some of the factors to be considered when designing shelters.

BASIC CONSIDERATIONS

Planning for expansion is particularly important because it seems that no matter how large an equipment shelter may be, soon after installation it will be found to be too small. Even when no expansions are foreseen, it is wise to build the shelter so that at least one wall can be removed to extend the building.

When transportable huts are used as equipment shelters, they should be placed on the site in a way that allows for additional huts to be added. Too often transportable huts are placed centrally on the site, mainly for cosmetic reasons. This can make future expansion awkward.

Switch rooms in particular should be built with ease of expansion in mind. It is almost impossible to relocate a switch once it is placed in service. As additional switches are added, the need to interconnect and monitor them is best met if the expansion allows good access to the Internal Distribution Frame (IDF) and the control room. High-level digital interconnections are cheaply and effectively achieved if they can be made without repeater equipment; but the switches must then be located within about 100 meters of one another.

There are a number of ways of providing for switch room expansion. In the example discussed here, a storeroom is incorporated in the switch room building with the plan that it will ultimately become a switch room itself.

Cellular operators, particularly new operators, should not underestimate the amount of space that may be needed for other services. In particular, space must be found for microwave, power and, sometimes, billing equipment. Other services, like paging, trunked radio, and voice messaging, can also consume considerable space.

BUILDING DESCRIPTION

The building will generally have only one floor and include the following areas:

- Equipment room
- Control and billing room
- Battery and power room
- Emergency plant room
- Uncrating area
- Storeroom
- Air-conditioning plant
- Toilets
- Staff facilities
- Cleaner's/janitor's room

EQUIPMENT ROOM

The equipment room houses the switch and microwave equipment. Because the equipment is often bulky, good access for delivery vehicles and movement of equipment is essential. The area must be air-conditioned and the humidity regulated.

Because demand is difficult to forecast, the room should be designed to expand in at least one direction (at least one wall should not be structurally supportive). Expansion can be facilitated by the use of steel-framed, non-load-bearing walls, which can readily be removed when additional area is required. The position of the equipment room with respect to the site should allow for this expansion.

An equipment room allowing for three suites of equipment should be about 7.5 x 5 meters. The actual dimensions will depend on the actual equipment purchased.

Floor tolerances for the equipment room should be precise. In the largest dimension of any floor area, the level should not vary by ± 12 mm; in any 3 meter length, the floor should not vary by more than ± 5 mm; in any 300 mm section, the floor should not vary by more than ± 2 mm.

CONTROL AND BILLING ROOM

A room about 5 x 10.8 meters is included to house the control and billing equipment. If necessary, billing functions can be handled remotely, and there are often good reasons for doing so. Real-time billing (hot billing), however, requires a data link between the switch and the billing computer.

BATTERY AND POWER ROOM

The battery and power room should be about 7 x 5 meters—large enough to accommodate two battery stands and the rectifiers. High ohmic distribution should be employed. The room should be fitted with a small washbasin and a handheld hose spray attachment, as well as an exhaust fan.

When lead acid batteries are used, the battery and power room should be physically isolated from the switch room. Today, however, sealed, rather than lead acid batteries are usually used, and these can be located in the switch room. In fact, with sealed batteries, it is common to place the batteries and rectifiers very close to the equipment to reduce copper costs (smaller bus bars), which can result in overall savings of 30 percent.

EMERGENCY-PLANT ROOM

The emergency-plant room should accommodate a diesel generator adequate for the total load of equipment, air-conditioning, and lighting. Typically, the load will be from 30–150 KVA (the upper limit being where a co-located base station is included). This room contains an electrical switchboard and must have an exhaust fan. Measures should be taken to contain any fuel spillage in this room, including surrounding the room with curbing 150 mm high (or making the floor 150 mm lower than the rest of building). Figure 22.1 shows a typical emergency-plant room. To allow easy access and safe passage, raised ramps should be provided for doorway access.

UNCRATING AREA

The uncrating area, a part of the switch room, is deliberately left vacant to provide space for additional equipment. The area serves as a workplace for installation and maintenance staff and is the last area of the switch room to be occupied. Good access should be available for vehicles transporting equipment, and the area should be fitted with doors 1.5 meters wide.

STOREROOM

The switch building will probably house its own spare parts. The value of spare parts is usually about 5–10 percent of the switch value, so good, safe storage is essential. A storeroom 3 x 4 meters would be adequate if storing only switch room spare parts. The switch room shown later in Figure 22.3 has a very substantial storeroom meant to hold base-station spare parts, too.

WALLS

The walls should be made of bricks or cavity blocks. Because cavity blocks cannot be cut to size as bricks can, when cavity blocks are used, room dimensions must be exact multiples of the block length. The internal walls to the power room/storeroom, equipment room, and emergency-plant room should be non-load bearing and surfaced with a fire-stop plasterboard or other fire-retardant material with a one-hour rating. The internal ceiling height should be 0.6 (minimum) to 0.9 meters higher than the switch rack or the minimum height specified by the local planning authority, whichever is the highest. Such considerations ordinarily yield a wall height of around

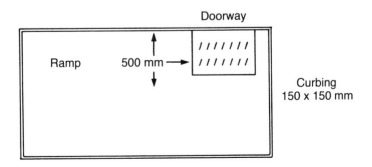

Figure 22.1   The emergency-plant room should be lower than the rest of the building so that fuel spillages will not seep into the main equipment room.

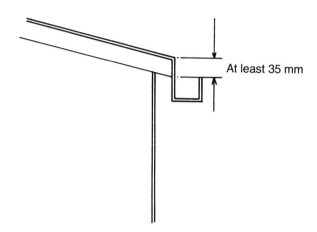

Figure 22.2   Because gutters can become clogged, it is recommended that the outside part of the gutter be lower than the wall side to prevent overflow down the walls and perhaps ultimately to the equipment.

3.4 meters (equipment height 2.5 meters), although suppliers have rack heights from 2.2 to 2.9 meters. Of course, these measurements do not apply if under-floor cabling is used.

When an existing building is used and the ceiling is significantly higher than the optimum, adding a false ceiling is worthwhile to substantially reduce the air-conditioning load.

ROOFING

The building must be totally waterproof because water on the equipment can cause total malfunctioning of the switch. The roof should have a minimum pitch of 5 degrees (assuming steel decking is to be used). The appropriate local high-wind code should be applied. Particular attention should be paid to the guttering and downpipes, which should be a "leaf-free" type to decrease the likelihood of blockage. The gutter sections should be designed to prevent overflow of blocked systems into the building, as shown in Figure 22.2.

INSULATION CLADDING

Table 22.1 shows the minimum insulation in fiberglass batting or equivalent that should be provided. Insulation should be used in walls and ceilings to reduce air-conditioning costs. All materials should be fire resistant.

CLIMATE ZONE	INSULATION THICKNESS (mm)
Temperate	100
Mediterrain	75
Sub tropical	50
Tropical	75

Table 22.1   Thermal insulation for equipment shelters

FLOOR LOADING AND CONSTRUCTION

The floor should be designed to carry live loads of 9.5 KPa throughout the building. Suspended floors should be used only where the site would require excessive fill.

Ideally, the whole building should be on level, consolidated ground. A reinforced concrete raft slab, incorporating edge beams to the perimeter and ground beams under the walls, should be used. Newly filled land requires time to compact and is not recommended for the switch room.

The raft slab should be placed on a consolidated base, leveled with sand, and covered with a waterproof membrane.

The floor should be elevated sufficiently above ground level to ensure against the entry of water under the worst flood conditions. (A flood-free site should always be selected).

CEILINGS

Ceilings should be of plasterboard or similar material and insulated to the recommended thickness.

WINDOWS

Windows should be of laminated glass. External windows, for security, should be high and covered with a metal security screen that also diffuses sunlight. To reduce air-conditioning costs, double glazing or thick glass (60 mm+) should be used for external windows. Adequate provision should be made for cleaning the windows (a sliding construction with a key lock facilitates cleaning).

APPEARANCE

Because the switch room will probably be located in a residential area, building design is important. The building should not look like a residence but it should blend in with the area and should not be conspicuous. The visual effect of the large equipment-access doors can be reduced by painting the top of the doors in a dark color.

TORNADO/HURRICANE AREAS

In tornado/hurricane areas, brick or blockwork should be reinforced with galvanized tie rods, from the footings to the roof beams, providing post-stressing.

STRUCTURAL STEELWORK

All structural steelwork should be painted to protect against corrosion. Galvanizing is not necessary except for lintels.

CABLE WINDOW

A cable window measuring about 1000 x 1000 mm provides cable access into the equipment room. This window will be positioned so that its bottom edge is rack height above the floor. The window should be made of galvanized, painted mild steel, with one plate on the outside of the wall and one inside. A cable tray (gantry) supports cables from the tower.

INTERNAL FINISHES

A dust-free environment, ease of cleaning, and hard wear are the main factors to consider when choosing internal finishes.

FLOORS

The toilet and entrance lobby should be floored with ceramic or vinyl tiles. All other areas can be finished with vinyl or similar tile.

WALLS

All brick walls should be cement, rendered with a rubber float tool to a fine-sand finish. Partitions to the emergency-plant and battery rooms should have a one-hour fire rating. Steel-studded plasterboard can be used in other areas except the toilets. The surface coating should be a gloss enamel finish to minimize maintenance.

DOORS

All internal doors should have a durable gloss enamel finish. Door frames in brickwork should be pressed steel and finished with the same coat as the adjoining walls. The doors to the uncrating area should be 1.5 meters wide and able to swing 180 degrees to be flat against the external walls. All external doors should be faced with sheet metal.

EXTERNAL FINISHES

External finishes should be maintenance-free; painting should be avoided. Aluminium frames should be anodized, and steel fixtures should be formed from material with an aluminium/zinc coating and finished with polyester.

All external doors should be fitted with dual heavy-duty mechanical locks vertically separated by 1 meter. An electric combination lock for use during normal hours of operation is also recommended. A combination lock allows the staff to move freely but keeps out others.

EXTERNAL SUPPLY

Three-phase, four-wire (or sometimes three-wire, single-phase for small-switches) power is required. Electrical utilities usually offer a range in price; select the option that is most economical.

The main switchboard should be located in the battery room. The power board should be fully enclosed with hinged doors for access to all components except main switches and changeover switches.

ELECTRICAL POWER OUTLETS

All equipment and control rooms should have adequate power outlets. In most locations, double outlets every two meters is adequate. The outlets should be located 150 mm above the floor.

EXTERNAL EMERGENCY PLANT

Because the emergency-plant room requires adequate access to a diesel generator, the doors to the room should be at least 1 meter wide.

Provision should be made to attach an external emergency generator. A manual changeover switch with three positions should be provided: Normally on, Off, External Emergency Plant.

A suitable three-phase external socket should be fitted to the building.

ESSENTIAL POWER

When using the emergency plant, only the essential power is provided. This power supports:

- Main switch rectifiers
- Power and battery room ventilation fans
- Control room
- One three-phase future expansion with fuses
- Emergency lights and outlets
- Essential air-conditioning

AIR-CONDITIONING

The whole building should be air-conditioned except for the battery and power room and emergency generator room. Separate air-conditioning for the switch room and control room should be powered from the emergency plant. It is probably not necessary to run the rest of the building from the emergency power.

Temperature should be kept in the range 20–30 degrees C, with a relative humidity of 20–60 percent (non-condensing) and dust to a maximum of 60 mg/28 cu meters of air by weight (5 micron diameter).

TYPICAL SWITCH ROOM

The switch room can take many forms, and maximum provision should be made for future expansion of the switch room. Figure 22.3 shows a typical switch room.

The switch room in Figure 22.3 could be used for a cellular switch. A billing computer room has been included. The billing need not be co-sited, but because the switch has emergency power and high security, it may be desirable to do so.

Figure 22.3 also shows a large storeroom. For a new cellular business (non-wireline), safe and secure storage is vital; normal office-type storage probably will not suffice. Locating storage in the switch room at first and then using that space as an expansion of the switch room in the future may prove cost-effective.

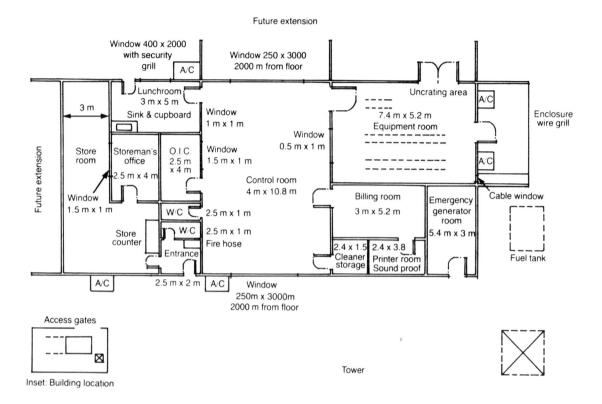

Figure 22.3   This typical switch room has a larger than normal storeroom (to be used also for base-station spare parts) and has the billing computer co-located with the switch.

BASE-STATION HOUSING

The base-station housing can take many forms—for example, existing buildings, transportable huts, and specially built structures (usually to meet town planning requirements). With the exception of very remote sites, all new structures or acquisitions should be adequate in size for ultimate expansion to full-base size. This full-base size depends on the system type, frequencies allocated, and the cell plan used. Due to intermodulation and frequency plan problems, it is generally not practical to place more than one fully equipped base (which uses the full cell-frequency allocation) at any one site. This problem does not occur in low-density regions.

The housing should be designed for ease of equipment-rack expansion, especially in installations where estimates of channel requirements at particular sites are, at best, guess work.

Base-station housing can present many challenges for the installation engineer. It is possible to use a standard shelter for new sites; for installations in existing buildings, however, it is necessary to conform to the space and layouts available. This constraint often also applies to building rooftop installations, where sufficient space for an optimum layout is not generally available. It is often necessary to use small spaces, often with awkward shapes, in order to conform with rooftops that were never designed to accommodate a base station.

Dead-load limitations can also restrict the type of installation that can be placed on a rooftop; this is particularly true if a tower structure is also required.

Shelters can be made of concrete, metal, or fiberglass. The cost of a well-designed shelter built with any of these materials is similar, and all materials can be expected to be suitable for about 20 years. Locally available materials and expertise may well determine the most cost-effective choice.

Concrete generally provides the most durable shelter and, where necessary, a virtually bullet-proof enclosure. The concrete must be sealed about every four years but otherwise requires very little maintenance. Concrete hardens with age, however, and can suffer cracking in extreme weather conditions.

Steel structures require a little more maintenance but can be expected to last about 20 to 25 years. Steel and fiberglass panels are easily moved and are most suitable for assembly in awkward places like rooftops.

Fiberglass, with careful upkeep, can be expected to last 15 to 20 years. Care should be taken to ensure that panel joints and sealing (caulking) are permanent and do not require annual maintenance. Fiberglass and steel are both less vandal-proof than concrete.

Transportable huts can be prefabricated and have all hardware installed before being moved onto the site. These huts can also be placed on top of office blocks and other high-rise buildings. When using high-rise buildings that are not the property of the cellular operator, locating equipment in prefabricated huts is recommended because the huts can be moved relatively easily in the event of future leasing disputes. It is often necessary to construct the huts on site because of access problems that make the placement of a complete hut on the rooftop impractical.

Using transportable huts reduces installation costs (because of standardization). It may be possible to assemble the huts and the base-station equipment before transporting them from a central workshop site. Attention should be paid to adequate thermal insulation to reduce air-conditioning costs. The huts should be of a type that is easily dismantled, or they should be designed for transport by helicopter, vehicle, or crane.

It is easier to meet deadlines if prefabricated huts are used, because they can be under construction even before the site is selected.

In areas where the water supply is undependable, the roof of the hut should have adequate gutters to collect water. And, if the hut contains lead-acid batteries, a secure water supply is needed for washing off acid spills. In addition, safety warnings should be posted around the shelter, and signs showing how to treat acid burns and, in particular, damage to the eye, should be prominently displayed.

Figure 22.4 through 22.7 show various base-site floor plans and structures.

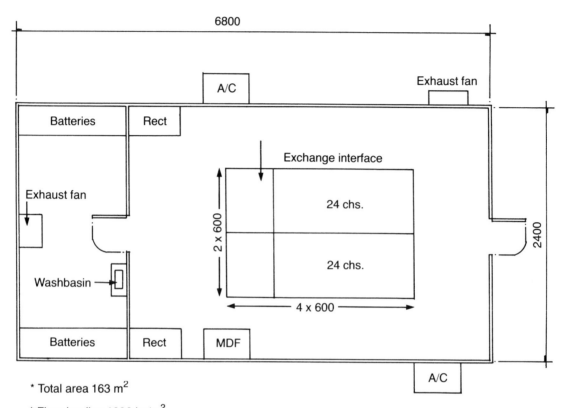

* Total area 163 m^2

* Floor loading 1000 kg/m^2

Figure 22.4   A minimum cell-site shelter designed for maximum economy of space; back-to-back equipment racks can be used.

Figure 22.5   A metal base-station shelter designed by the author and built in the Philippines

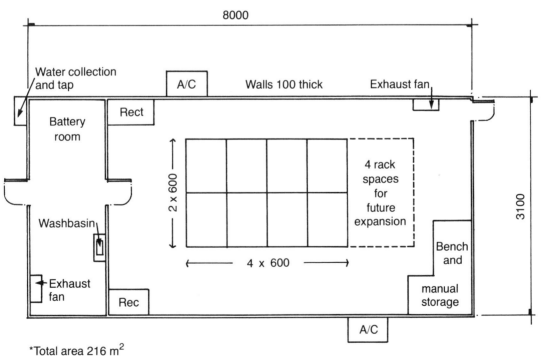

*Total area 216 m^2

*Floor loading 1000 kg/m^2

*Floor to ceiling 3100 mm

Figure 22.6   A preferred base-site floor plan where space for expansion is provided. This floor plan allows for some expansion and has room to allow ease of movement for installation and maintenance.

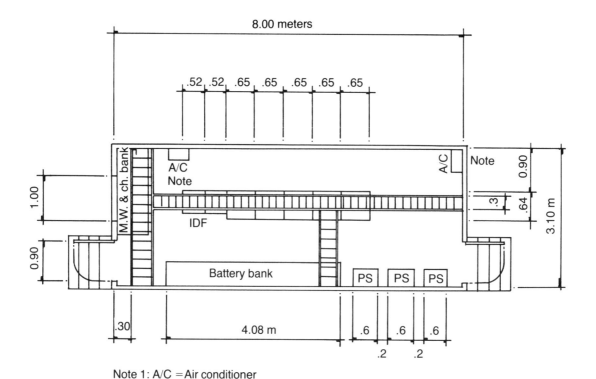

Note 1: A/C = Air conditioner

Figure 22.7   A base-station layout for single-sided equipment racks. Access is required to both the front and the back of the equipment racks. This drawing also shows cable trays.

FLOOR LOADING

A 48-channel radio base station will have approximately 4000 kg of equipment installed. Table 22.2 details the equipment.

The floor loading should be designed for the actual equipment loading, typically 700–800 kg/m^2.

ADDITIONAL AREA REQUIRED

Most cellular operators contemplate expanding into paging and possibly PMR, so three to four additional rack spaces should be provided. It is also desirable to leave some room for possible future digital expansion.

EQUIPMENT	WEIGHT (kg)
48 fully equipped radio channels	1500
Base-station controllers	900
2 rectifiers	300
Batteries (two-hour backup)	1300
	4000

Table 22.2   Equipment for a 48-channel radio base station

LIGHTING

Adequate lighting can be provided by eight (4 x 2 twin units), 40-watt fluorescent lights within the hut.

SECURITY

It is usually necessary to build a human-proof fence around the hut/tower installation to prevent vandalism. The antenna structure should have some means that prevent attempts to climb it. Serious legal liability could result should an intruder fall and be injured. All doors should be fitted with entry alarms that are monitored 24 hours a day at some central location.

An independent means of communicating to the main operations control center should be provided. A conventional telephone and land line, or a PMR (two-way radio) link, is adequate. This is needed if complete failure of the base station, its link, or the switch, occurs.

EQUIPMENT MOUNTING

The floor of the building should be constructed to allow the racks to be anchored by suitable bolts. The wall construction should be such that equipment (such as power boards and microwave links) can readily be attached. In the case of metal construction, mounting can use the supporting studs and the steel frames between them. For fiberglass construction, it may be necessary to include a plywood layer in the wall that can be used as a structural support. With fiberglass, it is important not to rupture the outside wall because a rupture can lead to problems with waterproofing.

INSULATION

To minimize the air-conditioning load, fiberglass-batting insulation should be placed in the walls and ceilings of the shelters. Table 22.1 on page 330 gives the proper insulation thickness.

The grade of insulation is a trade-off between cost and energy conservation due to heat losses. A thermal resistance rating in the range R-12 to R-22 covers the range of insulation normally used in such shelters.

CABLE WINDOW

It is necessary to bring RF cables for the base station and perhaps microwave antennas into the base-station housing. Because these ordinarily come from the same antenna support structure, a single cable entry of approximately 800 mm x 800 mm suffices. The cable window should be able to support additional cables as the base station expands.

For transportable shelters, the orientation of the shelter cannot always be determined by the cellular operator before construction (the shelter often must be made to fit into available space), so it is sometimes necessary to bring the cables into the shelter through any of the walls. In these cases, the location of the cable window must be flexible.

ELECTRICAL

The shelter should be wired for the following:

- Air conditioners
- Rectifiers
- Eight double, general-purpose power outlets

- An exhaust fan (when lead-acid unsealed batteries are used)
- Lights
- Emergency lights that can run off the equipment batteries
- An external power-inlet socket for an emergency portable generator
- General-purpose power outlets, one double unit every 2 meters, 150 mm above the floor level

Notice that the power consumption of a large base station is quite high (around 30 kw); if possible, it is a good idea to use a three-phase power supply (some authorities may require it). Approximately 200 watts per channel should be allowed for equipment power and a similar amount for air-conditioning.

The power supply may be single or three-phase. If a three-phase supply is used to power the base station, it can be used directly only if the rectifiers and air conditioners are three-phase units. Even in this case it is necessary to derive some single-phase supplies for auxiliary items such as the lights and power outlets.

Because three-phase power is not available at all locations, the decision to use three-phase equipment where appropriate means that the network will probably end up being a mix of three-phase and single-phase hardware. Consequently, spare rectifiers and air conditioners of both three-phase and single-phase-type must be kept.

Single-phase voltage feeds can be derived from three-phase supplies and the load shared between the phases. Figure 22.8 shows the relationship between the line and phase voltages of a three-phase supply from both star and delta transformers.

Providing a power inlet on the outside of the building so an emergency generator can be plugged in if required is a good idea. If an emergency generator is used, a suitable isolation switch should be provided at the switch board to enable the generator to be engaged and disengaged without danger to the operator or the equipment; a three-position switch that includes a neutral position will suffice.

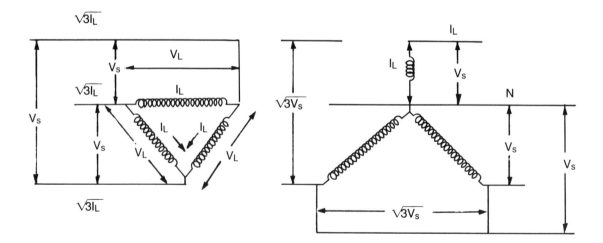

Figure 22.8 Power grid three-phase supply configurations. Star- and Delta-connected main power transformers showing the voltages from line-to-line and line-to-neutral as a function of V_S (the single-phase voltage, which is usually 110/220/240 volts) and the current relationship for three balanced (equal) loads.

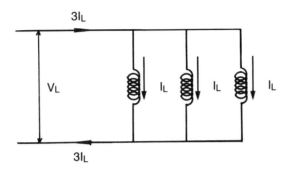

Figure 22.9   The single-phase equivalent of Figure 22.8

Power companies generally supply bulk power in a three-phase form because in this form transmission costs are lower than for a single supply. Consider the delta connected supply in Figure 22.8. If all the loads are equal, then the power supplied to the three loads = 3 x V_L x I_L x Pf (where Pf = power factor). The corresponding current carried by each of the three feeders is $\sqrt{3}I_L$.

If these loads were supplied by a single-phase line, the line current would be 3 x I_L, as shown in Figure 22.9. Hence, two conductors carrying $3I_L$ each would be necessary. The conductor size is a function of the current carried. If I_L = 10 amps, then the single-phase conductors would need to be AWG 6 or 4.13 mm. A 1-km feed length would require 237 kg of copper (two wires).

A three-phase conductor feeding the same load would need to carry $\sqrt{3}$ x 20 or 34 amps and would be AWG 10 or 2.59 mm. A 1-km feed length would require only 141 kg of copper.

GROUNDS AND PATHS

For both the switch and the base station, it is important that dirt and dust not get into the equipment rooms. The grounds surrounding the shelters should be paved and the rest of the ground should be free of dust, mud, or other potential pollutants.

Chapter 23

BUDGETS

Before undertaking any project, it is necessary to ensure that sufficient resources are available to complete it. Many cellular projects fail because of poor resource allocation. For budgetary purposes, some broad assumptions about costs and staff hours need to be made. These assumptions should be sufficiently general so that costs can be estimated before getting a detailed engineering study, and equipment orders can be placed before completing a detailed engineering design. Equipment orders generally require from 6–12 months of lead time because manufacturers produce equipment only upon receipt of an order. And, at any given time, equipment manufacturers normally have quite a few unfulfilled orders on hand. Note that the guidelines presented in this chapter provide a first-order approximation useful only for budgeting studies.

EQUIPMENT REQUIREMENTS

Before determining how much equipment is needed, you must first establish the level of demand and when service will need to be provided. This exercise is relatively simple if the equipment is for ongoing expansion only. Estimating quantities becomes more difficult, however, if the installation is for a new area. The estimates must consider the number of base stations, channels, switches, links, shelters, and towers, as well as the type of billing system to be used.

BASE STATIONS

In most new installations, assuming that high-density coverage is not required initially, the number of base stations depends more on the area to be covered than on any other factor. When the area to be covered is small, say less than 500 km^2, a first-order engineering approximation of base-station requirements is necessary in order to account for large variations in local terrain or other conditions. In larger areas where this variable tends to average out, the general rules discussed below are more applicable.

Table 23.1 gives the expected range of base stations in a city environment from a 30-meter

TERRAIN	AREA COVERED (km²)			EQUIVALENT RANGE (km)		
	AMPS NMT450	TACS	NMT900	AMPS	TACS	NMT900
High-density urban	12.5	10.2	4.5	2	1.8	1.2
Med.-density urban	50	41	18	4	3.6	2.4
Suburban	120	92	41	6	5.4	3.6
Outer suburban	200	163	72	8	7.2	4.8

High-density urban: Central business district (CBD) of major cities, such as London, Hong Kong, New York, and Sydney
Medium-density urban: CBD of smaller cities, such as Manila, Bangkok, Munich, and Singapore
Suburban: Area with only a small proportion of buildings higher than two stories, mainly residential
Outer suburban: Mixed residential, farm land, and open land

Table 23.1   Expected base station coverage for various systems from a 30-meter antenna elevation

structure. In rural areas, significantly larger towers and larger ranges can be achieved, so these areas should receive at least a map study before estimating coverage.

The number of bases required for a given installation can be estimated from Table 23.1.

Low-density areas permit the use of prominent sites, and coverage in these regions can extend to 20 km for NMT900 systems, and to 40 km for AMPS systems.

Additional bases may be required, depending on density (traffic-capacity) rather than coverage. The time to make this determination is when estimating the number of channels that will be needed. Also, some isolated, high-density areas may require dedicated bases. Whether your area requires dedicated bases can be determined later, when a detailed coverage study is done.

SWITCHES

The switch is usually a "stand-alone" unit (that is, it is not used for non-mobile purposes).

There are several factors that may lead to the decision to install a second switch: capacity, link costs, reload time, and security.

- Capacity. Usually a switch is limited to 10,000 to 50,000 customers (depending on the actual switch and software).
- Link costs. When the base stations become more and more remote, link costs rise. It may be more economical to provide a second switch. Generally, this is not a factor for links shorter than 100 km.
- Reload time. In the event of a system "crash," a reload at the switch occurs, and each base station must be downloaded from the switch sequentially, with its data and software. The time to reload one base is about 2 minutes, and as the number of bases increases, the total

NUMBER OF CHANNELS/CELL	FORMULA
1. Less than 10 channels/cell 2. More than 10 and less than 30 channels/cell 3. Greater than 30 channels/cell	K x 8 customers/channel K x (0.6 N + 2) K x 20
where	N = number of channels/cell K = 30/calling rate (mE)
Add 1 control channel/sector or 2 if standby exclusive control channel provided.	

Table 23.2   Budgetary estimate of customers/channel

reload time can become prohibitive. The reload time, however, is very dependent on the system architecture and can vary greatly from one manufacturer to another.

- Security. In the event of a major disaster, such as fire, earthquake, or war, the system integrity is greatly enhanced if two physically separate switch locations are used.

CHANNELS

This is an area of great uncertainty because the number of channels depends on the forecast demand, which experience shows is never very predictable. If actual traffic figures are available, they should be used, but for new service areas or even modified service areas (for example, a sectored omni cell), estimates must be made.

The sum of the traffic of individual cells (measured as time-consistent busy-hour traffic) will usually be 5–10 percent more than the traffic carried by the switch (measured in the same way).

This difference in traffic occurs because traffic peaks occur at different times for different cells and, in effect, measuring time-consistent busy-hour traffic at different bases is not the same thing as measuring time-consistent busy-hour at the switch.

Table 23.2 can be used to formulate channel-capacity requirements when traffic readings are not available.

When actual traffic figures are available, they are used with the Erlang B table to determine requirements (for more information, see Chapter 16, "Traffic Engineering Concepts").

Install sufficient channel capacity to allow for the lead time for the next expansion (typically one year) and, if the network is a new one, build in some overcapacity (for example, 30 percent) to allow for errors in forecasts.

TYPICAL WORK-HOUR REQUIREMENTS

Table 23.3 gives the approximate work-hour requirements for cellular radio installations.

COSTS

Table 23.4 shows the cost of switches, Table 23.5 shows the cost of switch sites, and Table 23.6 shows typical base-station costs.

INSTALLATION TASK	WORK HOURS

Design radio
(site selection and radio-path design)
        Up to 4 bases
        5 to 15 bases
        16 or more bases
	
	1000 hrs
	300 hrs/base
	400 hrs/base

Radio survey (includes survey of rejected sites)	300 hrs/base
Switching design	500 hrs
Installation of bases (includes links and installation design, and commissioning)	 1000 hrs/base
Installation of switch (includes installation design and com- missioning)	 3000 hrs/switch

Acceptance testing
        Per base
        Per switch
	
	40 hrs
	400 hrs

Moblie installation in vehicle	4 hrs/unit
Handheld sale (includes paperwork and customer in- struction and validation)	 2 hrs/unit
Changeover (i.e. from one vehicle to another)	4 hrs/unit
Maintenance	1 hr/customer

Table 23.3   Work-hour requirements

SWITCH SOFTWARE

Switch software comes in packages that cost from $500,000 to $1,000,000 and includes a guarantee of support and updates for a fixed period (usually about one year). Some manufacturers charge once only for software; others charge on a per-switch basis. Usually, there are additional charges for ongoing software support after the first year. The switch costs shown in Table 23.4 include an allowance for the software component.

MAXIMUM SWITCH SIZE	COST (US$)
500 lines	300,000
2,000 lines	500,000
10,000 lines	1,000,000
50,000 lines	1,600,000

Table 23.4   Switch costs

NUMBER OF SUBSCRIBERS	COST (US$)
0–5,000	100,000
10,000–20,000	300,000
More than 20,000	500,000

Table 23.5   Switch shelters (including land and site work)

ITEM	COST (US$)
Base station shelter	40,000
Tower/mast	1,000 per metre
Antennas and feeder (each)	2,000
30 metre pole (each)	10,000
Power supply for base stations incl. batteries	50,000
Emergency power plant	30,000
Base station radio equipment	13,000 per channel
Common equipment (incl. base-station controller)	230,000
Site and site work	40,000

Table 23.6   Base-station costs

COST MODELS FOR LINE TRANSMISSION SYSTEMS

Cellular bases are ordinarily connected to the switch by 2-Mbit, 8-Mbit, 34-Mbit, or T1 spans. Table 23.7 gives rules for determining the cost per circuit of the links. This model is based on a cost model of:

$$Cost\ (\$) = A + B \times L$$

where   A = fixed costs
B = length-dependent costs
L = length (km)

SYSTEM TYPE	$(A +B x L)$ COST COEFFICIENTS IN $1,000 PER VF CIRCUIT FOR CABLE TYPES					
	CABLE TYPES					
	CPFUT		SQC		SMOF	
	A	B/km	A	B/km	A	B/km
2 Mbits (protected)						
Existing (25 km)	0.19	0.08				
Existing (100 km)	4.63	0.08	0.54	0.08		
New (100 km)					0.55	0.16
2 +2 Mbits						
Existing (25 km)	0.13	0.062				
Existing (100 km)	2.47	0.062	0.27	0.039		
New (100 km)					0.34	0.081

* Costs of cable routes can be modeled as Cost = (A + B x L), where A = fixed costs, B = distance dependent costs, and L = distance.

CPFUT = Cellular Polyethylene Insulated Filled Core Unit Twin (cable)
SQC = Single Quad Carrier (cable)
SMOF = Single-Mode Optical Fiber

Table 23.7   Cable costs as a function of route distance

DIGITAL RADIO SYSTEMS (DRS SINGLE HOPS)

Digital radio systems have a very small marginal cost for an additional circuit and so costs are figured using a different model:

$$Cost\ (\$) = C + D \times N$$

where   C = fixed cost
D = cost per channel
N = number of channels

This model assumed that the existing cellular towers, power, and shelters are used for the microwave. These costs include installation. Table 23.8 shows the cost of microwave systems.

Wireline carriers have a good chance of having an existing plant, but if new links are provided over distances greater than 10 km, Digital Radio Systems (DRS) usually are the least expensive.

SYSTEM TYPE	C	D
2 Mbits	$56,000	$500/channel
8 + 34 Mbits	$100,000	$500/channel
* Microwave systems modeled as cost = (C + D x L), where C = fixed costs and D = additional costs/channel. For hot standby, add 40 percent.		

Table 23.8   Cost of microwave systems

BILLING SYSTEM (INCLUDING COMPUTER AND SOFTWARE)

Table 23.9 shows the cost of typical billing systems.

SYSTEM	COST
Simple system for 5000 subscribers, basic billing only	$ 80,000 + $ 10,000/year maintenance
Simple system for up to 10,000 subscribers	$ 120,000 + $ 50,000/year maintenance
System with on-line billing, direct switch validation for up to 10,000 subscribers	$ 250,000 + $ 25,000/year maintenance
Fully expandable system for up to 100,000 subscribers, with full-management information capability	$ 600,000 + $ 50,000 /year maintenance*
* Add $120,000 per additional 10,000 subscribers (mainly computer upgrade costs). Budget $250,000 for test equipment and tools.	

Table 23.9   Billing system costs (includes computer and software)

COSTS OF A "TYPICAL" CELLULAR SYSTEM

For an AMPS system, if you make some basic assumptions about start-up sizes and switch capacities, you can calculate the cost of a "typical" cellular system. Figure 23.1 plots costs against subscribers for a system that includes all infrastructure (shelters, links, towers) and also the marginal cost where these are not included (that is, the cellular-specific equipment only—a wireline operator may need to add this cost).

The switch in Figure 23.1 is assumed to be supplemented by a large (50,000 line) unit at 10,000

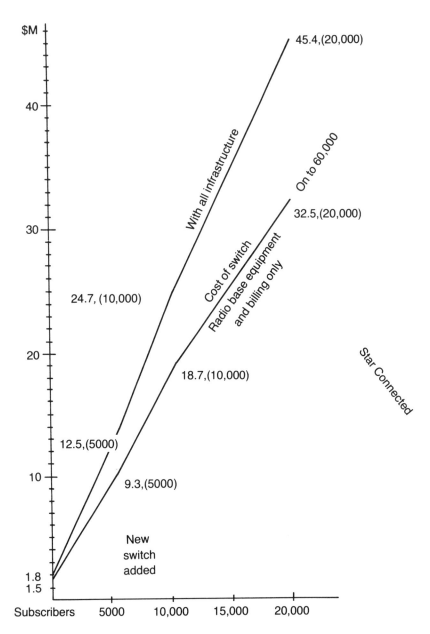

Figure 23.1   Costs of a cellular system against subscriber capacity (assumes most sites are sectored with N=7 pattern)

subscribers. Start-up costs, which are very operator-dependent and can include license fees, lobbying costs, advertising expenses, and other costs, have not been included here.

Some large operators of mature systems have reported the marginal cost of adding new subscribers at $1000. This figure probably has more to do with the method of calculation than economies of scale of large systems. For example, a wireline operator could easily claim that the marginal cost at 20,000 subscribers was around $1600 (based on Figure 23.1), but this budgets links, equipment space, and marginal switch costs as zero. Often, links are leased from the PSTN operator and therefore are not considered to be equipment costs.

AN EXERCISE

Estimate the cost of covering a city of a million people with an area of 600 km^2 and two CBDs separated by 5 km. Each CBD is approximately 2 km x 1 km in area and an AMPS A band (not extended) system is planned. Assume shelters are budgeted at zero (already owned) and all bases are omnidirectional. A 7-cell pattern is used, and microwave links must be provided. Note that this system needs some infrastructure (microwave only) and so should fall between the two extremes shown in Figure 23.1. Other infrastructure is assumed to be already owned or leased.

ASSUMPTIONS

You should base this example on the following assumptions:

Switch price[1] = $1.6 M (50,000 line switch)
Base stations = $230,000 and $13,000/channel
Work-hours = $60/hour (contract rate)
Calling rate = 30 m Erlangs/subscriber
First year's estimate of customer demand = 5000

BASE-STATION COSTS

To determine base-station costs, follow these steps:

First, determine the minimum number of base stations required for coverage only. Each CBD requires one base (total of two bases). An area of 600 km^2 (assuming it is suburban) requires 600/120, or five, bases, based on Table 23.1. Seven bases are required for coverage.

Second, determine the number of channels required for traffic. Assume that forecast customer demand equals 5000 and that there are 20 subscribers per channel. In this example, 5000/20, or 250 channels, would be required.

Because only the A band is available, it is reasonable to assume that each base has an average of 22 channels (50 percent of the 44 maximum) equipped (assuming a 7-cell pattern). The number of bases required is 250/22, or about 11.

---

1. Note these prices apply fairly well as of July 1989, but actual prices should be determined for each calculation.

Therefore, the cost of base stations is as follows:

$$(11 \times \$230 + 250 \times \$13)K = \$5780\ K$$

BASE-LINK COSTS

Assume, an average of one link/base (11). For this exercise, all links are considered to be 2-Mbit DRS, except the main link to the PSTN. Some sites will have 2 links (as more than 30 channels will be needed at some sites) so, say 15 links are needed.

Therefore, the cost of base links (from Table 23.8) is as follows:

$$(15 \times \$56 + 0.5 \times 250)K = \$965\ K$$

Add to this amount \$200 K for a 34-Mbit link to PSTN (with standby). Thus, the total link cost is \$1165 K.

Work-hours for this installation are shown in Table 23.10.

The cost of work-hours equals \$60 x 21,300 = \$1,278 K, assuming \$60 per work-hour. Therefore, the cost of this installation can now be calculated as shown in Table 23.11.

INSTALLATION TASK	WORK HOURS
Radio design	3,000
Survey	3,000
Switch design	500
Installation of bases = 11 x 1,000	11,000
Installation of switch	3,000
Acceptance testing of radio bases = 40 x 10	400
Acceptance testing of switch	400
Total	21,300

Table 23.10   Work-hours calculated from hours shown in Table 23.3

BUDGET COMPONENT	COST (in millions of dollars)
Work-hours	1.278
Links	1.165
Base stations	5.780
Switch	1.600
Ancillary costs:	
Training	0.100
Tools and test equipment	0.250
Billing system	0.135
For capacity of 5000 subscribers the total cost is	10.308

Table 23.11   Cost of typical cellular installation (assuming 5000 subscribers, 11 bases, and 250 channels)

Figure 23.1 (page 349) leads us to expect that the cost of this system should be between $9.3 million and $12.5 million, which it is ($10.3 million).

For budgeting purposes, this example gives a good description of the required system. The total work-hour requirements can also be used to estimate involvement by staff and consultants. A detailed engineering design may indicate some deviation from this estimate, but usually the results of this procedure are realistic both in terms of cost and base-station requirements.

Chapter 24

BILLING SYSTEMS

The billing system is an important but independent part of a cellular system. Most cellular system suppliers do not offer a billing system as part of their product line and recommend that this be purchased from an independent supplier. Wireline operators will almost certainly already have a billing system for their other operations, and this can be adapted for the cellular system. Some operators incorporate the billing so that the customer receives the cellular bill in the same format as if it were an additional wireline telephone; others prefer to bill separately.

The billing computer is physically separate from the switch; frequently, the only link is a tape. The raw billing information is collected by the switch but is not processed. The tape is then removed and sent to be read and analyzed by the billing computer.

This billing computer can be a minicomputer or a mainframe computer that has its own software to interpret the billing tape and then produce the subscriber's bill and the operator's ledger. Because various manufacturers have different proprietary operating systems, the billing (DAS) tape format is different for each equipment type. If the billing system provider has not previously worked with a particular switch, some software must be written to enable the tape to be read.

Charges vary widely from country to country and vary somewhat from operator to operator within the same country.

Charging can involve many parameters. These are some of the main parameters:

- Cost of local calls
- Cost of airtime
- Call rate variation with time of day
- Charges for long-distance calls
- Charges may vary with call length (for example, the first minute may be charged at a higher rate than subsequent minutes)

The following requirements should be built into the software:

- Frequency of billing (monthly, two-monthly, and so on)
- Charges applicable to roamers
- Billing format (what the bill looks like)
- Currency (this can present problems in countries where the exchange rate is thousands of units to the dollar so that existing billing formats may not be able to accommodate a sizable bill)
- Language (sometimes the bill may be required in languages other than English)

Real-time (or hot) billing requires that the billing computer can access the switch in real time and produce a bill on demand. The advantage of this is particularly evident in the case of roamers and rental units, where an instant bill is needed. To keep the demands on the processor within practical limitations, only some of the customers are flagged for real-time billing. Typically, a real-time billing package may allow up to 1000 customers at a time to be marked for real-time access. Such a system could use an RS232 link between the billing computer and the switch. Figure 24.1 illustrates this.

Billing systems can perform billing and ledger functions only, or they can perform validation and produce Management Information Systems (MIS). There are considerable advantages in the more capable systems, but they are also considerably more costly.

Billing systems that can perform validation allow customer data to be entered manually only once into the billing system. If validation is not available, the dual entries of customer data into the switch and to the billing system are sure to leave considerable room for false and missed entries. Once the system is bigger than a few thousand customers, it is almost impractical to use two parallel data entry systems.

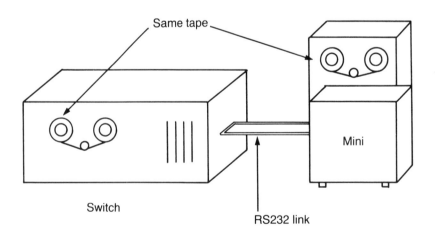

Figure 24.1   The link between the switch and the billing computer for real-time billing may use an RS232 link. The bulk of the billing information, however, is transferred via tape.

RESELLERS

The concept of a reseller evolved in countries that do not allow the cellular operator also to be a mobile retailer (notably the US and UK). This concept, however, can apply to all cellular operators. It involves giving limited access to third parties to the validation and/or the billing system. In this way, a reseller can validate and bill a subgroup of customers; in effect, the reseller becomes a de facto cellular operator. Figure 24.2 shows a billing system for resellers.

Most billing systems were developed with the major US or UK markets in mind, so this concept is generally available on most commercial billing systems.

BILLING HOUSES

In the US, it is possible to make use of the services of a number of centralized-billing systems and do without an in-house billing system. There are several advantages to this approach, particularly for small and medium-sized operators.

Billing systems cost upward of $100,000 (perhaps as much as $1,000,000). Because these need to be in place on day one, the billing system represents considerable capital outlay. It also requires space and staff to operate it. Having a third party undertake this responsibility, particularly at

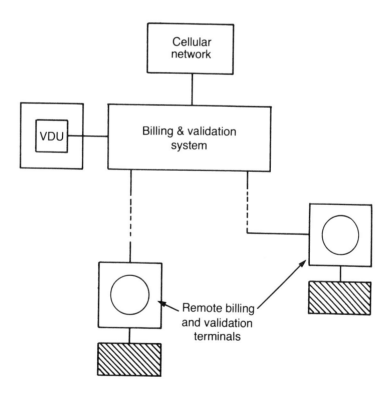

Figure 24.2   The billing system can be configured for resellers so that the reseller effectively becomes a de facto operator.

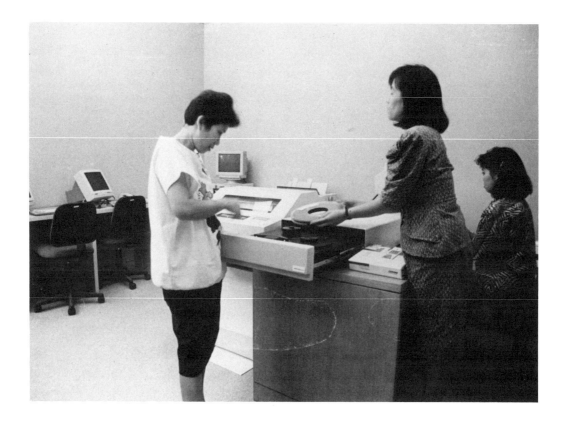

Figure 24.3   The billing room of Extelcom, in Manila, comprising four workstations, one minicomputer, a magnetic tape drive, and a high-speed line printer. Photo courtesy Extelcom.

start-up, can make good business sense. A medium-sized minicomputer billing system with four VDU terminals is shown in Figure 24.3.

Billing houses also have greater access to credit checking, roamer databases, and invalid roamer databases and information on rogue users. They can also provide sophisticated packages for system management to suit a wide variety of user demands.

The billing house can simply process the operator's tapes or, in some instances, can offer on-line facilities via a data link. The option of using a billing house is rarely available outside the US.

DO-IT-YOURSELF BILLING

It is, of course, possible for a cellular operator to write dedicated software to read and process the billing tape, and a number of operators have chosen to do this. In principle, the concept of reading the billing tape and processing a bill is simple. This can make the "do-it-yourself" approach look attractive, particularly when billing systems can cost hundreds of thousands of dollars. But commercial billing systems do a lot more than just write bills; they offer numerous system-management reports that can be very expensive to incorporate into a "do-it-yourself" system.

Software always takes a lot longer to develop and debug than seems reasonable, and adding on new subsystems can be very costly. Software experts consider 5 lines of debugged code per day to be a good average rate and 20 lines to be exceptional. Thus, a reasonably complex program of 2000 lines can be expected to take 2 years for one person to write.

Of course, writing and debugging smaller programs is much easier, and it may be possible to write a 100-line program and debug it in a single day (but not every day).

In general, the "do-it-yourself" approach is really only practical for the very smallest (who don't need sophisticated outputs) and the very largest (who can afford anything) operators.

BILLING SERVICE

The billing system should contain the following customer information:

- Age
- Occupation
- Sex
- Business/work address
- Vehicle (for fixed units)
- Average revenue (that can be subdivided into similar revenue groups)
- Areas generating most revenue
- Revenue as a function of time of day (or week)

This information tells a good deal about the types of customers that are connected, but it can never yield the profile of the mythical "typical customer."

A marketing manager can use this information effectively. If it is used intelligently, this information can give a company a competitive edge. The goal, of course, is to get maximum return for minimum investment. In a mature network, achieving this goal means identifying and exploiting unused capacity and using existing capacity more efficiently.

For example, if certain days show under-utilization (usually weekends), then perhaps an incentives scheme can be worked out. Similarly, after business hours, the traffic drops dramatically. A two-tier charge rate that offers incentives for night-time calls could be considered.

Customers who have high calling rates (and hence generate high revenue) can be identified and marked for special attention to ensure that they are retained. To gain customer loyalty, perhaps package deals that offer reduced charges for those users can be arranged. The system could alert customers whose calling rates might entitle them to a better deal if they choose a package. A small drop in revenue is more than compensated for by gaining long-term, satisfied customers.

Incentive packages are often aimed at high-calling-rate customers who, although they generate the most revenue, also use the system most and have a higher system cost per customer. In a fixed network, a customer who is connected but has a low calling rate costs the provider just as much as a high-revenue customer. Such customers are therefore undesirable to the operator. In cellular systems, however, a connected customer who pays the monthly fee and doesn't use the system costs only the price of the billing services. In cellular, therefore, a low-usage-rate customer normally generates the best return/cost ratio to the network operator.

Sensitivity to pricing can be gauged by correlating new customers with price variations. Where the option exists (that is, it is not government controlled), this information could be very useful in determining the revenue mix from monthly access fees versus call rates. It may be found that particular locations are not generating their share of revenue and that a sales promotion in those areas is warranted. Or it may be determined that an area is intrinsically one of low-demand. Knowing why demands are low (or high) can be extremely valuable in future expansion plans. But even if the cause cannot be identified, some generalizations on the structure of the areas with abnormal demand can help to focus expansion.

Information from widely disparate sources reveals at least an order-of-magnitude variation in customer sign-up rates from similar sized cities with similar per capita incomes in different parts of the world. Within the same country, there are still large variations in what are otherwise similar markets. The best information, therefore, is obtained from the operator's own records; the marketing manager who realizes this is the one who will succeed. In most Western markets, the average monthly bill per customer ranges from $80 to $120.

MANAGEMENT INFORMATION SYSTEMS (MIS)

Management Information Systems (MIS) are software packages designed to organize customer billing and provide further information about such things as roamers, equipment inventory, traffic distribution and analysis, and customer profiles.

Some billing systems allow, as an input, a field that describes how a potential customer came to inquire about the service. Then these people can be asked how they first heard of the company. This information can then be sorted by the MIS to measure the effectiveness of the marketing strategy.

Advertising can be related directly to sales performance and, by carefully identifying the advertising (for example, TV advertising as opposed to newspapers), the most effective media can determined.

MIS services are available as specialized software packages with numerous user options. Faced with many choices, it is often very difficult for the cellular operator to define clearly what facilities are needed. Because of this uncertainty, some operators have resorted to using more than one MIS because of their comparative advantages in different areas such as customer information, marketing, planning, and, not of least importance, billing. MIS is not cheap, so most operators must settle for a single package. One US company, GTE Data Services' Cellular Business Group, for example, has 150 different reports available from its MIS.

In the US, it is possible to have either an in-house MIS (that is, one solely operated by the cellular operator) or a MIS provided at various centralized locations. Although there is a definite trend towards in-house service, some operators cite the large database facilities, lower start-up costs, and expertise of the central MIS systems as adequate reasons to use centralized services. This option is generally not available to operators outside the US.

Chapter 25

MARKETING

The ultimate goal of a cellular radio system is to sell or rent the end product (a mobile phone) to the customer. Marketing is an integral part of any successful business, and in cellular radio it must be coordinated with all other aspects of the business. Marketing should not be confused with sales. Marketing is the means by which the demand is estimated, the target customer is identified, and the sales strategy (advertising, pricing, and incentives) is determined. The marketing department should also set standards for customer presentation, interaction, and service standards. More than with most products, there is a need in cellular radio for the technical, marketing, and sales staff to establish clear lines of communication.

The cellular customer often is a relatively high-profile and demanding person, who has paid a considerable sum for the mobile unit and expects service of a commensurate standard. The salesperson, on the other hand, is usually just that, a salesperson. As good salespeople, they generally believe that they can sell anything, and that all products are just products.

Salespeople are usually proud of their ability to sell anything regardless of their knowledge (or lack thereof) of the product itself. They like to tell customers what they want to hear. In cellular radio, however, this attitude, if not properly regulated, can be disastrous. It is the responsibility of the marketing department, in consultation with engineering, to ensure that the salesperson does not promise the customer anything that the system operator cannot deliver.

CUSTOMER DISILLUSIONMENT

Provided the system has been well thought out, most customers are prepared to accept the known limitations of the system. They become very disillusioned, however, if they have been promised more than can be delivered. Some common areas of potential disillusionment are coverage, call charges, handheld battery life, handheld coverage, battery drain, and transportables.

Coverage should be clearly defined before the sale is made. In particular, any doubt about coverage in areas considered important to the potential purchaser should be clarified. Sometimes a field test is necessary.

Each salesperson should also do the following:

*   Provide a reasonably definitive coverage map
*   Define as accurately as possible the difference between the coverage area for handheld and "in-vehicle" mobiles
*   Stress that radio systems will always have some areas of poor coverage and that marginal areas may work one day and not the next
*   Point out that the mobile antenna can contribute to limited range in fringe areas if a suitable antenna and mounting system is not deployed

Most customers have only the vaguest idea of how a radio telephone system works; unfortunately, some salespeople contribute to this lack of information. For example, customers often have difficulty with the concept of a limited coverage area, reasoning that if the area covered is, say, 40 km from the center of the city, then calls can only be made to or from fixed subscribers in the same zone.

I recall one customer who was quite upset at the idea of getting a hole drilled in his new car for an antenna, arguing, "Well, I don't have an antenna on my phone at home and I don't see why I should have one in my car."

Most customers initially expect nationwide coverage and are a little disappointed when they first realize that it is not generally available. In addition, call charges are often confusing to the customer. Usually, mobile and rental call charges are much higher than the normal fixed network. Some customers have difficulty accepting these higher rates, and they will have even more difficulty if they first become aware of them with their first bill.

Handheld battery life varies from one-half hour to about one and one-half hours and can be a source of annoyance to many users. Customers should be made aware of this limitation and told of the options available for additional batteries and charger packs. Some chargers are very slow (taking up to 15 hours per charge) and not suitable for heavy users. Some vehicle adapters for handhelds allow the handheld to run on the vehicle battery, but will not concurrently recharge the handheld unit battery. Such accessories can be a major source of profit for the mobile phone seller.

Handheld coverage can be severely limited in some local areas (for example, in elevator shafts and within metal structures). If the system has not been designed for handheld coverage, it is usually inadequate in most high-rise buildings, especially on the lower floors.

Vehicle battery drain is also a problem. If a vehicle-mounted mobile is left switched on overnight and the vehicle battery is a few years old, the battery will probably be dead in the morning. Some installers wire the mobile to the ignition switch to avoid this problem. But this can lead to voltage regulation problems that, in turn, lead to impaired performance when the battery voltage is low. To save the battery, some mobiles have an automatic turn-off facility that takes effect if the mobile has not been activated for eight hours.

Transportables are usually very heavy (especially if the charger is carried as well) and not very portable. Further, they often perform little better than handhelds (due to the relative height from which they are operated). If transportables are carried only short distances, for example, from the car to a desk, it is very useful to customers such as real estate salespeople or building supervisors because of its longer talk-time (typically 3–4 hours).

WHAT CUSTOMERS WANT

It has been well established from marketing surveys that customers look for the following five things from the cellular operator (in order of importance):

- Coverage
- Price
- Service
- System reliability
- System performance

In certain local areas, the relative importance of these factors can change.

Coverage includes both the extent (what areas are covered) and quality (how well the area is covered). Operators naturally want to cover the greater part of major cities before expanding into more distant rural areas. Operators usually decide (quite correctly) that the demand is mostly in larger cities and expansion brings only marginal returns. What is often not clearly understood (and is difficult to quantify) is that the city customer's decision to buy can be quite strongly influenced by the total service area available (including the area outside the main city), even though in practice, the customer may rarely use the extended coverage.

Experience has shown that, when extended coverage is provided (for example, covering the roads between major cities), the local buy rate is generally low, and the transit traffic does not seem to indicate that the extended coverage is heavily used. Despite this, there is considerable evidence that the availability of extended coverage enhances customers' perception of the quality of the service offered even if they rarely use the coverage.

There will, of course, be isolated low-density regions with high traffic, particularly when a smaller town is a satellite to a larger one and is interdependent for trade, or where there is a good deal of two-way traffic between the towns and local traffic may be high. Knowing the importance that customers attach to coverage, the operator may be tempted to overstate the coverage, particularly by failing to mention areas of marginal coverage. This "oversight" often leads to disappointed and disillusioned customers.

It is probably better to understate rather than overstate coverage, so that if marginal coverage is available, customers can "discover it" for themselves; they may then feel as though they've received a "bonus." Calls from elated customers describing good-quality coverage from areas up to 50 km away from published boundaries are often received. These calls are usually made from hilltops or at the coast where an over-water path to the base station is available. Nevertheless, customers seem to feel that they have received more than they paid for and are very satisfied.

Poor and patchy coverage is often a source of customer complaint. This can occur because of poor design (base-station location), failure of the operator to provide an adequate number of base stations, or system overload or excessive blocking. In each case, the result is the same: noisy channels on which it can be difficult or impossible to hold a normal conversation. These areas, when they are within the stated coverage area, should be clearly identified (in the form of a map); and the customer should be aware of these areas before a service is sold.

The price sensitivity of sales of cellular mobiles is fairly well established and upward discontinuities in sales graphs corresponding to drops in purchase price are a common phenomenon. Note, however, that as prices plunge below the $1000 mark, this sensitivity decreases rapidly. It does appear, however, that when telephones are above $1000, the ongoing operating costs are far less important to potential customers than the start-up costs. When call charge or airtime rates increase, a few customers will seek disconnection, but these customers are often those who were making marginal use of the system and were looking for an excuse to cancel.

Below the $1000 threshold, further price reductions are likely to attract bargain-basement buyers who like the idea of a cheap telephone. The first monthly bill, however, often dampens this enthusiasm somewhat and such buyers frequently cancel or even default. Buyers of the "below cost" bargains sometimes offered in the US and United Kingdom are particularly prone to canceling. (Many "bargain basement" mobiles are beginning to reappear in various parts of the world, having been purchased in the US, validated locally, and then re-exported.)

Service is a key to good customer relations in all industries, but especially in cellular radio, whose users are, as a group, more demanding than the average customer. Customers generally feel that they have paid a premium price for a premium product, and for that they expect good, efficient service. In practice, such service means personal attention at the installation center, a minimum of paperwork for connection, and prompt and efficient response to complaints about system problems.

When systems approach their capacity-call drop outs, the frequency of crosstalk, blocking, and failed handoffs increases, and the result is a degradation of service. Poorly designed and maintained systems suffer the most from this malady, but all systems with frequency reuse will have some problems. Customers will then experience a gradual but continuous degradation of the quality of the service as the systems grow. As a rule the bigger the system, the worse the service.

INITIAL MARKETING SURVEYS

Most cellular radio operators initially commission some type of marketing survey to estimate demand and to identify customer profiles. Nearly all such surveys have been spectacularly unsuccessful. Usually, very little correlation is found between demand for cellular phones and other telecommunications equipment already in service (telephones, PABXs, FAX machines, TELEX, and pagers).

Generally, almost no relationship is found between the answers given to market survey questions and public demand for the product. This discrepancy is not surprising when the questions asked are examined (they usually indicate a limited understanding on the part of the market researcher). Furthermore, the answers usually indicate a very vague comprehension of the question by the customer.

Today, it is usually unnecessary to spend large amounts of money on futile market research because most big cities have an existing market from which the necessary projections can be determined. If no market exists, cities elsewhere, preferably in the same country or in those with a similar GDP, can be used as benchmarks. Although this approach may not be very scientific, it is much cheaper and likely to produce results at least as good as market surveys. (Note, however, that expensive, beautifully presented, feasibility and marketing studies still tend to impress bankers and financiers, and so may still serve a purpose.)

Cities with populations larger than 500,000 in developed countries are generally considered capable of supporting two operators. Use caution when interpolating the results of big city sales to small ones, however, even when they have the same per capita income. There is evidence that small cities have a lower per capita demand rate than larger ones, and that for relatively small cities (less than 100,000), this demand rate can be very small. It is still not clear how to assess the market potential of smaller cities, but the experience in the US and Australia over the next few years may shed some light on this market.

The situation becomes more complex when assessing the potential of developing and under-developed countries. The initial reaction was to assume that, because the land-line network was (generally) quite bad, business would flock to cellular and that the calling rates would be correspondingly high. This does not appear to be the case, and calling rates in developing countries are, at least, similar to those in developed countries. There is also no strong evidence that there is any trend to regard cellular phones as a serious alternative to the fixed network. It does appear that in developing countries it is still the business/official user who makes up most of the market, and that usage is similar in nature to usage in developed countries.

Most big developing cities (pop. 5–10 million) seem to have a demand on the order of 1000 units per month. On a GNP per-capita basis, this rate is about three to four times higher than in developed countries, which probably only confirms the very uneven distribution of wealth in those countries. Developed cities of one million or more can expect about 1000 customers per month per million persons.

MOBILE TERMINAL POLICY

The cellular operator can adopt one of these two distinct mobile terminal policies: a monopoly or limited monopoly (the network provider basically controls the mobile terminal market), and an open market for mobiles.

The operator may have no choice of which policy to follow; government legislation may dictate the choice. There are some essential differences between these two policies. These differences are examined in the following sections.

MONOPOLY

By choosing a monopoly, the following may occur:

- Capital investment is much higher; as about 30 percent of the investment in a cellular network is for mobiles and all advertising and promotion is now basically with one party.
- The connection rate is lower; regardless of the determination of the supplier, a monopoly can never achieve sales figures as high as the free market (this also means that network revenue is lower).
- The number of models offered is limited. (In 1989, production quantities are such that an order for 10,000 units is still a big order to a manufacturer, so in order to get the benefits of quantity buying, a monopoly market should limit the number of models carried.)
- The network supplier can control or tailor the demand as necessary.

Often, in small markets with small sales volumes, a monopoly is the cheapest way to provide a service. Because the cellular market is relatively lucrative, any monopoly will be difficult to keep indefinitely, so marketing plans should include the contingency of ultimately having that monopoly broken.

OPEN MARKET

An open-market policy normally reduces the average sales price of a mobile unit and stimulates demand. In small markets, however, the large diversity of models available (there are over 17 manufacturers with an average of at least three models each) may mean that the benefits of bulk purchase cannot be realized and that costs actually rise as a consequence.

Results of implementing an open-market policy are as follows:

- Greater sales volume and more network revenue due to the accumulative individual efforts of a large number of relatively small, competitive suppliers
- The ability to support a very diverse number of models
- Inability of the network supplier to regulate demand (which may be necessary due to network constraints)
- A large diversity in installation and customer awareness standards
- A need to revalidate blocks of numbers for each supplier (or alternatively have on-line validation)
- Margins on mobiles sales are often low

It is generally a good idea for network operators to also have installation and sales sections so that they can get direct customer feedback and also be familiar with installation practices. In some major markets, however, including the US and UK, network operators ar not permitted to be directly involved in the supply of mobiles.

An open market can result in very low profitability, with fierce competition between suppliers. There is little or nothing that the network supplier can do to prevent such competition. The network supplier can contribute to the problem by competing at very low margins, knowing that losses will be recouped on the call charges and rentals or through incentives offered to mobile sales groups. Below-cost mobile phones are now common as the retailers have become dependent on "connect fees" to be profitable.

MOBILE INSURANCE

Very few vehicle or household insurance policies automatically cover a cellular phone, but coverage is often possible under riders written to cover specific items. Because the cost of additional coverage on existing policies is sometimes unattractive, specialized cellular policies may have advantages. Specialized policies can also have other benefits, such as a 24- or 48-hour replacement period compared with a month for conventional policies.

To avoid future disappointment, operators should inform prospective customers of the need to make sure that the cellular phones have adequate insurance coverage. Many people either think they are automatically insured or do not even think of the possibility of loss. Thieves, often very ill-informed, sometimes just take the control head. But organized theft rings that are very efficient both at stealing and disposing of the mobile units, are known to exist in many countries. Mobiles can also be damaged in vehicle accidents or become waterlogged in convertibles with the roof down. Uninsured customers will be unhappy customers and are bad for business.

Salespeople may be reluctant to push insurance because of the extra price or because the implication that the mobile phone makes the car more attractive to thieves may deter to the customer. (As the cost of mobile units drops, however, the relevance of insurance decreases.) Leased units, of course, should almost certainly be insured (either directly or indirectly).

Aside from advising the customer to take out insurance on the car telephone, the salesperson should inform the customer of some simple ways to protect the device. For example, locking the vehicle is essential. Most stolen vehicles are not locked and the mobile is more likely to be stolen with the vehicle rather than the vehicle being stolen because it has a mobile. When transportables and handhelds are used, it is best to stow them out of sight (not on the front seat in full view). It is also wise to use the lock facility on the mobile to prevent unauthorized use.

Although these warnings will probably go unheeded, the operator should point them out to the customer, preferably in writing.

SAFETY

In some countries, it is illegal to use a mobile radio or car phone while driving. Only hands-free units can be used legally while driving. In these countries, the driver's insurance may be revoked if an accident occurs while using a handset, regardless of whether the driver was at fault. Studies have shown that using a handset while driving measurably increases stress on the driver, whereas using a hands-free option or a boom microphone causes little stress.

Dialing requires a high degree of concentration and is probably the one action most likely to cause an accident. The driver normally needs to look away from the road to dialing. Dialing is done most safely when the vehicle is stationary or by using the telephone's memory. Voice-activated dialing (using voice recognition to associate a name with the required number) is an elaborate and expensive way to overcome the hazard of dialing while in motion. Positioning the keyboard so that it can be seen without looking away from the road can also help. But this is usually difficult to achieve. It is in the interest of the operator to bring these safety matters to the attention of the customer.

DISTRIBUTION OF MOBILE UNITS

The early trend in the cellular market showed that the biggest portion of sales came from the direct-sales sector. As customer awareness grew, so did indirect sales through such outlets as electronics and auto-sound specialists. This trend is predicted to increase significantly, with some market reports suggesting that direct sales will account for as little as 25 percent of the total sales volume by the end of 1990.

Auto accessory dealers, department stores, computer stores, direct mail sellers, mass merchandisers, and customized sellers will take advantage of this opportunity. Together with this diversification, the markets are forecast to stabilize while prices continue to drop. Although nearly all early sales were made by personal sales calls, there is a general public dislike for such sales; consumers report that they prefer visiting retail outlets. The cellular operator must be prepared for this change and must be open to new and diversified outlet opportunities. Among the most popular of the new outlets are retailers, particularly those who already have a high profile in electronics and radio consumer goods.

The need to maintain standards of installation and maintenance in the industry is apparent, and this presents a challenge with diversification. Some operators have introduced a certification scheme, in which all agents are required to undertake prescribed training and satisfactorily complete the course. Providing training is relatively easy for operators who have their own installation centers and can conduct the training in-house. Minimum standards of premises, equipment, and facilities should also be set, and all prospective agents should be subject to inspection. Spot checks are also a good idea to ensure that standards are maintained.

Special attention should be paid to major customers (corporate and fleet sales) and frequently an "account executive," an experienced salaried professional, is appointed to handle this market sector. Agents and junior sales personnel, on the other hand, are usually geared for commission sales.

When an operator or reseller has both direct and indirect sales outlets, conflicts can occur between the two; it is a good idea to have one sales manager responsible for both outlets. The sales manager should be responsible for ensuring fair competition between outlets. In the current environment of heavy discounting, however, this job can be most difficult.

A survey in late 1988 by *Cellular Marketing* revealed that only 10 percent of cellular salespeople are totally dependent on cellular sales; the average salesperson is only 35 percent dependent on cellular for income. The same survey revealed that 53 percent of salespeople were self taught (a very dangerous condition), and those salespeople generally considered customers to be rather vague about cellular radio. A case of "the blind leading the blind" emerges from this survey.

COMPETITIVE NETWORKS

Very few network providers do not have competition. Almost invariably, the competition uses the same system but uses the other half of the frequency spectrum. The three most common bands in use are the following:

- 420–590 MHz (2 x 5.4 MHz for NMT450 and NAMTS)
- 824–894 MHz (AMPS)
- 890–960 MHz (TACS, NMT900, NTT)

In most developed countries only one of the last two bands is available. (For example, in Europe, the 824–896 MHz spectrum is used for UHF TV; in the US 890–960 MHz is used for various mobile and paging systems.)

Most cellular systems have the built-in capacity to operate as two competitive systems using one half of the spectrum each (an exception to this is the NMT systems, which have no provision for a second operator). Subdividing the band to allow competition between suppliers in the one location was not anticipated by the original NMT equipment specification and causes difficulties with roaming due to non-standard control-channel allocation. Channel efficiency also drops as base-station maximum capacities decrease. Thus, in most developed countries there is a natural limit of two providers per city using existing technologies. New technologies, however, offer the prospect of further de-monopolization of cellular. This particularly may apply to the proposed 1 + GHz systems and other "microcell" systems under development.

At this point, it is interesting to look at the experience of Hong Kong, where this limit does not apply. Hong Kong has three suppliers using four different systems. They are:

- An AMPS system using one-half the AMPS band
- An ACS system using the other half of the AMPS band
- A TACS system (with three users using the A, B, and ETACS bands)
- An NEC (NTT) 400-MHz system

Despite competition, the network charges for these systems are the same.

Hong Kong, like a good number of countries outside Europe and the US, has been able to make use of both cellular bands. Although doing so solves a short-term frequency problem and provides more "competition," the result is that now Hong Kong, like Europe and the US, has little spectrum reserved for a future digital system and will need to recover spectrum and possibly mobile telephone hardware early in the 1990s. In general, the plan in the US is to use the extended portion of the A and B bands for the introduction of digital. If this band is not already in use for analog, the introduction of digital will be relatively painless.

Another problem has to do with interference between competing cellular systems. In Hong Kong, for example, problems have been reported where an AMPS handheld can be desensitized by a TACS handheld operating a few meters away.

"Competitive" network suppliers rarely have very different network charges. This is either because their charges are subject to government approval or regulation, or because they form a cartel to regulate them. Competitive networks, however, do have the incentive to provide enhanced customer services such as improved coverage, roaming, message services, network reliability, and customer facilities.

EFFICIENCY AND OPERATION OF
TWO OR MORE NETWORKS

Splitting the circuits for two providers incurs only minor penalties in circuit efficiency in high-traffic regions. In the rural areas, however, where traffic is low and base stations are sited mainly to provide coverage rather than to carry substantial traffic, circuit splitting entails heavier penalties for operators.

Because the market is more diffuse in rural areas, it is in the interest of competing operators to minimize expenditures by cooperating whenever possible. And because it is normal for both operators to use the same system, roaming (from one system to the other) can be provided. Differences in hardware can cause difficulties when any form of automatic roaming is proposed. Thus, in rural areas (or along roads between major cities), it is practical to allow mutual roaming in order to avoid duplication of facilities. This roaming can be limited only to certain specified bases. A method of cross-charging and cross-validation will be necessary, but this should not prove to be a major problem.

Where different cities have different network providers (as in the US), the multiple bilateral agreements can become very complex. The complexity is increased if the systems are not operated identically. Another major problem in the US is the size of the potential roamer database (approximately 2,200,000 in mid-1989) and the fact that most agreements are bilateral.

There is, to date, no central universal database. And because numerous types of hardware and software are used, it is difficult to transfer information from one switch to another. A universal roaming switch that will connect all switches to each other should be available around 1990.

Operators contemplating an environment similar to that in the US are advised to consider future roaming options before purchasing hardware or determining operating procedures. Standardization can be of great value.

CUSTOMER RELATIONS

Customers will generally expect prompt attention to mobile problems. Customers have a strong tendency to equate the people to whom they take their complaints with the company itself. As a result, the people who work directly with the customer can make or break the company image. To distinguish operator errors from problems with the mobile unit, it is important that the customer's first contact be a person who is knowledgeable about both the operation and the limitations of a mobile telephone. This need is particularly critical because the customer often has great difficulty communicating the problem.

Moreover, because customer education is often necessary, the contact person also needs skills in handling people and their complaints. Technicians, generally, are not ideally suited to handling customer problems.

SERVICE POLICY

The customer probably expects same-day attention to problems. Such service is expensive because the operator must provide the capacity to handle unforeseen peaks in demand. The cost of providing this service has to be built into the sales price of a mobile (in the case of a warranty) or into an ongoing service charge.

Operators must also formulate an after-hours service policy. The provider must decide if after-hours service is to be provided and, if so, by whom and at what cost. A minimum of two hours is usually necessary to enable the serviceperson to reach the service center and attend to the problem. About one hour should be allowed for service. Some operators provide fully equipped mobile service vans for after-hours service. A number of mobile supply centers offer service policies that provide post warranty service (and sometimes also insurance).

CONSUMER AWARENESS

Ensuring customer satisfaction is very important because the customer is the greatest source of new sales. About half of all new customers first become aware of cellular radio by word of mouth. Some operators use the billing system to foster company awareness, and some operators follow up after two to six weeks with a personal call by a service adviser to check on customer satisfaction.

In a market where many suppliers are offering many types and brands of equipment, product-orientated marketing can be difficult; marketing that focuses on the supplying company's corporate image can be more efficient. Mobile brand names and image have a relatively low influence on purchasing decisions.

Despite its widespread popularity (mainly because of its low cost and low overhead), tele-marketing (direct telephone marketing) accounts for much less than 1 percent of initial customer awareness in the US overall. In isolated instances in the US and Australia, however, it has proved most successful.

ADVERTISING

Advertising is an essential ingredient in any marketing policy, but it must be monitored. Designing an ad campaign is best left to the experts, but the operator should monitor sales against advertising outlays. Advertising can also "spill over," and major campaigns that are not well-directed toward the particular advertiser can cause increased sales in mobile phones from other suppliers. This "spill over" can sometimes be monitored.

It is a good idea to concentrate on different media campaigns sequentially, so that the most effective form, in terms of sales per dollar invested, can be evaluated. The effect of an advertising campaign is usually felt as increased sales that last from one to three months.

An advertising budget should be set at the beginning of each financial period as a percentage of gross sales. Allocating between 2 and 5 percent of sales is typical but consideration must be given to actual demand and, more importantly, to what the competition is doing.

Too many unsatisfied customers can be as big a problem as too few customers. Consequently, there is often a need to regulate demand downward due to supply or system capacity problems, and not advertising is often the best way to do this. In this respect, the marketing and engineering departments should be well-coordinated, because advertising requires a lead time of four months (from placing an advertisement to customer response).

QUALITY OF SERVICE OF THE CELLULAR SYSTEM

A cellular operator who has underestimated demand (and most operators have been caught doing this at least once in their careers) is sometimes tempted to put on "a few more services" than engineering recommends. This practice is very dangerous.

The number of subscribers who can be carried without congestion (system overload) can be quite accurately calculated. As the number is increased beyond that point, the first thing the customer notices is that calls get noisier. This is because fewer channels are available and customers are getting second and third choice base stations when the first (nearest) choice is full.

Increasing the numbers a bit more (further increasing congestion) causes calls to fail (due to no available channels) or drop out (when no channels are available for handoff). By this time the operator has a reputation for providing second-rate service. Even if the situation is promptly rectified, it may take years to live down the bad image generated by the expediency of putting on a few extra services.

PRODUCT LAUNCH

Because of the very long lead times in setting up a cellular system (usually about two years—the first devoted to preliminary preparation and the second to installation), marketing people can easily be tempted to send off a series of press releases detailing progress as it occurs. Premature publicity, however, can be counterproductive; information, for example, that an approval to operate has been given, that an equipment supplier has been named, that construction has begun, and so on, is neither interesting to the press nor to potential customers. Premature releases can even dull the impact of the product launch when it actually does occur. Old news does not attract media attention. Therefore, the operator who takes a low-key approach to the set-up phase and then has a spectacular launch is able to get the undivided attention of the press at the most critical time. The product launch can include some quite spectacular stunts when a handheld and a bit of imagination are combined.

Here are some examples of spectacular product launches:

- A live telecast of a parachutist in free-fall using a handheld to receive a call from some well-known personality who also has a handheld in some distant city with a well-known feature (the Eiffel Tower, the Sydney Opera House, or the Golden Gate Bridge) in the background
- A call made to the top of a high tower, building, or flag pole from a prominent person using a handheld at a well-known site
- A call made by a prominent person to a person in an awkward place (a lion's cage or a crocodile pond)

Both ends of the call would be broadcast live and the call could be billed as "the first official call." It is preferable if the stuntperson is shown by zoom camera in a place where the telephone is the only valid voice link.

This stunt could coincide with the official launch, to which the company invites potential customers, prominent first customers (whose telephone number can be activated on the network during the launch), and others as appropriate.

This official public product launch should be preceded by a press launch at which a large number of telephones (different types) are made available for free calls. Long-distance calls should be allowed because they make a more lasting impression. A vehicle fitted with a working telephone and brought into the function area is also quite effective at such launches.

PROMOTIONS

The giant cellular telephone seems to be a must for all cellular operators. There are available inflatable models up to about six meters high. Some operators have even stripped down cars and replaced the car body with a huge functional mobile telephone.

Figure 25.1 shows a giant Motorola Micro-TAC mobile telephone used for displays by Extelcom. (The Micro-TAC is the world's smallest cellular phone.) The unit lights up when the power button is pressed and can be used to make calls because there is a conventional hands-free unit inside.

Giveaway gimmicks can include hexagon-shaped key rings, plates, clocks, pens, and coasters. Some of these items are shown in Figures 25.2 and 25.3.

OUTDOOR (BILLBOARD) ADVERTISING

Because a large portion of the cellular market is directed at motor vehicles, it makes sense to use advertising aimed at motorists. Billboard, taxi, and perhaps even bus advertising can appeal to the target audience. Operator experience with billboards generally has been positive. Some operators change their message and sites every few months to give the impression of being in more locations than they have paid for.

Since the attention time for billboards is small (8 to 10 seconds), the message must be brief; most advertising concentrates on projecting the company name along with a straightforward message like "phones for people on the go" or "your car phone company." Billboards, however, rate very low in customer awareness surveys, accounting for only about 2 percent of "first awareness" reports, according to *Cellular Marketing*.

INFORMATION SOURCES

Probably the best source of marketing information available to an operator can be extracted from the company's own files. Once a system is mature (usually defined as being above 2000 subscribers), clear customer patterns begin to emerge. The two best sources of marketing data come from the mobile installation centers and the billing records.

One valuable source of customer information is in the files on handheld units. Surveys in the US reveal that about 50 percent of handheld users come from a two-car phone family. These people can be followed up from warranty cards or other records.

MOBILE INSTALLATION CENTER

The mobile installation center can provide very valuable information, including the following:

- Portion of handhelds/transportables/fixed-in-car services
- Vehicle make and type
- Customer home and business address
- Monthly fault statistics

Figure 25.1   A giant handheld is often used by cellular operators for display. Photo courtesy Extelcom.

Figure 25.2   A promotional clock. Photo courtesy Extelcom.

Figure 25.3   A promotional brass plate. Photo courtesy Extelcom.

- Customer-attrition rate
- Customer-complaint rate
- Preferred options

Apart from these direct sources, indirect information from the installation center staff on perception of service quality and customer attitudes can provide valuable feedback.

MOBILE TELEPHONE RENTING, LEASING, AND BUYING

Although most customers prefer to buy their telephones outright, others, for cash-flow or sometimes tax reasons, prefer a time-payment method. Most operators prefer outright sales or leasing because they then do not have to deal with "used" phones.

RENTING

The renting option appeals to many users. In particular, prospective customers may choose to rent for a few months before committing themselves to a purchase. Rental can be arranged on a daily, weekly, or monthly basis, with short rentals requiring that the operator has real-time (on-line) billing facilities. Rates charged for rentals are usually relatively high and typically enable the operator to write-off the mobile after 12 months. Rental units lead a hard life and cannot be expected to last as long as privately owned units.

A particular problem with rental equipment is battery life, which is often reduced due to poor charging practices. Portables have nickel-cadmuim (NICAD) batteries, which need to be completely discharged and then fully charged in order to achieve full battery capacity.

The mobile dealer can make renting available at airports, hotels, and through rental-car operators. A credit card or substantial deposit is usually used for security. Real-time billing makes renting an attractive option.

LEASING

Although more expensive in the long run, many customers prefer to lease for cash flow or security reasons. Leasing often involves a package that includes the following:

- Installation fees
- Extended warranty on parts and labor
- Loss and theft insurance
- Loan of replacement phone during repairs
- A discount package for call charges (depending on arrangements with the financier)

The leasing finance is generally provided by a specialized leasing company assigned by the operator.

BUYING

Outright purchase is the option preferred by the operator, whose only remaining obligation on the sale is the warranty.

SALES PROMOTIONS

There are, in most countries, many competing suppliers of mobile telephones. Price cutting is inevitable in such an environment, with some operators even selling below cost in order to get customers into their systems.

Bargain-basement mobiles are sometimes available. These mobiles normally are very basic and offer few extras. Some of these extras which are not really optional, such as the antenna and a warranty, must be paid for at the time of purchase. These often cost more than the mobile.

Ultimately, there are no beneficiaries of a price war; non-wireline operators, in particular, should beware that against the resources of the wireline operators, they are particularly vulnerable to chronic cash flow shortages during such a price war.

End of model sales can result in temporary bargain prices, but because the main advantage of the newer models is usually that they are smaller and cheaper to produce, these "sales" are short lived.

EMERGENCY USERS

Local emergencies and disasters can often be handled more effectively if good communications are in place. By moving quickly at such times and offering an adequate supply of free handhelds, the cellular operator can be seen as a good corporate citizen, while graphically (and publicly) illustrating the value of the cellular product to the community.

CHARGES

Charging packages are usually fairly complex and vary considerably from operator to operator both in magnitude and type. In 1989, the operator's gross annual income per subscriber (without installation, connection, sales, and other once-only charges) ranged typically from $800 to $2,000. Most operators are interested only in the bottom line, so the large variation between monthly access charges and airtime costs, not surprisingly, leads to similar total costs per subscriber per month.

Once-only charges include installation fee, connection fee, deposit on rental or leased equipment, special-number charge, special facilities fees, access fee, connect-time charge, and special-package fees.

INSTALLATION FEE

Installation charges vary from a low $15 in Asia to more than $200 in Australia and the US, and probably span the full range in between. Vehicle installations take an average of four hours (including the customer service component), and the cost to the installer is very dependent on the hourly wages.

CONNECTION FEE

The connection fee covers the cost of validating a new user on the system and the associated paper work. Typically it is about $50.

DEPOSIT ON RENTAL OR LEASED EQUIPMENT

Deposits are required on rented and leased equipment, and sometimes they are required even when the customer buys the unit. This deposit may be held as a security on future call charges or it may cover the first few months' access charges. This charge is typically around $250.

SPECIAL-NUMBER CHARGE

Some operators make special numbers available at a premium. Particularly in countries like Hong Kong, where people are somewhat superstitious, numbers like 888888 can fetch a very high price. The cost and inconvenience to the operator of reserving and providing such numbers is considerable. Most operators either have a policy not to provide such numbers or to do so only at a considerable fee ($100 or more). Special numbers can be considered an additional resource, so it is well worth considering selling them.

SPECIAL-FACILITIES FEES

There are many special facilities available on the average cellular switch, and some operators capitalize on this by charging a fee (either once only, or recurring) for features such as call forwarding, call waiting, conference calling, and call restriction. The charges levied for these features are usually modest.

ACCESS FEE

The access fee is charged monthly for access to the system. The fee ranges from about $10 to $60 per month.

CONNECT-TIME CHARGES

Connect-time charges are a grey area, where the number of possibilities is limited only by the imagination of the marketing department. The first major distinction between connect charges is whether the call is charged with airtime added (as in the US), or only the call is charged but at a premium rate (as in most other countries). The second method is self-explanatory, but the airtime charge needs elaboration.

In general, airtime charge is based on the user-pays concept. That is, as soon as the mobile begins to occupy a channel, the charging starts. The channel is occupied as soon as the customer presses the send button and continues until the customer hangs up the phone or presses the END button. A variation on this approach is to commence charging when the called party answers. The actual call is charged separately, usually at the same rates as a land-line telephone, so the subscriber is billed separately for call charges and airtime. This scheme is often used when the operator is not a wireline carrier. These two connect-time charging methods can, of course, also be applied when a land subscriber calls a mobile telephone.

A philosophical difference does occur, even here. In the US, the mobile subscriber pays for a call *whether or not* the subscriber originates a call or receives one. In many other countries, particularly where a premium call charge only is applied, incoming calls to a mobile are paid (at the premium rate) by the caller. Mobiles using this scheme usually have an area code prefix (a prefix beginning with 0 or 1) so that the caller would normally expect the call to be a "trunk rate" call.

SPECIAL-PACKAGE FEES

Many operators have multiple charging packages that may offer cheap off-peak rates, special accounts for high calling rate customers, and packages that include a number of "free" local calls (for example, the first 100 calls per month or first 200 calls per month).

These packages are usually budgeted with the access fee, so the more attractive packages are associated with a higher access fee. Unless these are carefully budgeted, they can prove to be unprofitable for the operator. For example, if the operator offers big discounts for high-usage customers, the total return can be calculated by comparing the package cost with the income expected. High-usage customers, however, make greater use of the network; and the marginal cost of network expansion must also be considered.

This line of reassoning suggests that the low-usage customer is the one producing the greatest return on investment and therefore is perhaps the one to be encouraged. (This reasoning, of course, does not usually apply to wireline customers.) For example, imagine an operator who has 1000 customers who never use the system. They pay their access fee (the lowest one), for example, $20 per month. The revenue from these customers is thus $240,000 per year and their cost to the operator is billing only; because they do not use the infrastructure (base stations, switch, and so on), these customers represent zero capital investment costs. These are indeed prime customers! Unfortunately, they are also very rare.

LAND-LINE AND CELLULAR BOUNDARIES

In analog systems at least, it is impossible to precisely control coverage. Therefore, cellular radio systems usually have service areas that do not conform with established toll-charging boundaries. This discrepancy can, and often does, lead to the situation where a mobile call to certain destinations can be cheaper than the same call via a land line. Some operators are philosophically opposed to such outcomes and devise elaborate schemes to circumvent them. Because of the unpredictability of radio propagation, however, it is difficult, if not impossible, to ensure that charges will be consistent with land-line charges.

Where base stations overlap the land-line charge zones, a decision needs to be made whether to charge according to:

* The location of the base-station accessed
* The lesser of the charges that could apply, depending on the actual location of the mobile
* The greater of the charges that could apply, depending on the actual location of the mobile

When possible, it is therefore advisable to treat any continuous radio coverage as a single zone or, at least, as a small number of zones. Some "natural" radio boundaries (such as hills) can often be found and using these, rather than the old land-line boundaries, should be contemplated. Marketers should ask the design engineers to locate these boundaries.

Chapter 26

CHOICE OF ANALOG MODULATION/ DEMODULATION METHODS

Most analog cellular systems use FM for speech and FSK (Frequency Shift Keying) for data. Other modulation systems could be used, but this combination gives the best performance when signal-to-noise (S/N) and simpler modulation methods are the main considerations.

Because of the threshold effect (discussed in the next chapter), the mathematics of noise performance above and below the threshold are very different. It is assumed that in cellular radio applications, all transmissions occur above the threshold level (approximately where the S/N of the off-air carrier is 17 dB). Therefore, the information in this chapter is valid only for values of S/N above 17 dB.

RECEIVER PROCESSING GAIN

The baseband reference gain or processing gain (G_B) is defined as the S/N obtainable at the detector output compared to the S/N at the receiver input if the noise is considered to have the bandwidth of the baseband only.

Thus, if:

> The received power = P_r
> Noise/(Hz) = N_d
> Modulating signal bandwidth (Hz) = B_m

then:

Equation A

$$SNR_R = \frac{P_r}{N_d \times B_M}$$

where SNR_R = reference signal-to-noise ratio

The signal-to-noise improvement factor is then:

Equation B

$$G_B = \frac{SNR_A}{SNR_R}$$

where SNR_A = actual signal-to-noise ratio

This ratio becomes more meaningful when the results for various systems are tabulated, as shown in Table 26.1.

For SSB – SC and DSB – SC, $G_B = 1$, then, as a relative measure of noise performance, the S/N performance of any modulation mode compared to SSB – SC or DSB – SC can be used. The values of G_B for the major cellular systems are shown in Table 26.2, comparing results with other modulation systems.

Clearly, FM is superior in all cases, with the difference between the various systems being a function of their peak deviations. For this reason, FM was chosen for the speech channels.

The data channels generally use FSK, which can yield excellent S/N performance particularly at low data rates. FSK is used frequently for signaling in noisy mobile environments. At the time most of the existing cellular systems were developed, techniques for good digital performance over reasonable bandwidths for speech channels had not emerged, and so only analog systems were considered for speech. Today, techniques are available that transmit good-quality speech over bandwidths less than the base bandwidth.

THRESHOLD EFFECT IN FM SYSTEMS

All angle-modulation techniques exhibit a threshold effect; the result of this effect is that a received signal moves from acceptable quality to unacceptable very rapidly as the signal level drops below a certain critical value. FM systems have a processing gain in S/N performance that is a function of their modulation index, which explains why FM is also chosen as the modulation method for high-quality commercial broadcasts. FM stations use 75-KHz deviation, which ensures a very high processing gain. The consequent lower threshold, however, requires a high input signal.

SYSTEM	$G_B = \dfrac{SNR_A}{SNR_R}$
SSB - SC	1
DSB - SC	1
DSB	$\dfrac{m^2}{1+m^2}$
AM	$\dfrac{m^2}{2+m^2}$
PM	$(A\emptyset)^2$
FM	$\dfrac{3}{2}\beta^2$
Additional gain with FM pre-emphasis	$\dfrac{\pi}{6}\left[\dfrac{W}{f_1}\right]$
Compander	G_B is a function of level and the compression ratio being maximum at low levels of modulation
SSB = single sideband DSB = double sideband SC = suppressed carrier AM = amplitude modulation PM = phase modulation FM = frequency modulation *and* m = modulation index $0 \le m \le 1$ $A\emptyset$= maximum phase deviation for PM deviation β = modulation index = $\dfrac{\text{deviation}}{\text{audio bandwidth}}$ W = baseband bandwidth of modulating signal f_1 = 3-dB point for pre-emphasis and de-emphasis	

Table 26.1   Processing gain of different modulation systems

A very noticeable improvement in commercial FM-received S/N can be noted with increased input signal level, especially if the receiver is operating in a poor reception area.

At some level, L_{it}, the baseband reference gain, drops off sharply, and within a few dB of input

SYSTEM G_B	OTHER MODES	NMT 450/900 D = 4.7 KHz	TACS D = 9.5 KHz	AMPS D = 12 KHz
SSB	1			
DSB	1			
AM (m=1)	1/3			
PM*		2.4 (3.8 dB)	10 (10.0 dB)	16 (12.0 dB)
FM		3.6 (5.6 dB)	15 (11.8 dB)	24 (13.8 dB)
* PM using same bandwidth.				
All cellular systems are assumed to have 3-KHz baseband.				

Table 26.2   The relative noise performance or processing gains of different modulation systems as signal-to-noise ratios.

level can drop to less than unity. L_{it} can be shown (by somewhat arduous theory or by practical measurement) to be in the range of 10–20 dB for practical FM systems. This effect is not present in SSB or DSB systems, but it is present in AM detectors that envelope detectors. But because AM systems have processing gains less than unity, this effect is not so important. SSB systems that use synchronous product detectors do not exhibit this threshold phenomenon, so this mode can be used as a reference.

Figure 26.1 illustrates the threshold effect. Figure 26.2 shows a synchronous product detector.

All types of linear modulation (DSB, AM, SSB, VSB) can be detected by a synchronous product demodulator. For low-level data detection in AM systems, using a synchronous product detector can avoid the threshold effect. There is, however, little practical value in doing this for non-data circuits because the advantages are only realized at very low S/N.

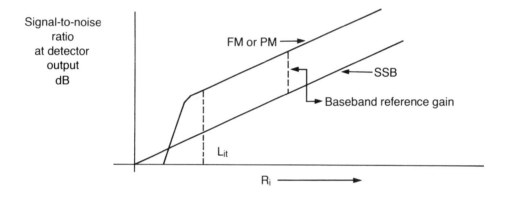

R_i = Receive input signal-to-noise ratio (dB)

Figure 26.1   Process gain of PM or FM over a linear system such as SSB

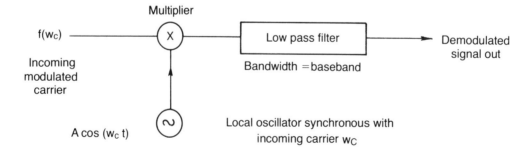

Figure 26.2   Synchronous product detector

The f(w_c) for an AM signal can be represented as:

$K \cos w_m t \cdot \cos w_c t$

where w_m = modulation frequency
          w_c = carrier frequency

If the detector product,

$$P = K \cos w_m t \cdot \cos w_c t \ x \ A \cos (w_c t) = AK \cos^2 (w_c t) \cdot \cos w_m t$$

is taken, we can expand:

$$\left[using \ \cos^2 (w_c t) = \frac{\cos (2w_c t) + 1}{2} \right]$$

$$P = \frac{AK}{2} (\cos 2w_c t + 1) + \cos w_m t$$

Equation C

$$P = \frac{AK}{2} (\cos 2w_c t \cdot \cos w_m t + \cos w_m t)$$

Because the term $\cos 2w_c t$ equals the second harmonic of the carrier frequency, a low pass filter can easily remove this product, leaving only:

$$\frac{AK}{2} \cos w_m t$$

(the original modulation).

For a reasonably good-quality signal (acceptable S/N performance), the receiver must operate at an output S/N level of about 35 dB (30-dB S/N is considered the lowest level at which the noise is not obviously intrusive). This is well above threshold.

Because S/N performance determines range, the coverage from the same site of the three main cellular systems can vary significantly. Table 26.3 shows the relative gains of the three systems at S/N output levels of more than 35 dB compared to the lowest, NMT450/900.

An advantage of 8 dB results in an improved coverage factor of about 1.7 times in range. There are, however, as always, trade-offs; in this instance, they are bandwidth and immunity to adjacent-channel interference.

SYSTEM	GAIN
NMT450/900	0 dB
TACS	6 dB
AMPS	8 dB

Table 26.3   Relative processing gains at S/N output of more than 35 dB for the three main cellular systems

BANDWIDTH

The bandwidth required for an FM system is given by Carson's Rule, which states that 98 percent of the power in the sidebands is transmitted if the bandwidth of the system is such that:

Equation D

$$B_T = 2 \left(\Delta F + f_m \right)$$

where B_T = bandwidth
ΔF = maximum deviation
f_m = maximum modulation frequency

Table 26.4 shows bandwidth for the three systems, and also lists the channel spacing. Notice that only NMT systems have any margin between the channel spacing and the bandwidth required. Because filters have less than ideal response, the channel filters need to be somewhat wider than 30 KHz or 25 KHz, respectively for AMPS and TACS. As a result, adjacent-channel interference is a potential problem in AMPS and TACS systems.

SYSTEM	*BT (BANDWIDTH KHz)	CHANNEL SPACING
NMT450/900	15.6	25
TACS	25.2	25
AMPS	30.2	30
* Assumes f_m = 3.1 KHz (audio speech bandwidth)		

Table 26.4   Bandwidth versus channel spacing for cellular systems

PRE-EMPHASIS AND DE-EMPHASIS

The noise at the output of an FM detector has the following density function:

Equation E

$$S_{(f)} = \frac{N_i \times f^2}{2P_r}$$

where $S_{(f)}$ = noise density function
N_i = noise power density at the receiver input
f = frequency
P_r = power received

From this equation, it can be seen that the noise power is inversely proportional to the input power, and that this is responsible for the FM quieting effect (that is, the noise level decreases as the input carrier level increases). Figure 26.3 shows the noise power output of FM as a function of deviation. Also, the noise power has a parabolic spectrum as it is proportional to f^2.

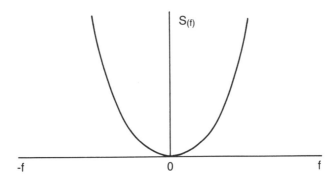

Figure 26.3   Noise power output of an FM detector as a function of instantaneous deviation

The detected S/N, to input S/N can be shown as:

Equation F

$$SNR_{(gain)} = \frac{3}{2} \, x \, D^2 \, x \left(\frac{B}{W} \right)$$

where   D = peak deviation ratio
          B = bandwidth
          W = base bandwidth

Thus, the noise power increases rapidly as the bandwidth increases. Fortunately, this can be partially compensated for by pre-emphasis. This involves boosting the transmitted signals in proportion to their frequency.

This is usually achieved using the transfer function:

Equation G

$$P_E = K \, (1 + j \frac{f}{f_1})$$

where P_E = pre-emphasis
          f = baseband frequency
          f_1 = cutoff frequency

The f_1 is the point at which the modulating signal is boosted by 3 dB. The slope of this curve approaches 6 dB per octave, which is the inverse of the noise spectral density function $S_{(f)}$. In practice, a 6-dB emphasis has been shown to be both readily realizable and capable of producing good results. Increasing the boost has not been found worthwhile. Naturally de-emphasis must be applied at the receive end which has an inverse form:

Equation H

$$D_E = \frac{S_o}{(1 + jf/f_1)}$$

where D_E = de-emphasis transfer function
f = frequency
f_1 = 3-dB point
S_o = input signal to de-emphasis circuit

It can be shown that the improvement in S/N performance due to pre-emphasis and de-emphasis is given approximately by this equation:

Equation I

$$\text{Improvement} \approx \frac{\pi}{6} \times \frac{W}{f_1}$$

where   W = base bandwidth
f_1 = 3-dB point

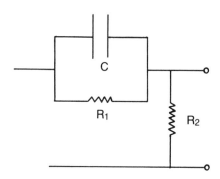

Figure 26.4   A simple RC pre-emphasis network for FM

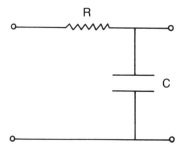

Figure 26.5   De-emphasis circuit

Nearly all FM systems employ some form of emphasis and de-emphasis, including commercial FM and cellular radio systems. Thus, a pre-emphasis network can be a simple RC network, as shown in Figure 26.4.

It is normal that $R_1 >> R_2$ so that the time constant of this network is R_1 x C for the low-frequency cutoff point and R_2 x C determines the high-frequency cutoff.

Figure 26.5 shows the de-emphasis circuit. This circuit is sometimes described by its time constant R_1 x C with values of 50, 75, and 100 µsec being common.

SIGNAL-TO-NOISE IMPROVEMENTS WITH A PHASE-LOCKED LOOP

A Phase-Locked Loop (PLL), which can be used in FM receivers, can be designed to improve S/N by 2.5 to 3 dB in the region below the threshold by incorporating a loop response that has spike suppression. Above the threshold, the processing gain of a PLL is the same as a conventional discriminator.

Figure 26.6 shows a simple PLL demodulator. For the equilibrium of this loop it is required that:

$$\frac{d\phi}{dt} = \frac{d}{dt}\left(G \int_{-\infty}^{t} V(\lambda) \cdot dt\right)$$

and if:

$$w = \frac{d\phi}{dt}$$

then:

Equation J

$$w = G \cdot V(t) \ or \ V(t) = \frac{G}{w}$$

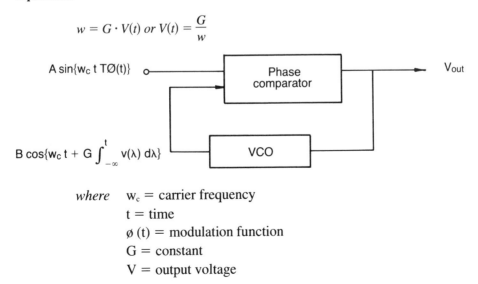

where w_c = carrier frequency
t = time
ø (t) = modulation function
G = constant
V = output voltage

Figure 26.6   Simple phase-locked-loop detector

This demodulator is a first-order (or single pole) device; a further improvement can be obtained by using a second-order transfer function.

Such a PLL can yield a 2.5- to 3-dB threshold improvement. More elaborate PLL detectors are available (with second- or third-order transfer functions); one such device is known as an FM Demodulator using Feedback (FMFB) and is used for greater improvements in threshold extension. These techniques are widely used in satellite receivers.

COMPANDING

Companding is the process of compressing the signal before transmission and expanding it at the receiving end. Figure 26.7 illustrates this process.

A compander, as illustrated in Figure 26.7, reduces the distortion generated at the transmit end by reducing the voltage excursions of the modulating signal. It also improves the S/N of an analog linear system by boosting the level of the low-level signals. The transmission noise (N_t) is added (logarithmically) to the signal on the transmission channel but is reduced by the compander ratio at the receiver. In FM systems, a compander also improves the S/N performance by increasing the deviation of the low-level signal components. Figure 26.8 shows a typical transfer characteristic of a compander.

This same compression technique is used in some hi-fi tape recorders to reduce the relative S/N and is known commercially as HX noise reduction in some tape recorder systems.

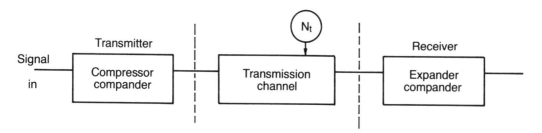

Figure 26.7   A compander system. The terms "compressor" and "expander" are somewhat misleading as the compressor compander compresses signals above the mean level and expands those below it, whereas the expander compander does the opposite.

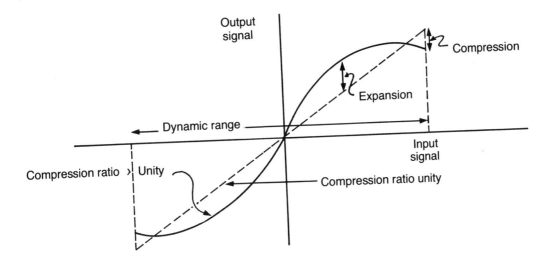

Figure 26.8   Typical transfer characteristic of a compander

Chapter 27

NOISE AND NOISE PERFORMANCE

All communications systems are noise-limited in the strict mathematical sense that noise determines the maximum data rate that can be transmitted over any fixed bandwidth and with any arbitrary error rate. The bit rate that can be communicated at an arbitrarily low error rate in a Gausian noise environment was calculated by Claude Shannon in the early 1940s to be:

Equation A

$$C = W \log_2 (1 + S/N)$$

where C = the baud (or bit) rate
W = the bandwidth in Hz
S/N = signal-to-noise ratio

This equation is true for all signal-to-noise (S/N) ratios (including those < 1), provided a suitable encoding (modulation) system is used.

All analog speech-encoding techniques are very inefficient because speech itself uses very high redundancy and makes poor use of the channel bandwidth.

From Equation A, it can be seen that the baud (or information) rate is improved with any increase in bandwidth and, for reasonably high S/N ratios, the improvement is almost directly proportional to the log of the S/N ratio. FM was an early and successful attempt to exploit this relationship. In FM broadcasting, a wide bandwidth is used to gain an improved signal quality. Commercial FM broadcasting uses a bandwidth five times broader than the baseband (the modulation frequency).

For HF (High Frequency, 3–30 MHz) the limiting noise is mainly manmade (for example, electrical and ignition systems), but also includes storms, solar flares, and, more importantly, other transmissions on the same frequency (if they are considered to be noise). Thus, with HF, the improvements in receiver technology cannot achieve much improvement in transmission over any single channel. Frequency-hopping techniques, however, can achieve a marked improvement.

Noise can have many origins, including galactic and extra-galactic, thermal, manmade, and even quantum mechanical. Modern radio receivers in the VHF/UHF bands operate close to the theoretical limits of sensitivity imposed by these noise sources.

At VHF and UHF frequencies (30–3000 MHz), and particularly above about 300 MHz (where all cellular radio operates), the background noise is relatively low, and the limits of performance are set by the equipment. Galactic noise can be significant in the region of 40–250 MHz.

At these frequencies, the predominant source of noise is thermal. Modern receivers operate at very low-noise factors (levels of introduced noise) and, short of using very expensive technology (such as masers), few improvements in the basic sensitivity of these receivers are possible.

GALACTIC AND EXTRA-GALACTIC BACKGROUND NOISE

Galactic background noise (which originates within the galaxy) and extra-galactic background noise (which includes the microwave emissions left over from the Big Bang some 16 billion years ago) are significant radio noise sources.

Galactic noise has its origin in stars, supernovas, neutron stars, black holes, quasars, and other noise sources that are scattered throughout the galaxy. Particular noise sources include the sun, Cygnes A (thought to be a black hole that emits high levels of X-rays and radio but only low-intensity light radiation), and Cassiopeia A (a particularly noisy star at radio frequencies). In addition, because the universe is composed of some 10^{12} galaxies, it is not surprising that some of the noise emanates from outside our local galaxy.

Below 1 GHz, the maximum levels of noise are for the beam pointed at the galactic poles. At higher frequencies, the maximum levels are for a beam just above the horizon, and the minimum levels are for the zenith. A low-noise region between 1 and 10 GHz is most amenable to application of special, low-noise antennas.

Antenna noise temperatures are used to define the background noise levels and have nothing to do with the actual temperature of the antenna itself. The noise power is directly proportional to the bandwidth of the receiver. It is important to note that the noise power is independent of antenna gain.

The quietest area is from 1–10 GHz, where galactic noise is at a minimum. Cellular radio systems are usually at 800–900 MHz and operate at antenna noise temperatures of about 2.5 degrees K–100 degrees K. The radio noise is very much a function of frequency, as the Earth's atmosphere acts as an attenuator at high frequencies, while the ionosphere attenuates the lower frequencies. The large variation in antenna temperatures is due to the differences in how the measurements are made. At lower frequencies (below 1 GHz), the maximum temperature occurs when the antenna is pointed at the galactic poles. At higher frequencies, the maximum temperature occurs when the antenna is pointed at the horizon.

THERMAL NOISE

Because of random molecular movement caused by thermal energy, all passive and active components at any temperature above absolute zero generate a certain amount of wide-band energy. In electronic devices, this energy manifests itself as system noise and imposes fundamental limits on usable sensitivity of receiving and detecting systems.

In any conductor the available noise power can be determined by this relationship:

$P = KTB$ *watts*

where   P = available noise power
T = temperature in degrees absolute (Kelvin)
K = Boltzmans constant = 1.38×10^{-23} joule/Kelvin
B = the bandwidth in Hz

Notice that the noise power depends only on temperature and bandwidth and is not dependent on resistance.

For a conductor with a resistance R (see Figure 27.1), it can easily be shown that:

Equation B

$E^2 = 4\,RKTB$

where   R = the equivalent resistance
N = the equivalent noise voltage generator

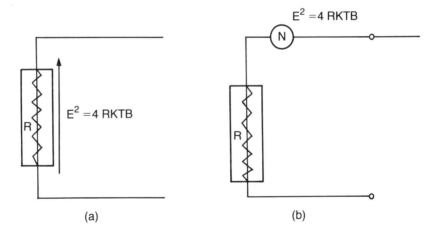

R = The equivalent resistance

N = The equivalent noise voltage generator

(a) Voltage across a resistor

(b) Equivalent voltage generating circuit

Figure 27.1 Equivalent voltage of a resistive noise source, where (a) is the voltage across a resistor and (b) is the equivalent voltage generating circuit

Notice that the voltage is represented as E^2. The square of the actual value of voltage (that is, a value proportional to energy) is used in lieu of E, the actual voltage, because the average voltage is zero (that is, noise with negative-going pulses is just as likely to occur as noise with positive-going pulses).

The RMS value of voltage is:

Equation C

$$E_{RMS} = \sqrt{4\,RKTB}$$

In practice, amplifiers are usually cascaded (that is, they are used in series). The nature of amplifier noise factors dictates that the first stage is the most critical in determining the overall noise performance of a system.

ATMOSPHERIC NOISE

Atmospheric noise is largely due to lightning discharges and is consequently very seasonal. It predominates in the frequency range up to about 20 MHz. Atmospheric noise is not generally a factor at cellular frequencies except in abnormal circumstances.

MANMADE NOISE

Due mainly to low-frequency devices such as motors, neon signs, power lines, and ignition, manmade noise sources tend to decrease rapidly in intensity with increasing frequency. Typically, suburban areas are about 15 dB quieter than city centers, and rural areas are about 15 dB quieter than suburban areas. This noise source is significant up to about 1 GHz, but it is generally not a serious problem above 500 MHz.

SUBJECTIVE EVALUATION OF NOISE

The ultimate determination of S/N performance is the perception of the user. In the audio environment, it is possible to classify S/N in terms of quality. Table 27.1 shows some categories of everyday experience of S/N.

Another method that was developed by the mobile-radio and amateur-radio community is to evaluate S/N on a scale of 1 to 5. Table 27.2 shows this method and its approximate S/N equivalents.

TYPE OF SIGNAL	S/N RATIO (dB)
Limit of operation of 5 tone sequential pager	0–3
Barely readable two-way radio	5–10
Telephone voice quality	25–40
Hi-fi analog recording	55–65
Compact disc	80+

Table 27.1   Some common S/N levels in everyday systems

SIGNAL QUALITY	APPROXIMATE S/N (dB)	SIGNAL NUMBER
Broken and unreadable	5	1
Broken and just readable	10	2
Readable with some difficulty	15	3
Readable with noise	20	4
Clearly readable	25+	5

Table 27.2   Signal quality as a function of S/N ratio

The fact that this table ends with "clearly readable" is indicative of mobile two-way standards, where a high-quality signal is not generally sought. An S/N of 20 dB would be a low limit of acceptability for cellular subscribers and would be acceptable in fringe areas only.

NOISE FACTOR

In order to look more closely at the noise performance of cellular receivers, it is necessary to introduce the concept of noise factor. Noise factor can be defined as:

Equation D

$$F = \frac{available\ S\,/\,N\ power\ ratio\ at\ input}{available\ S\,/\,N\ power\ ratio\ at\ output}$$

where   F = noise factor

Figure 27.2 shows the noise factor.
From Equation D, you can see that:

Equation E

$$F = \frac{P_{si}}{P_{ni}} \times \frac{P_{no}}{P_{so}}$$

For linear amplifiers, F is always greater than 1 (that is, noise will be added).

$$G = \frac{P_{so}}{P_{si}}$$

where   G = the gain of the amplifier

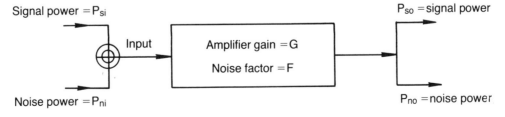

Figure 27.2   A single-stage amplifier, where the noise contribution of the noise factor F results in the amplifier adding to the output noise

Similarly, from Equation E:

Equation F

$$F = \frac{P_{no}}{G\,P_{ni}}$$

THE AMPLIFIER'S CONTRIBUTION TO NOISE (REFERRED TO THE INPUT LEVEL)

It is often useful to determine the contribution of the amplifier to the overall noise of a system. The output noise referred to the input is P_{no}/G. So, the noise contributed by the amplifier is:

$$\text{AMP Noise} = \frac{P_{no}}{G} - P_{ni}$$

$$= F P_{ni} - P_{ni}$$

Equation G

$$\text{Noise contributed by amplifier (referred to input level)} = P_{ni}\,(F - 1)$$

CASCADED AMPLIFIERS

Amplifiers connected in cascade (series) result in an overall noise factor that includes the contributions of each stage. Figure 27.3 shows cascaded amplifiers.

From Equation G, it can be seen that:

Equivalent noise at the input of amplifier 2 = noise input to amplifier 2 by amplifier 1 plus contribution to noise of amplifier 2 referred to the input (from Equation G)

$$= G_1 \text{ x } P_{ni1} \text{ x } F_1 + P_{ni2} \text{ x } (F_2 - 1)$$
(from Equation G and EquationE)

Hence:

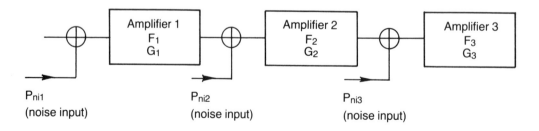

Figure 27.3 Amplifiers connected in cascade can be regarded as having inputs that are the sum of the previous stage outputs, regarded as NF = 1 stage and the net noise contribution.

Noise output power of amplifier 2 $= G_2$ x the total noise input power

Therefore, the noise output power is:

Equation H

$$\text{Noise output of stage 2}, P_{no2} = G_2 G_1 \text{ x } P_{ni1} F_1 + P_{ni2} (F_2 - 1) G_2$$

If the amplifier input and output impedances are matched, and amplifier 2 is at the same temperature as amplifier 1, then:

$$P_{ni1} = P_{ni2}$$

$$= KTB \text{ watts}$$

But, from Equation F:

$$F = \frac{P_{no}}{G P_{ni}}$$

which results in the overall gain for series amplifiers 1 and 2 being:

$$G = G_1 \text{ x } G_2$$

From Equation H:

$$P_{no2} = G_2 G_1 \text{ x } P_{ni1} F_1 + P_{ni2} (F_2 - 1) G_2$$

Then, from Equation F:

$$F_o = \frac{G_2 G_1 \text{ x } P_{ni1} F_1 + P_{ni2} (F_2 - 1) G_2}{G_1 G_2 P_{ni1}}$$

But $P_{ni1} = P_{ni2}$ for impedance matched amplifier stages, so the cascaded noise factor F_o is:

Equation I

$$F_o = F_1 + \frac{F_2 - 1}{G_1}$$

It can similarly be shown that for additional cascaded amplifiers:

Equation J

$$F_o = F_1 + \frac{F_2 - 1}{G_1} + \frac{F_3 - 1}{G_2 G_1} + \dots$$

Two important conclusions are now drawn from these equations:

- In the cellular mobile environment, the first RF stage virtually determines the noise performance of the receiver. A typical mobile RF stage has the following specification:

    Gain = 15 dB (ratio = 31.62)

    F = 1.5 dB (ratio = 1.41)

    The next stage may have a gain of 30 dB (1000) and a noise factor of 10 dB (ratio 10). So, the overall performance, denoted F_c is:

    $$F_c = 1.41 + \frac{10 - 1}{31.62}$$

    $$= 1.41 + 0.284$$

    $$= 1.694$$

    The overall noise factor in dB is 2.28. In other words, the noise factor has not been significantly increased by the addition of a noisy (10 dB) record stage.

    It can be further shown that additional amplifiers can be of progressively lower quality (with a higher noise factor and therefore cheaper) without significant degradation in the overall system performance. This fact is most fortunate because the mixer stage in a superheterodyne is very noisy indeed and so, in all high-performance receivers, the mixer is preceded by a low-noise amplifier.

- The noise factor of a typical UHF receiver can be calculated from S/N ratios using Equation B, once measurement details are known. Assume the following:

    - S/N ratio = 12 dB
    - Audio (measured) bandwidth = 3500 Hz
    - Modulation is 1 KHz at 1 KHz deviation
    - Measured sensitivity = 0.2 microvolts

Then:

$$\text{Actual input noise} = \frac{1}{2} \sqrt{4\ RKTB}\ volts\ RMS$$

$$= \sqrt{50 \times 1.38 \times 10^{-23} \times 290 \times 3500}$$

$$= 0.02646\ \mu V$$

where   T = 290 degrees K or 17 degrees C (an accepted standard room temperature for noise calculations; in practice, this temperature may be somewhat high or low, depending on location)

Hence:

$$S / N \text{ at the input terminals} = 20 \log \frac{0.2}{0.02646} = 17.57$$

Thus, the noise factor is 5.57 dB.

Some measurements of signal-to-noise ratio may yield negative noise factors (or less than unity if ratios are used). Negative noise factors can occur when using deviation ratios higher than unity and noise-reduction techniques such as emphasis and companding. The lesson is that S/N ratios mean very little unless the conditions of measurement are clearly stated.

Further, when the measurements are made at very low S/N levels (less than 15 dB at the receiver input), threshold effects can mask the true nature of the S/N performance.

NOISE FIGURE OF AN ATTENUATOR

The mathematics of the noise contribution of an attenuator, such as a transmission line or coaxial cable, are a little complex and involve thermodynamic considerations. Only the result is listed here. Figure 27.4 shows the noise factor contributed by cable loss.

The noise factor of an attenuator F is:

Equation K

$$F = 1 + (L - 1)\frac{T_c}{T_o}$$

where   L = the attenuation factor

so that $L = \dfrac{power\ out\ of\ attenuator}{power\ into\ attenuator}$

T_o = 290 degrees K

T_c = cable temperature

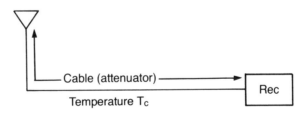

Figure 27.4   Noise factor contributed by cable loss. The feeder cable to an amplifier is equivalent to a series attenuator.

When $T_c \approx T_o$, then $F \approx L$ (the attenuator). Hence, the noise factor introduced by the feeder cable will be similar to the attenuation of the cable itself.

As an example, these calculations can be used to determine whether a low-noise mast-head amplifier would be of value in the mobile environment, given the need to operate in a fringe area. Assume the following:

- The mobile receiver has a noise factor of 6 dB.
- The cable loss is 3 dB.
- A mast-head amplifier of 15 dB gain and 4 dB noise factor is available. (This is a low-grade wide-band amplifier.)
- Receiver gain is 70 dB.

In the case of no mast-head amplifier, the overall noise figure is shown in Figure 27.5.
    For $T_c = T_o$:

$$F_{overall} = F_c + \frac{F_r - 1}{G_c} \text{ (from Equation K)}$$

$$= 1.99 + \frac{3.98 - 1}{0.502}$$

$$= 7.92 \text{ or } 9 \text{ dB}$$

Now, consider the use of a mast-head amplifier, as shown in Figure 27.6:

$$F = 2.51 + \frac{1.99 - 1}{31.6} + \frac{3.98 - 1}{31.6 \times 0.502}$$

$$= 2.51 + 0.031 + 0.187$$

$$= 2.72 \text{ or } 4.35 \text{ dB}$$

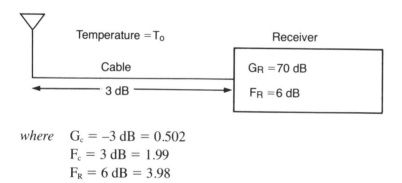

$$\textit{where} \quad G_c = -3 \text{ dB} = 0.502$$
$$F_c = 3 \text{ dB} = 1.99$$
$$F_R = 6 \text{ dB} = 3.98$$

Figure 27.5   Noise factor of a receiver and feeder cable. The values shown are for a practical receiver with a 3-dB cable and a 70-dB receiver gain with a noise factor of 6 db.

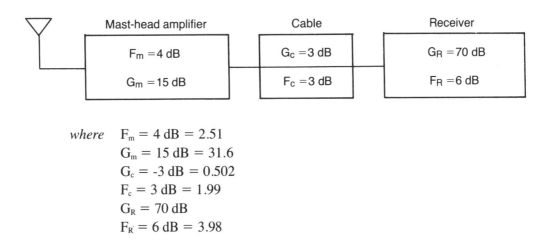

$$\begin{array}{ll}
\textit{where} & F_m = 4 \text{ dB} = 2.51 \\
& G_m = 15 \text{ dB} = 31.6 \\
& G_c = -3 \text{ dB} = 0.502 \\
& F_c = 3 \text{ dB} = 1.99 \\
& G_R = 70 \text{ dB} \\
& F_{R} = 6 \text{ dB} = 3.98
\end{array}$$

Figure 27.6   A typical mast-head amplifier is placed as close as possible to the receiving antenna.

So, an improvement of 4.65 dB in S/N performance would result in this example and, in marginal areas, could well be worth the effort.

It should be noticed here that the "mast-head" amplifier used is of better quality than the receiver, resulting in an improvement that exceeds the cable loss. When the amplifiers are of similar quality, (that is, about a 4-dB noise figure on the first stage of the receiver), the improvement will be similar to the cable loss, because the contributions to the overall noise factor (F), after the first term, tend to be small. This system is not necessarily what would result in practice because mast-head amplifiers with noise factors in the range 0.7–2 dB are available.

This suggests that mast-head amplifiers are useful for survey purposes, particularly when low-powered transmitters are used as a source and measurements are done near the limits of the noise performance of the receiver. Mast-head amplifiers are difficult to install and are very prone to lightning strikes. In a mobile environment, because they are usually wideband devices, they are somewhat prone to intermodulation, which can limit their utility. They cannot be used on antennas that are connected to a duplexer.

PROCESSING GAIN AND NOISE

The processing gain, shown in Table 20.1 on page 289, for an FM system is:

$$G_B = \frac{3}{2} \beta^2$$

$$\textit{where} \quad \beta = \Delta f_d / \Delta f_m = \frac{\text{deviation frequency}}{\text{modulation frequency}}$$

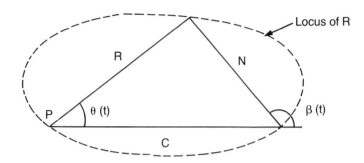

Figure 27.7   The locus of an FM noise pulse. The received signal is the vector sum of the carrier level (C) and noise level (N). This figure shows the locus of the resultant (received) vector (R), which causes the phase reversal associated with FM "picket fencing."

However, this formula is valid only at high input S/N levels, where the noise levels are not high enough to cause the noise spikes familiar in FM systems operating in the threshold region. At very high noise levels (at the receiver-input port), noise spikes are generated by the noise components, which are of sufficient level to cause a phase reversal in the incident wave form. Figure 27.7 illustrates the locus of the vector sum of the carrier and noise signals.

The signal out of the receiver, S_o, is such that:

$$S_o \propto d\theta/dt$$

where θ = the phase of the resultant of the incident signal and noise
                 t = time

From Figure 27.7, it can be seen that a phase reversal produces a pulse. If a plot is made of $\beta(t)$, $\theta(t)$, and $d\theta/d(t)$, the form of S_o can be determined. Here, we assume a steady carrier and a noise signal of approximately constant amplitude (but larger than C) that is rotating uniformly with respect to C.

Figure 27.8 shows a typical FM discriminator output filter.

Input = R, where R = C + N

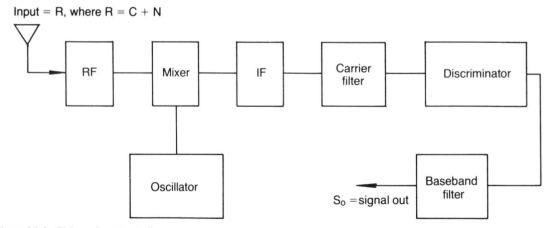

Figure 27.8   FM receiver block diagram

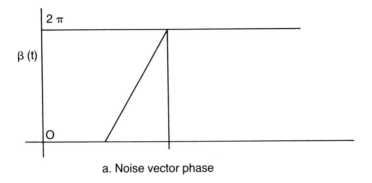

a. Noise vector phase

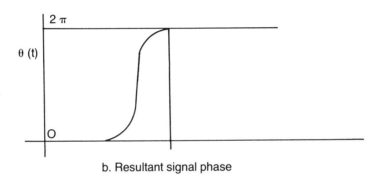

b. Resultant signal phase

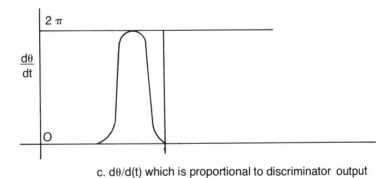

c. dθ/d(t) which is proportional to discriminator output

Figure 27.9   Noise output pulse generated by a noise pulse that causes the resultant locus to pass through 2π radians

Figure 27.9 shows that this phase reversal produces a pulse. Similarly, it can be shown that when the locus of the resultant does not encircle the point P (as shown in Figure 27.7), a different pulse form arises (as seen in Figure 27.10 on the next page).

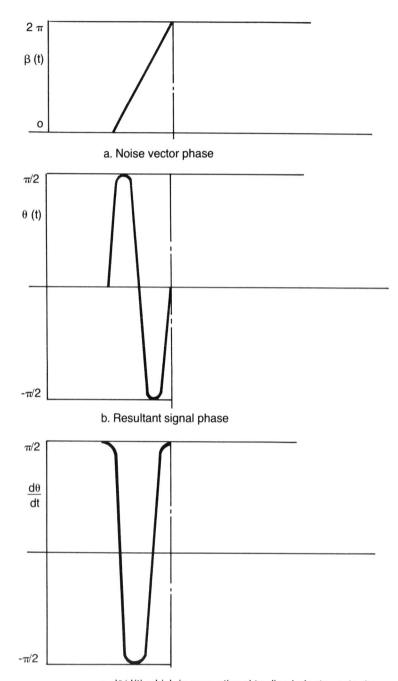

a. Noise vector phase

b. Resultant signal phase

c. dθ/d(t) which is proportional to discriminator output

Figure 27.10 Noise output pulse from a noise vector N, which is smaller than the carrier and rotates through 2π radians

The noise pulse in Figure 27.10 can be shown to be less energetic than the noise pulse in Figure 27.9. Thus, the effect of noise increases rapidly with deterioration of the incident signal. The actual S/N under these conditions can be shown to be:

Equation L

$$(S/N)_o = \frac{(3/2)\beta^2 (S/N)_i}{1 + (12\,\beta/\pi)(S/N)_i \exp\left[-\frac{1}{2}\{1/(\beta+1)\}(S/N)_i\right]}$$

The term in the denominator of Equation L can be seen to tend to 1 when $(S/N)_i$ is large. To determine the levels at which this term becomes effective, it is traditional to define the threshold as the point where the effect of these terms is to reduce the $(S/N)_o$ by 1 dB, as shown in the following equation:

$$(S/N)_o = \frac{1}{1 + (12\,\beta/\pi)(S/N)_i \exp\left[-1/2\{1/(\beta+1)\}(S/N)_i\right]} = 10^{-0.1} = 0.7943$$

This function can now be plotted for the three major systems (AMPS, TACS, and NMT900). The values for commercial FM are also tabulated. Table 27.3 on the next page shows the S/N performance of various systems.

Using Equation L, the S/N, as measured at the discriminator output of systems with various modulation indexes, can be determined. This function was plotted for AMPS and NMT systems and for commercial FM in Figure 27.11 on the next page.

A plot for TACS would be very similar to the AMPS plot but 2 dB down. Notice that the AMPS system has a significant relative processing gain up to the threshold (18 dB $(S/N)_i$), but this advantage decreases rapidly as the $(S/N)_i$ decreases below the threshold. Above the threshold, all the systems have significant gains over a zero processing-gain system such as SSB. However, at very low levels of $(S/N)_i$ (for example, below 12 dB), the SSB system will out-perform the FM systems.

Notice that the threshold for the NMT system is 14 dB or 4 dB below that of the AMPS system.

The various systems have a processing gain that tends to unity in the range 12–13 dB S/N output. For this reason, it is usual to measure S/N at 12 dB SINAD, because this measure reflects the quality of the hardware (noise figure) and will be the same for equal-quality systems regardless of their processing gains.

In all real receivers (as opposed to the theoretical receivers considered previously), the processing gain does not improve the S/N indefinitely, and a threshold is reached where increases in S/N input do not result in increased S/N output.

S/N PERFORMANCE IN PRACTICE

Because of the lower operating frequency, an AMPS antenna has a slightly larger aperture than a TACS/NMT900 antenna of the same gain, and this contributes an additional 0.6-dB system gain to the AMPS system. The effect of this gain in practice is, however, minor. For modern cellular systems, the requirements of good handheld coverage and high S/N levels mean that those systems will always be operating above the threshold. Handhelds are the weak link in the cellular chain and are usually assumed to be operated from stationary positions inside buildings. Thus, handhelds do not ordinarily operate in high multipath environments. In general, the full benefits of the processing gain of the FM systems is usable in modern cellular systems.

The situation is similar, but a little different, if only mobile units are considered. With mobile units, there is some significant multipath, and because of occasional excursions below the threshold (contributing relatively high levels of noise), there is some reduction in the relative processing gain. The other requirement of good S/N performance, however, minimizes this effect.

(S/N)i	COMMERCIAL FM ß=5	AMPS	TACS	NMT900
50	66	63	61	55
40	56	53	51	45
30	45	43	41	35
25	41	38	36	30
20	34	33	31	25
19	30	31	29.6	24
18	25	28	28	23
17	20	24	25	21
16	17.3	20	21.6	20.3
15	14.3	16.3	17.8	18.7
14	12	13.3	14.4	16.3
13	10	10.8	11.5	13.4
12	8.6	8.9	9.3	10.6
11	7.4	7.4	7.4	8
10	6.5	6.1	6	5.9

Table 27.3   S/N performance of various systems illustrating the relative processing gains

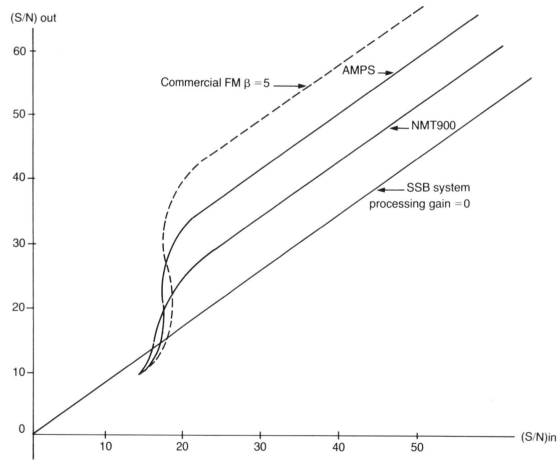

Figure 27.11 Plots of the processing gains of various FM systems. This graph shows S/N performance versus carrier signal to noise of the AMPS and NMT900 systems from Equation L.

A very different situation occurs in mobile communications systems where voice quality is secondary to range (such as CB and amateur transactions). In this instance, the systems are often operated below the FM threshold, and the performance depends only marginally on deviation. Such systems are usually designed to have low deviations and narrow channel bandwidths.

ABSOLUTE QUANTUM NOISE LIMITS

Without going into extensive theory, quantum mechanics demand that radio noise will exist in a vacuum even at 0 degrees K, and even though the vacuum is completely shielded from outside radio influences. In essence, the theory states that a condition of "absolute nothingness," free of any noise, is not achievable, even theoretically, in a perfect vacuum.

Some very sensitive measurements (like those used to measure gravity waves) are now approaching the limits of accuracy permitted by quantum mechanics. These noise effects are different from thermal effects (which are predictable by classical physics), and even though much smaller, they may one day limit the speed of future high-technology data communications by placing a fundamental limit on error rates.

Chapter 28

ROAMING

Roaming occurs when a mobile subscriber makes a call on a switch other than the usual home switch. This usually means that the roamer has left the normal city of residence, but when there are two operators, roaming can occur within the normal service area where coverage is not provided by the "home" operator.

Roaming is available only between systems that have roaming agreements and that use the same system (AMPS, TACS, or NMT900). Roaming cannot occur between different systems because of their technical incompatibility. (A roamer can change mobile units in a foreign service area, but this is not true roaming.) All new systems should have national roaming as an essential component, so that the difficulties now being experienced in the US, due to the lack of automatic roaming, can be avoided.

Roaming between countries is beginning to occur with bilateral agreements. Perhaps the biggest problem to be overcome before international roaming can be considered is customs clearance. Until it is possible to freely move mobile units (and, in particular, handhelds) between countries, there is little point in considering roaming. Countries that practice protectionism have made roaming practically impossible by requiring a formidable mountain of paperwork before roaming can occur; this negates the convenience of roaming.

ROAMING METHODS

International and national roaming have been proposed in a number of ways. The following sections describe these schemes.

RENTING A LOCAL MOBILE

This method is not true roaming; rather, it is a rental agreement. Even so, it can be useful for travelers staying in a foreign area for long periods, or for the user who wants additional phones for special occasions. Renting a mobile, however, is not much value for the business travelers who spend short periods in foreign areas.

It is worthwhile to look at the disadvantages of this scheme from the view point of the frequent traveler. Of course, this method assumes the user can conveniently pick up a mobile on arrival at the airport. Upon arriving at the foreign area (presumably by aircraft), the traveler must first find the office where the mobile unit can be obtained. Presumably, this office would be at the airport. For the operator, this means that the airport office must be staffed during the hours of operation (typically 16 hours per day, assuming a night curfew), which can be expensive. If the operator works through an agent (for example, a rental car company), then the operator takes a risk on the competence of the representative, particularly when the user is not familiar with portable phones and requires operating instructions.

So for short-term users on the move, the rental method is neither convenient nor hassle-free; rather, it is almost unworkable. Although there is a place for rental phones within the market, they are not a substitute for true roaming.

ROAMING PROVIDED BY BILATERAL AGREEMENT

Roaming provided by bilateral agreement between the user and the operator is an option used both nationally and internationally. The roamer must either contact the operator on arrival in a new area to make the necessary payments, or make the necessary arrangements before leaving home.

This method does lead to true roaming, but it also has limitations in convenience. First, the method assumes that the roamer is aware of the agreements and knows what is required for roaming. As the number of bilateral agreements increases, so does the number of forms they may take. The customer may not be aware of the agreement variations.

Reports are discouraging from the US for travelers who want full roaming while driving across the county. The diversity of agreement types, office hours, and acceptable forms of payment is formidable. Very few people have managed to make all agreements in time for their trip. The confusion caused by these different agreements detracts from the value of the agreement in the first place.

INTER-OPERATOR AGREEMENTS

Inter-operator agreements are another option. With this method, each operator agrees to honor the roaming charges incurred by the customer (or some subset of customers) on another operator's network.

From the customer viewpoint this arrangement is by far the most satisfactory option. But it is also the one most vulnerable to fraud. To cover possible losses, operators can charge those customers who want to be regarded as roamers.

Operators generally have a special (higher) roamer charge borne by the user. This charge does not necessarily need to be the same as that of other operators. In addition, the home operator may also levy a surcharge on the roaming costs.

ROAMING WITH A PAGER

This is a variation on semi-automatic roaming, where a wide-area paging system is used to locate the particular roamer from the fixed network. With semi-automatic roaming, a PSTN caller needs to know the whereabouts of the mobile unit sought. To be effective, the pager used should be at least numeric and preferably alphanumeric.

The roamer in this case is alerted by the pager that someone is calling. It is not necessary for the caller to know the whereabouts of the traveler. This is the only method of calling a subscriber in the CT2 systems, where subscribers have outgoing automatic access but no incoming alert.

The lack of a truly international paging system, however, does generally limit this method to national coverage. Efforts to establish an international paging network have, to date, met with only limited success. Generally, paging systems are even more diversified than cellular systems.

FULLY AUTOMATIC ROAMING

All major mobile telephone systems are capable of full automatic roaming. NMT900 and TACS were both implemented with full automatic roaming and, in 1986, a fully automatic, nationwide roaming AMPS system was introduced in Australia. Roaming was later extended to include New Zealand.

The NMT900 system has a non-proprietary signaling system so it can be readily interconnected. Unfortunately, the AMPS/TACS system is not specified to the same level. Although a number of manufacturers can now provide automatic roaming among their own switches, the ability to interconnect different switches has only recently become feasible.

The US is one of the few places that has a large number of manufacturers providing architecturally different switches. This is because most other countries have only one or two operators, who have chosen to stay with only one supplier. Most other countries have national operators and not regional operators (as in the US), so this problem generally does not arise.

In the US, it was realized at a very early stage that bilateral agreements between the many manufacturers (at last count, nine in AMPS) would not be particularly productive, so the Cellular System Operation Technical Committee, TR-45.2, a committee of the Electronic Industry Association (EIA), has been working on switch-to-switch compatibility standards. An early subset of standards has been released so that development work can begin.

ACCOUNT SETTLEMENT

Accounts between operators can be settled in full (that is, cross-billing according to use) or settled as a net sum only. While the accounts are small and not very diversified, direct cross-billing is the simplest and most workable method. If an operator has a large number of roaming agreements, and if roaming income becomes a significant portion of the total income, some refinements may be desirable.

An obvious alternative to direct cross-billing is net settlement only. Using this method, only the net difference due between two operators is paid. The operators collect all charges due for a period of, say, one month and then settle with one another by paying the net difference. For example, if operator A has 10 roamers for the month with charges of $1000, and operator B has 20 roamers with charges of $1500, operator B pays operator A the net difference of $500. Operators charge their own customers. Net settlement may require a slightly more sophisticated accounting system, but it does mean that large sums of money do not need to be moved.

As the diversity of roaming destinations increases, the attractiveness of a central billing system grows, particularly for smaller operators, who may not have sophisticated accounting systems, and who may be anxious for prompt payments to ensure an adequate cash flow. (Of course, cash flow can be a problem for operators of any size.)

ROAMING IN THE A OR B BAND

The roaming subscriber must first find out whether to use the A or B band. Suppose a subscriber arrives in a foreign switch area to use the system operated on the A band (because the necessary agreements are in place) but cannot use the B band system. In this case, unless the subscriber locks the mobile onto the A band, there can be multiple call failures because the mobile may instead lock onto the B-band control channels (on which the subscriber is not valid).

From a user's perspective, it is easier to roam to an area where both the A and B bands are available. Otherwise, the user must know which one is available and set the mobile accordingly. This requirement makes "exclusive" roaming agreements a little less attractive than block agreements with the foreign operators. Even so, the need to know which band to use is really the only potential technical complication for a roamer in an automatic roaming network.

To understand automatic roaming it is useful to first look at the trunking of a home call to a mobile. Assuming that the mobile telephone switch is connected at the trunk level, the mobile telephone number will consist of two parts: the trunk access code, and the mobile phone ID number. For example, a car telephone may have the access code 007-714789. In this example the digits 007 are used by the local switch to connect the call through the trunk switch to the mobile telephone switch. Figure 28.1 shows this process. The digits following 007 (714789) identify the actual mobile phone sought. Hence, the path to the mobile is fully defined.

Now, suppose that the mobile user is in a foreign mobile telephone area that has automatic

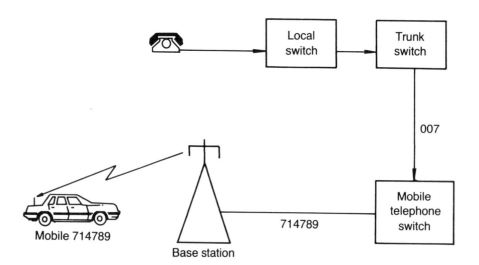

Figure 28.1   Trunking of a home call to a mobile telephone switch

roaming with the home switch. When a call is made by the fixed subscriber, it still attempts to connect as before. It can only be rerouted if, when the call arrives at the mobile switch, it is redirected to the new zone. This redirection can be done only if the home switch has information that the subscriber has roamed, and where the subscriber has roamed.

Roamers can announce their presence in a foreign area by attempting to make a call, by using system options, or by automatically "registering" when the mobile is turned on (the roamer sees a foreign zone identification, SID, being transmitted on the control channel).

In any case the foreign switch must recognize the roamer and then contact the roamer's home switch to inform the home switch of the presence of the roamer and obtain the subscriber's file (containing such details as the ESN, subscriber's category, options and roaming status) from the home switch. The foreign switch must be able to determine the location of the roamer's home switch, which can readily be done by having the first few digits of the subscriber's number identify the home switch (for example, <u>71</u> 4789). The two mobile switches can then be connected by a common-channel data bus to exchange the necessary information. Figure 28.2 illustrates this exchange.

Once registration occurs, it is possible for a fixed subscriber in the home network to call the mobile in the foreign network without the need to know that roaming has occurred.

The rerouting of a roamer call is shown in Figure 28.3. When the call request arrives at the mobile switch, it is rerouted via the trunk network to the foreign area where the mobile was last registered. This particular scheme assumes that all mobiles in the network have a unique identity

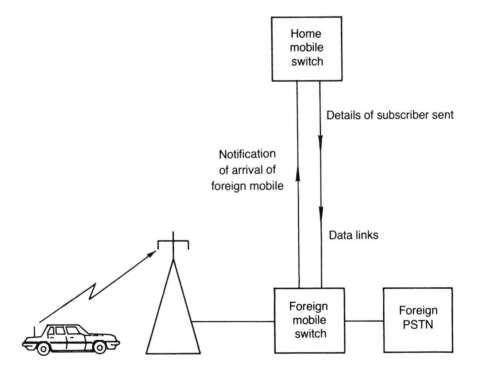

Figure 28.2 Notification and information exchange on an automatic roamer

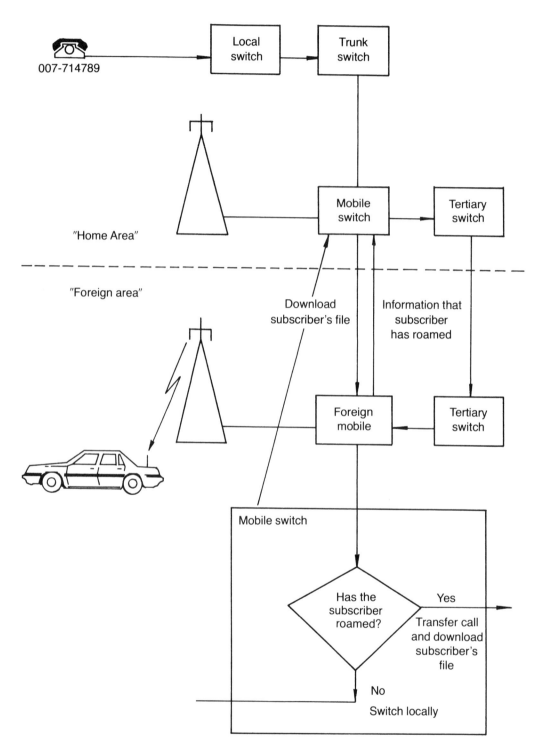

Figure 28.3   Automatic rerouting of a call to a mobile subscriber

that enables the mobile switches to determine their home areas from their mobile IDs. This identification can be readily achieved in modern cellular systems if careful numbering schemes are maintained. In addition, the use of unique country codes is necessary for international and some inter-area roaming.

Automatic roaming means that subscribers do not need to advise callers that they have moved. The interesting question, of course, is who should pay the extra call charges from the home switch to the foreign one, the customer or the land-line caller (who is not aware of the call diversion)? An outgoing call from the mobile unit in the foreign area is made in much the same way as a call in the home area. The call is usually charged locally (that is, in the foreign switch area), with billing reconciliation occurring later.

The only indication that a roamer in an automatic network has about being in a foreign area is that the ROAM light is illuminated (the mobile automatically detects that it is a foreign area). Otherwise, the subscriber makes and receives calls exactly as in the home service area.

An alternative method of automatic roaming is to have a central database of all valid roamers and have each switch connected to that database for roaming validation. This "star network" can provide considerable advantages as the network increases in size. Figure 28.4 shows the star network method.

The star network allows each base station to access the database at the baud rate of the data lines. A ring network (as may be used for a large number of switch databases) requires time-sharing. Although this is not a problem for small networks, time-sharing becomes increasingly slow as the networks grow. Therefore, future, large-scale roaming databases are most likely to be centralized.

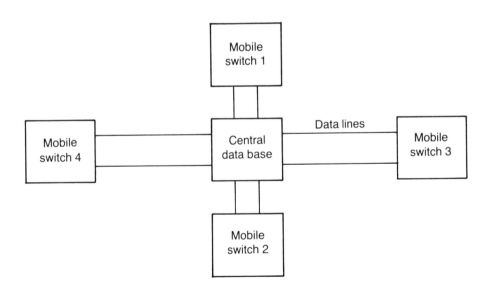

Figure 28.4   Using a star network to validate roamers. All switches can validate via a central data base.

SEMI-AUTOMATIC ROAMING

The attraction of automatic roaming to a user is its simplicity. The drawback is that it is a somewhat more complex network.

Semi-automatic roaming, however, allows the roamer to make outgoing calls without complication, but requires the land-line party not only to know that the desired mobile has roamed, but also to direct the switched network to cause the rerouting. The roamer is validated manually in this instance. That is, access to the foreign network depends on the manual transfer of the roamer's customer records to the foreign switch.

For simplicity, the customer can be part of a group of customers who are validated as a block (that is, all subscribers from one operator can be regarded as valid on a second operator's switch). This technique, while simple, has implications for roaming fraud (discussed later).

Making an outgoing call is not very difficult on a semi-automatic system, where it is made in the usual way. Some operators automatically assume that, if no access code is used, the caller is making a call to the home service area (not the area the caller is actually in). Thus, a local call to the area in which the roamer is operating must have an area code. Requiring all roamers to *always* use access codes is often necessary to reduce confusion.

Making a call to the mobile unit, however, is somewhat more complicated. Because the system is only semi-automatic, the land-line caller must know where the mobile is operating and dial a number that routes the call to the appropriate foreign switch.

The roamer access codes used in the US illustrate how this works. For example, if a mobile with the number ABCDEF from South Dakota is roaming in Oklahoma, a caller would need to follow these steps:

1. Dial the area code for Oklahoma (405)
2. Dial a seven-digit access code (627 ROAM). (Note that ROAM is made up of the digits that spell ROAM in the telephone keypad.)
3. Pause and wait for the continue tone
4. Dial the South Dakota area code (605)
5. Dial the called mobile's number (ABCDEF)

In other words, the caller must dial 405-607-ROAM-605-ABCDEF.

Figure 28.5 shows the routing of the semi-automatic roaming call. The first part of the number dialed (405-627-ROAM) directs the call to a switch inlet in Oklahoma in exactly the same way as a call would be directed to a normal subscriber in Oklahoma. This number, however, has a switch inlet that expects to receive further digits that identify a roamer to be called. This switch may or may not be the same switch as the Oklahoma mobile switch; the number 627-ROAM ordinarily identifies not a single inlet but an in-dial group (that is, multiple lines on the same number). When the switch is ready to receive further digits, it sends a continuous tone (much like a PABX does), and the digits 605-ABCDEF tell the switch that car number ABCDEF from South Dakota (605) is wanted.

Charges are usually levied on roamers in the same way as home subscribers but often at a higher rate and with a higher *daily* access fee. Within any one switch or a group of switches with fully automatic roaming, a positive validation system minimizes fraud. Positive validation means that

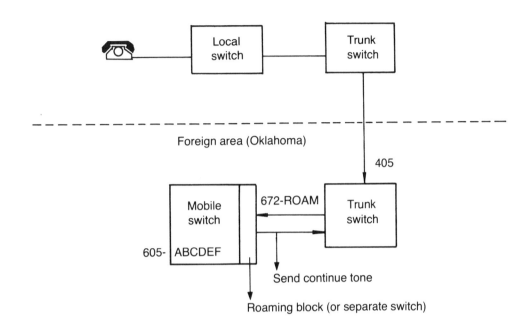

Figure 28.5 Semi-automatic roaming occurs when the land-line caller must know the roaming area in which the called party is operating in order to place the call.

a valid number is permanently filed and is checked before each call is permitted. This situation is complicated when roaming from foreign systems is contemplated.

GROWING ACCEPTANCE

The original AMPS system as specified by AT&T was designed to be operated, by them, as a monopoly and as a nationwide service. AT&T designed a network of interconnected switches that would have allowed nationwide positive automatic validation, such as is available today in Australia (AMPS), UK (TACS), or Scandinavia (NMT). With the Modified Final Judgement that broke up the AT&T monopoly, however, this plan was no longer valid, and individual operators, eager to start their own local operations, gave little thought to roaming.

The need for roaming became increasingly evident, however, as the systems grew and more service areas opened. Worldwide, there is now a current recognition of the value of roaming, and numerous bilateral agreements exist between operators. Bilateral agreements are particularly common in the Asia-Pacific region, where AMPS systems are used in most countries. In the US, such agreements are largely internal, but other countries, which generally have nationwide roaming, are exploring international agreements.

Roaming has been generally well accepted, and operators can project their service areas as being very extensive (at very little cost to themselves) when they have good roaming agreements in place.

ROAMING FRAUD

Both within countries (like the US) and between countries, roaming is becoming increasingly popular as a means of extending the effective coverage of an individual operator. In either case, the total number of potential roaming customers is very large. The size of the customer base presents a major problem of validation.

For simplicity, early roaming agreements between operators allowed all potentially valid subscribers from one operator's system to have access in the other operator's system. This method quickly led to abuses. For example, a mobile installer could illegally provide a mobile programmed to appear as a valid roaming mobile and could then use the mobile until detected. And detection often took four to seven weeks. Once detected, the ESN (Electronic Serial Number) could be registered as invalid and details of that number could be entered into a negative file (a "black list" of invalid users).

As a natural progression, particularly in the US where the opportunity for abuse was greatest, the concept of a negative file was introduced as part of the normal system files. This file contained the serial numbers of all known invalid mobiles from other operators with whom roaming agreements were held.

Outside the US, roaming fraud is not yet a problem and a negative file is probably adequate, particularly where trans-border traffic is not high. In many cases, positive validation, by manual means such as telex or FAX, for roamers between countries, is practical for the time being. Frequent roamers can be validated for long periods. This method is slow and cumbersome but adequate when the number of users is small. Because the US has had the most fraud, it is instructive to look in detail at the American experience.

As the total number of systems grew and the need for roaming became more evident, a number of solutions emerged. The operators, of course, did not standardize on one switching system (there are presently nine manufacturers), so developing any direct interface between switches would be very costly.

The first safeguard attempted was the negative file, but this quickly proved inadequate. Manual validation is not practical on the scale of roaming in the US, which has 2,190,000 users and a growth rate of 800,000 per year (at June 1989). Apart from the problem of needing constant updating, the negative files quickly filled up.

GTE Data Services Clearing House, a service that offers validation to a number of operators, offered a solution. GTE had developed a central negative file of fraudulent users, but the limitations of negative files soon became apparent and GTE realized a better method was needed. Total positive validation, although not foolproof, proved to be the answer. GTE therefore introduced Positive Validation Systems (PVS), which uses a "post-first-call" method for validating roamers. In this system, a roamer is allowed to "post-the-first-call" as though the caller were valid. Full details of that call are recorded by the cellular switch, and then those details are sent, usually by tape or disk, to the PVS clearing house. Here, the call is examined in detail, first against the negative file (a file of invalid ESNs); then a positive validation occurs (where the clearing house computer checks, either against its own files or by contacting the home switch, that the customer is valid).

Assuming the validity of the roamer is verified, the switch validates the roamer on its own temporary "recent roamer" file. This recent roamer file is used to validate future calls and is purged periodically to ensure the historical accuracy of the roamer status.

Positive validation started in the US in late 1986; the discovery of 3000 invalid users in New York and Boston in the first six weeks of operation reveals the extent of fraud. GTE Data System Clearing House reports an 8 percent rejection rate of validation queries.

UNDERSTANDING FRAUD

One reason for the proliferation of fraud is that most people do not consider getting free telephone calls a crime. Most people have used a public telephone that gave free calls, and very few have felt any obligation to repay the call costs or even report the faulty telephone. A wide cross-section of the community can readily be enticed into telephone fraud, far more than just the subpopulation of conventional criminals.

Unless a positive validation system is operating, an installer can easily program in a foreign MIN (Mobile Identification Number) and other data to identify the mobile as a roamer. The installer does, however, risk having the mobile ESN placed in a negative file. Disconnected customers may find that their units work in foreign areas without PVS and may be tempted to use them. These calls are at least more traceable, but the chances of collecting the outstanding revenue are slight.

The most difficult fraud to prevent is done by deliberate illegal customers who can change the ESN. The original concept of the ESN was to incorporate it in such a way as to make alteration impossible. This, of course, is fanciful thinking; at best, it is possible only to make alteration of the ESN very difficult. Some manufacturers, however, have made it less difficult than others, so in some instances it is not particularly difficult to change the ESN, making negative files almost useless.

REDUCING FRAUD

Clearly, some form of positive validation is the most effective way of reducing roamer fraud. If negative files are used, they should be regularly updated with some daily analysis. Because fraudulent roamers make about twice as many calls as legitimate users, flagging all high-usage roamers and checking their validity on a daily basis can be worthwhile. Placing limits on toll calls for roamers can also reduce the risk considerably, but at the same time, it reduces the value of roaming to the customer.

Agreements with other operators should acknowledge the potential for fraudulent use and clearly spell out the responsibility for lost revenue. Time limits should be placed on the notification of invalid ESNs. Once roaming numbers become significant, an automatic PVS should be considered.

Finally, it is important to choose staff carefully and inform them that information about user identification numbers should be treated as confidential, both by themselves and by external installers.

INTERNATIONAL ROAMING

As handhelds get smaller and businesspeople get used to the idea of always carrying a handheld phone, the demand for international roaming will increase. Manufacturers are working on even smaller handhelds, and it is reported that truly compact, pocket-sized phones already exist in the development workshops of a number of manufacturers.

Although international roaming is still in its infancy, it will mature over the next few years. Many countries already have some international bilateral roaming agreements and many other agreements are under discussion.

Europe will have to wait for GSM because of its incompatible systems, but the countries of North and South America, East Asia, and Oceania, which predominantly have AMPS, are quietly introducing roaming today, the basis of which will serve them well into the future.

International roaming has some unique problems, including customs clearance and type approval. Customs requirements vary considerably from country to country, but most countries have a bond system whereby that makes it possible to temporarily import certain goods that otherwise would incur a duty. The condition is always that they will later be re-exported (they must be taken out of the country when the roamer leaves). It is perhaps unfortunate, but predictable, that mobile phones, which are largely imported, are subject to duty. The bond system, even though it works, is clumsy.

Mobile telephones have an additional disadvantage when it comes to clearing customs: They are also transmitters, and many countries particularly target transmitting equipment for customs attention. This leads to type approval and the problems caused by each country insisting on local type approvals. Despite years of debate, the world still appears to be a long way from a uniform type-approval system.

To enable fully flexible roaming, however, it is highly desirable that all mobile units be universally type-approved. Lack of universal improvement is a considerable impediment to roaming at this time, particularly to countries that have government approval on only a small number of models.

PROTECTIONISM

Added to the problems discussed earlier, some countries have introduced forms of protectionism that are aimed at supporting local content in mobile production. Some of these countries have a blanket ban on the operation of any mobiles not manufactured in the country.

VALIDATION

There is, at present, no centralized validation center for international roaming, which means that bilateral arrangements for validation must be made between countries that have roaming agreements between them. Although bilateral validation presents no problems when there are only a few participating countries, it becomes quite cumbersome as roaming spreads to more and more countries.

Manual validation inevitably causes problems. Numbers that should be validated for roaming will be missed, and canceled roamers will remain validated. Although validation problems must be anticipated where manual methods are used, participating companies need to implement detailed screening of the roaming files. Some form of central clearinghouse system similar to the ones now operating in the US will soon be required. A "post-first-call" system would probably be adequate, with the PSTN being used to confirm the validity of the user.

CHARGING

There does not appear to be a need for uniform charging for roamers. The simplest method is that each operator determines prices and bills accordingly. It would thus be the responsibility of the home operator to warn customers that roaming charges vary and to indicate the charges in roamer areas.

Another way to recoup costs may be to treat roamers as a special category of subscriber who attract additional rental charges in the home service area. This additional charge can be justified by the extra bookwork required to validate the subscriber and the extra software required on the switch.

Problems can arise when roaming is automatic and subscribers, making what they think are local calls to mobile subscribers, is, in fact, routed to some distant point on the globe. In these cases, it is probably most ethical to charge the roamer for the cost of the diversion, but doing so may be technically complex, particularly in networks without digital switching.

NUMBERING

There is no international coordination on network numbering for mobile phones, and the lack of it can confuse some users. Because of the way that telephone traffic is routed, it is usually necessary to give the roaming subscriber a new telephone number at the foreign switch, particularly if the foreign mobile network does not use a cellular access code. Mobile telephone systems that do not use an access code must use the first digits of the mobile subscriber's number to identify that the call is to a mobile switch.

Consider the case of a mobile user whose home code is 007-714789, and who then roams to a foreign country where all mobile telephones have the phone number 818xxxx (818 identifies that the call is for a mobile phone). The simplest way for the foreign operator to allow access is to assign one of the numbers in the local number range to the roamer and translate the number in the switch. For example, the roamer may be assigned the number 8181234. When this number is dialed, the call is routed to the mobile switch and then translated to 714789 (the number the mobile responds to).

The roamer, of course, has the probably difficult job of explaining to his callers that the number changes while the roamer travels. There are ways, however, of preserving the original number.

Consider the same call mentioned earlier, where a special access code, say 099, existed for international roamers in the foreign country. Now, all incoming calls to a roamer (whether local or international) are sent to the mobile telephone switch on a special route. Figure 28.6 shows this trunking of mobile roamers.

The special roamer trunk route can be programmed to accept a a format that differs from the normal 818 route. For example, if it is assumed that the roamer comes from a country with the cellular code 636, then the number sent to the mobile switch in the foreign area could be 99-636-714789.

This method allows the foreign switch to determine the country of origin and the normal home telephone of the desired mobile. The desired mobile can be called without using a number translation.

Notice that the "cellular" country code is used here and that it is different from the normal IDD country code. (For more information on the numbering scheme, see Chapter 18, "Numbering Schemes.") This method, if adopted, would simplify procedures for users by providing an international standard roamer access code (99). These methods still require the caller to know which country the roamer is in, because automatic roaming is not considered here.

If such a mobile roamer code is to be adopted internationally, it must be a little more flexible than the above code would allow. A useful enhancement would be to add one more digit, D, to indicate which mobile system the call is directed to. This is needed because many countries have more than one mobile telephone network.

The code described earlier could become 99 - D - 636 - 714789.

> *where*   D = 0 means AMPS A
> D = 1 means AMPS B
> D = 2 means TACS A
> D = 3 means TACS B
> D = 4 means TACS C
> D = 5 means GSM
> D = 6 means AMPS A digital
> D = 7 means AMPS B digital
> D = 8–9 as required

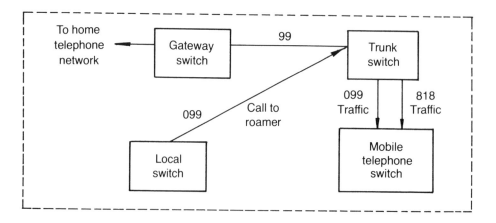

Figure 28.6   Marked trunking of mobile roamers

Note that expanding *D* to 2 digits in the future would be easy if required.

Of course, the actual access code (99 is used here only as an example) must to be carefully selected from codes available in each participating country. Notice that, although a worldwide uniform code would not be essential, it would be useful.

ROAMING AND GSM

Roaming has been identified as such an integral part of the new Pan-European GSM that it has been agreed that international roaming will be *mandatory*. Table 28.1 shows a number of restrictive categories for intra-country working have been allowed. As a logical consequence, operators must offer service to all roamers.

In Table 28.1, Category 2 allows access to only one national carrier, and Category 3 may restrict the home operation to only one city. In all cases, however, international roaming must be permitted. The purpose of these categories is to allow lower-priced, and limited-coverage home service areas if required.

The method of roaming in GSM is essentially the same as for the analog systems described earlier, except that GSM regards any active mobile as a roamer. Figure 28.7 on the next page shows roaming in GSM.

The mobile is registered in a Home Location Register (HLR), which can be part of the subscriber's switch or can be a centralized national register. The mobile uses a separate international mobile subscriber identity (IMSI) in addition to its mobile station roaming number (MSRN). The first number is fixed and the latter is allocated dynamically. Although this system requires additional signaling overheads, it leads to significant number conservation.

Thus, a PSTN call is registered in the local mobile switch which then checks the subscriber's home location register (HCR) to determine the subscriber's current address. The call is then switched in much the same way as current analog roaming networks. Because the PSTN must switch all GSM calls to the GMSC, the fixed network must be able to recognize all GSM codes; the numbering scheme must be coordinated.

CATEGORY	SUBSCRIPTION CATEGORIES ACCESS AREA
1	All GSM areas
2	One national plus all foreign areas
3	Regionally restricted home switch area plus all other areas

Table 28.1   Subscription categories

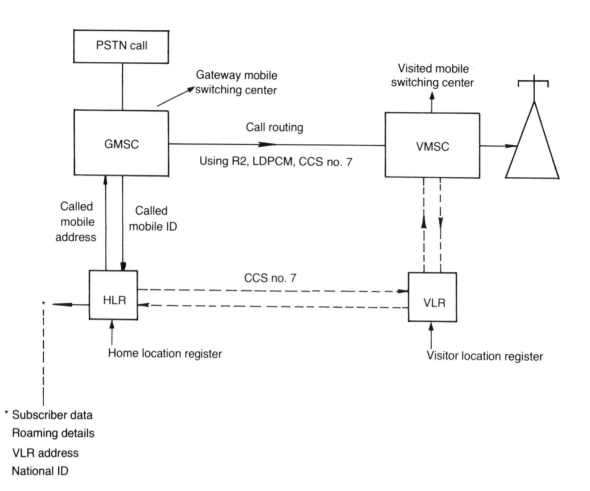

Figure 28.7   GSM roaming call establishment

Chapter 29

DATA OVER CELLULAR

Increasingly, data service is becoming a standard add-on to cellular mobile systems. This market, like the cellular market itself, has been largely underestimated both in its customer base and scope of application. The growth of the data-service market has caught many suppliers ill-equipped and, with the increasingly information-driven economy, a little embarrassed, both at their inability to provide services and their knowledge of the product.

The users of data services include construction, insurance, field sales, public services, real estate, service, delivery, and many other groups who require real-time access to databases in the mobile environment. As a rule of thumb, information can be transmitted in the data format at 10 times the rate of voice communications.

Because of the harsh environment, transmitting data over a cellular system differs from transmitting data over the conventional telephone network in two ways. First, the transmission medium is much noisier; second, during handoffs, disruption of the data stream can occur for up to hundreds of milliseconds. As a result, a more elaborate code for error detection and correction is needed for cellular systems.

Data is typically sent over analog media (such as telephone wires) via a modem. A modem is simply a device that converts the ones and zeros that form the language of a computer into an analog counterpart that can be transmitted over an analog medium. If the ones and zeros were sent directly down the telephone line, the distortion that would occur would make them unreadable at the far end. A simple modem uses a Frequency Shift Keying (FSK) mode to generate two tones, one corresponding to a one and the other to a zero.

Because errors can occur during transmission (due to noise pulses and non-linear transmission characteristics), a modem usually has some error-correction codes. These codes insert extra bits of information (redundant code) that have a definable mathematical relationship to the code bits containing the information. At the receiving end, the code bits are read, and then the mathematics done to check the integrity of the information received. The code is arranged so that not only can errors be detected but the original (correct) code can be reconstructed.

In a noisy environment like cellular radio, the error-correction codes need to be rather sophisticated, because there will be many errors. A resend capacity is also needed for data blocks missed during handovers. For this reason, normal modems do not perform adequately on a cellular mobile.

There are three common ways to transfer data over a cellular system:

• A dedicated modem can be fitted to the mobile and another to the computer. This method is adequate when only one host computer is involved but does mean buying two modems.
• A system of back-to-back modems, as shown in Figure 29.1, can connect a mobile personal computer with a home computer. This setup can be used either by private users or offered as a service by the cellular operator.
• A date transfer facility can be provided as an integral part of the cellular switching system.

In the system shown in Figure 29.1, the cellular modem on the land-line end can be shared by

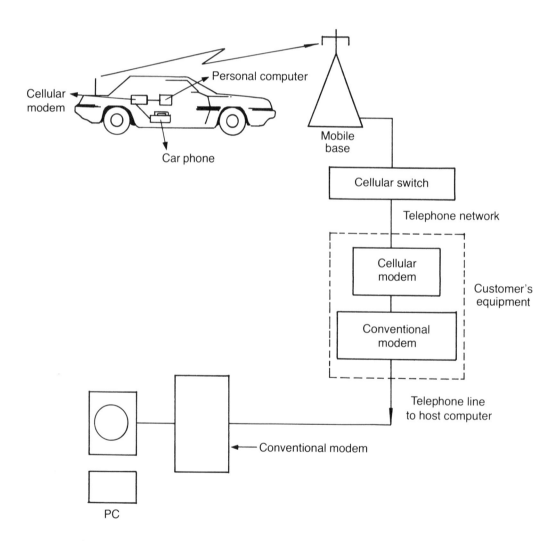

Figure 29.1   Back-to-back system modems can be used for data transfer.

a number of users. Each user can telephone the modem in the normal manner and, when it answers, the mobile dials through to the user's destination computer. The advantage of this system is that the host computer can have a conventional modem which, of course, can be used for a variety of other purposes. The so-called "SMART" system is a variation of this system. In a SMART system, data transfer is provided at the mobile switch and includes "smart" facilities such as password protection and user-friendly access to the host computer. Data transfer can be an integral part of the cellular switch which can automatically recognize a data call and decode and route it accordingly to the host computer.

Because different modems use different error-correcting codes, as well as different speeds and signaling formats, it is necessary to ensure the compatibility of modems before attempting to use them together. The modem industry is still relatively young, and standardization is only on the horizon. For this reason, cellular operators who become involved in data systems must focus on and become familiar with only one or two products.

Chapter 30

PRIVACY

Privacy is not an inherent feature of analog cellular systems. Although some operators claim a degree of privacy due to the difficulty of locating or following a particular call, the fact remains that analog systems do not inherently provide the sort of privacy that is associated (sometimes unduly) with the Plain Old Telephone System (POTS). Even so, it is widely rumored within the industry that many "shady deals" are transacted over the mobile telephone system rather than the POTS simply because of the difficulty of tracing a particular call. It is true that it is difficult to trace particular calls using a radio scanner, but full call accounting at the cellular switch ensures that details of the call are available to the proper authorities.

Radio scanners are wideband radio receivers that, under microprocessor control, can scan through a preset group of channels and stop when a carrier (an active channel) is encountered. The receiver stops on that channel indefinitely (unless programmed otherwise) while there is a carrier present. Once the carrier has gone (in the cellular instance, the call terminates) the scanner moves on through its list of programmed channels until a new carrier is detected, at which time it stops and proceeds as before.

For a few hundred dollars, anyone can purchase a scanner. There are a remarkable number of people who have not only done so, but who are content to spend hours every day listening to an interminable number of conversations on the off-chance of picking up some scrap of information, perhaps of an industrial, political, or even personal nature. The more organized "snoops" may have tape recorders attached to the scanners, and some may even have commercial outlets for various kinds of information. For this reason, it is wise to warn all users to avoid talking about secret business transactions, or at least to do so in a circumspect manner.

This may not be a problem for some people, as in the case of one rather well-known Australian politician who, when asked if the lack of privacy on his car phone was a problem, replied, "Heavens no, I never say anything that makes any sense anyhow."

Some cellular operators claim to provide reasonable privacy based on the fact that it is difficult to locate any particular mobile in the network and that handoffs produce a discontinuity in the conversation. It is true that it is difficult to lock onto a particular mobile, but it should be remembered that a snoop needs only to have a mobile at a high location to receive most base stations. Because some modern scanners are bus controllable and the signaling protocols are public knowledge, it would not be too difficult to build a scanner programmed to "track" a particular mobile. At least one instance of this, in the Netherlands, has been reported.

Handoffs provide some immunity to scanning, but they are not usually available in small systems, where there is nothing to hand off to, or when the mobile is a stationary handheld. In very dense networks, however, the claim is reasonably valid because handoffs do occur quite frequently. Still, a scanner programmed to follow a particular mobile could probably continue to do so.

There are basically two types of scanners on the market. Currently the cheapest type is usually limited to 512 MHz and therefore is unable to receive most cellular bands (and certainly not AMPS, TACS, or NMT900).

Other, more expensive scanners cover the spectrum up to about 1 GHz (1000 MHz) and do cover the modern cellular bands. These receivers are usually not particularly sensitive and are therefore limited to relatively nearby base stations. However, there are still usually plenty of channels within range (10–15 km) to keep snoops, both amateur and professional, quite contented.

The way in which conversations are received warrants some attention. All conversations between a mobile and the base station are full-duplex (that is, from the mobile, the transmission occurs on one frequency and the return signal, to the mobile, occurs on a different frequency, usually 45 MHz away). This fact has led many people to conclude that a scanner can therefore hear only half of a conversation, a conclusion that is, unfortunately, wrong. In fact, as shown in Figure 30.1, scanners hear both sides of the conversation.

As seen in Figure 30.1, although the two conversation paths appear to be separate, the path that transmits to the vehicle F_2 has a very wide range AGC (Automatic Gain Control) amplifier, which

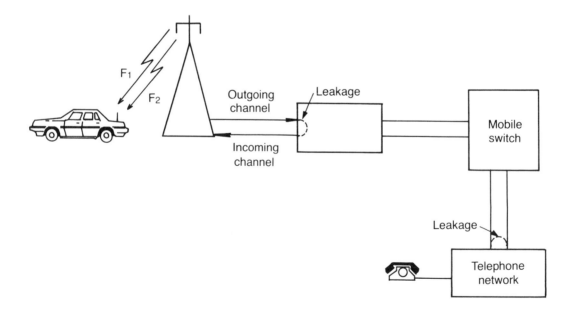

Figure 30.1 Leakage and internal automatic level controls enable scanners to hear both ends of a conversation even though it takes place over different duplex channels.

has the function of ensuring that constant audio levels are transmitted. This system works so well that side tone (or leakage from transmission to reception) is amplified to the level of normal audio.

Side tone is deliberately introduced into telephones (including cellular telephones) so that some response to the user's speech is heard in the receiver of the speaker. When this tone is not provided, the earpiece is silent during speech and this silence is interpreted (usually) as a transmission failure (the telephone sounds "dead"). The leakage may or may not be related to the generation of side tone but it is invariably present; the mobile transmission as well as reception is transmitted by the mobile base at approximately equal levels. The clarity of the leaked path as received on a scanner varies from poor to very good, depending on the system, but is usually intelligible.

PRIVACY METHODS (SCRAMBLING, ENCRYPTION)

Privacy has rarely been added as a standard feature to analog cellular systems for one simple reason—cost. To see how this cost rises, the simplest method of privacy or scrambling (frequency inversion) is examined here. Figure 30.2 shows how frequency inversion causes the speech spectrum to be inverted.

In Figure 30.2, the normal speech band of 300–3400 Hz has been inverted by a simple modulation technique so that the content of the high frequencies (3400 Hz) now appears as the low-frequency (300 Hz) part of the inverted band. Similarly, the 300 Hz component now appears as the high-frequency 3400 Hz component. This renders the speech virtually unintelligible.

The inversion can be achieved by a simple mixer and filter combination, as shown in Figure 30.3. The mixer produces the sum and difference of the input signal frequency and the 3700 Hz oscillator. If the mixer is followed by a low pass filter that cuts off at around 3700 Hz, then all the sum components are filtered out and only the difference frequency is passed. Thus, if the input signal is 300 Hz and the oscillator is 3700 Hz, the output will be 3400 Hz (3700 – 300 Hz). Similarly, if the input is 3400 Hz, the output will be 300 Hz. These frequencies are now said to be inverted. Demodulation consists simply of "re-inverting" and uses the same circuitry as shown in Figure 30.3.

This method has two disadvantages. First, it is simple and can therefore easily be decoded (unscrambled). This disadvantage can be overcome by splitting the audio band into a number of

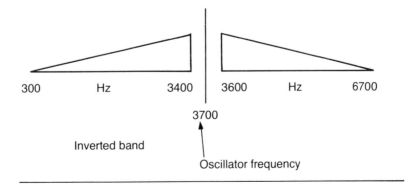

Figure 30.2   A simple frequency inverter

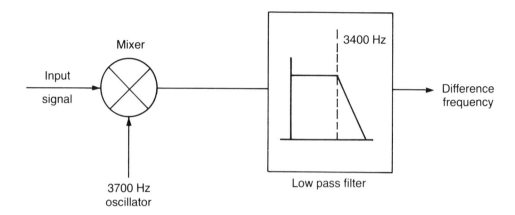

Figure 30.3   A double sideband A.M. modulator and filter produces frequency inversion.

components and inverting them separately; this is commonly done. More sophisticated scramblers use this method, but go even further. They divide the speech band into many smaller segments, which are not only inverted separately but sometimes transposed according to a user settable pattern. Clip-on, acoustic-coupled devices with thousands of combinations are commercially available.

The second disadvantage—the inherent increase in signal-to-noise performance—is more fundamental and has no easy solution. In its simplest form, noise necessarily introduced by the scrambling and unscrambling of this signal degrades the signal path by approximately 9 dB in S/N performance. This degradation reduces the useful range of a base station to about 60 percent of a base without scrambling. More elaborate decoders exist with filters that eliminate most of the noise caused by the inversion process. When time-dependent inversion is used, then the filters must be switchable and follow the pattern of the encoder. With such devices, the S/N degradation is reduced to about 1 dB and so the range reduction is not so serious.

Unless scrambling capability is provided as a feature of the cellular switch, it should be noted that any "add-on" scrambler requires a decoder/encoder at each telephone (whether mobile or not) where scrambling and descrambling are to occur. Some switches, however, do incorporate a built-in scrambler. Most switch-based scramblers are proprietary devices and can therefore be used only with the supplier's mobiles. These devices usually use elaborate algorithms to scramble the inversion points in manners that are difficult to replicate. Typically, a random number sequence, which is "seeded" by the serial number of the mobile (which need never be transmitted), and a second number, supplied by the switch, determine the inversion sequence. Such systems are difficult for all but professional interceptors to decode.

Other methods to discourage eavesdropping are in place. In most countries, it is either illegal to listen to cellular and other telecommunications, or it is illegal to use any information gathered by such eavesdropping. These laws are largely "paper tigers," however, because they are virtually unenforceable. There is little value in legislating that listening to the cellular bands is illegal unless that legislation can be enforced. Some administrations have gone a little further and prohibited import or manufacture of scanners that cover certain bands. This measure results in the marketing of scanners with discontinuous coverage. This method is also not particularly effective, because there are usually plenty of radio shops willing to "modify" these scanners for a relatively small fee so that full coverage is once again available.

One interesting way of discouraging eavesdroppers takes into account the way that the scanners actually work and turns scanner use into a disincentive in its own right. As noted before, scanners skip over unused channels and stop only on those that transmit a carrier. If all channels always transmitted, then the scanner could function only manually. By having all channels always transmitting, and by having them transmit loud and unpleasant sounds, the eavesdropper soon finds that the hours spent at the scanner are not particularly pleasant. This discourages a number of potential snoops.

On the negative side, however, there are quite a few costs. All channels must be turned on at all times. This significantly increases the power consumption of the base station and decreases the life of the power amplifier stage of the base-station channels. Unless some interlocking device stops this mode of operation during power failures, the time that the batteries can support a base station is significantly reduced.

For this method, base-station antennas must operate at all times at maximum power; this contributes to corrosion and antenna faults. Finally, because all channels are always transmitting, co-channel and adjacent-channel interference always operates at its maximum level, and contributes to a net decrease in the traffic capacity and availability of the system.

In general, one can see that, where attempts are made to add privacy to analog systems, the costs are high. This, of course, does not apply to digital systems, where very high levels of encryption can be applied at little or no cost. Digital systems can thus be expected to provide very high levels of privacy.

Chapter 31

RURAL APPLICATIONS OF CELLULAR RADIO

The application of cellular radio to rural areas is a relatively new concept, both as an alternative to the fixed telephone network and as a means of providing mobile radio. Some information suggests that the economics of rural cellular are, in many cases, quite attractive. However, as developed countries expand their cellular systems into smaller cities, a trend is emerging that indicates that per capita demand decreases as city size decreases. As the per capita demand decreases, so does the financial return. There appears to be a "critical mass," below which the population is too small to support a thriving cellular system. Early indications are that this population is about 100,000.

The reasons for this phenomenon are not altogether clear, but shorter traveling times and, hence, a smaller proportion of time spent in a car in rural areas, appear to be the major factors. The same low demand is evident in paging and other mobile systems (on a per capita basis). The exceptions to this trend are usually relatively large towns close to or along highways leading to major cities, although even these towns often exhibit a low per capita demand.

The economics of cellular radio, when compared to that of other telephone solutions, depend very much on the density of the subscribers. In large cities or towns the capacity limitations of cellular in the long term mean that a fixed land-line network must be used for most fixed subscribers. In medium-sized towns, cellular radio may be a complete solution for all telephone requirements. In very low-density areas, the high basic infrastructure cost of a cellular base and its links ultimately limit the usefulness of cellular radio.

Depending on the terrain, a lower limit of telephone density for economic rural application is probably in the range 0.05 to 0.5 customers per square kilometer. Within the boundaries of these extremes, the potential of cellular radio should be considered and budgeted against competing alternatives as part of the network planning.

ENVIRONMENTAL LIMITATIONS

In rural areas with little or no infrastructure, the environmental limitations can be daunting. The equipment must be able to perform while installed in areas with a wide range of temperatures, humidities, and altitudes. The reliability of power supplies (including brownouts) must also be considered. In some locations, vandalism (including deliberate attacks with high-powered rifles) may be encountered. Rural areas do not have good access to qualified maintenance staff or quality test equipment. Reliability is a vital factor and the installation should be a low-maintenance design.

SMALL SWITCHES AND REPEATERS IN RURAL AREAS

For rural areas that are remote from major switching centers, there are a number of possible ways to establish a cellular system. Trombone trunking (or backhauling) can be used to operate the remote bases from a distant switch, although doing so rapidly becomes less economical as the distance to the remote switch increases. Trombone trunking involves the use of a link from the centralized switch to the rural area for all calls (even local ones). In many rural areas, it may be economical to use repeaters to effectively extend the coverage area of a larger network to include the rural area.

Small switches, with capacities from hundreds to a few thousand customers, are most economical in remote areas where backhaul costs are prohibitive. An extension of this concept is the "remote switching unit," which is a small cellular switch with most of its "intelligence" (processing power) back at some central location. This last approach, however, has only limited application. Figure 31.1 shows a small cellular switch.

CELLULAR MOBILES AS "FIXED" TELEPHONES

There are a number of companies that specialize in "fixed" cellular telephones. A "fixed" cellular telephone system can operate either on a stand-alone basis or as part of a mobile network. Because there are no cables or overhead wires, the maintenance can be relatively cheap (compared to maintaining a hard-wired network), although in remote areas maintaining the mobiles can present problems.

There are two philosophies that can lead to very different maintenance workloads: One considers the fixed mobile to be permanently fixed; the other treats the "fixed" subscriber much like a mobile subscriber.

"FIXED" CELLULAR MOBILES

Cellular mobiles that are designed for permanent fixed installation and use a phone much like a conventional hard-wired phone are commercially available. Fixed cellular mobiles, as they are called, use a conventional mobile mounted in a custom package that includes the electronics to convert the cellular handset and ring interface into one fully compatible with a POT (Plain Old Telephone). In this instance, customers need not be aware that they are using a cellular system.

Figure 31.1   A small-capacity cellular switch. A specially designed small-capacity switch is the EMX 100, which is designed for a subscriber capacity of around 7000 (maximum). Notice that it consists of only three 600-mm racks. Photo courtesy Motorola Communications and Electronics, Inc.

There is an extra workload implied in this approach: The mobile must be installed and hard-wired in the customer's premises; and this extra installation must also be maintained. If the mobile has an external elevated antenna, additional installation and maintenance costs can be incurred.

The effective range of the mobile can, however, be increased dramatically by the use of external antennas. External antennas decrease the base-station infrastructure costs by reducing base-station antenna heights (and perhaps the total number of base stations). To further complicate the trade-off, in areas where external receive antennas must be used, base stations that provide continuous "fixed" coverage cannot do so for mobiles, so patchy mobile coverage results.

MOVABLE "FIXED" MOBILES

If the customer is simply given a conventional transportable or handheld telephone, then the customer-end costs are a lot lower. It is relatively easy to distinguish the movable "fixed" mobile from a normal mobile and have different tariff rates. Additionally, it is relatively easy to confine the movable unit to its home cell only. This last requirement prevents the customer from buying a cheap "fixed mobile" and then using it as a conventional mobile.

It is a good idea to have an additional mobile category that can cross-subsidize the rural customer. The most desirable arrangement would be for an adjacent metropolitan cellular operator to provide roaming into the area. This type of arrangement can easily be accomplished if one company provides both the rural and urban facilities. Even if more than one company is involved, the arrangement is not difficult, provided that both companies use the same hardware.

CELLULAR PAY PHONES

Cellular pay phones are a practical way to provide a public service in remote areas such as islands, mountains, and freeways (emergency) where a nearby cellular system is already installed. Commercial units exist (some complete with attractive cabinets) that can meter calls from the remote site and report back when the coin box is becoming full or a power failure occurs. The units usually contain a "look-up table" that enables the meter rate (call charge rate) to be read from the digits dialed. These phones, being portable, can be moved into place quickly for emergencies or for catering to large temporary gatherings. This is especially true if the phones are solar powered.

The meter pulses are generated locally so that a conventional pay phone can be attached. This point is significant because pay phones are subjected to a great deal of vandalism, and if the handsets can be replaced by a standard telephone unit, maintenance becomes easier and cheaper.

Table 31.1 lists the specifications of a typical pay phone unit.

Voltage	115/220 VAC 50/60 Hz
Dialing	Tone or pulse (10 to 20 PPS)
Power	15 V.DC @ 5.3 amps
Battery	12 V 10 amp hours
Operational temperature	-30° to 60° C

Table 31.1   Specifications for a typical cellular pay phone

RIGIDLY MOUNTED VERSUS MOBILE RURAL UNITS

There are thus two basic fixed cellular rural structures: Type I and Type II. In Type I, the customer units are conventional portables; in Type II, the customer's equipment is rigidly mounted, and external fixed antennas can be used.

Type I, which involves no installation costs, generally appears to be cheaper than the fixed installations, however, the fixed units almost certainly require less maintenance because they are not subject to the same environmental hazards. In Type I, the customer can deliver faulty mobiles to a specialist service center; Type II requires a visit by a service personnel. It is difficult to determine which type has the lowest overall maintenance costs.

The external antenna mount available with a Type II system leads to a greater usable range. It is shown later in this chapter that path-loss decreases of 18 dB can be achieved, but given the extra costs involved, this solution is not, on the face of it, economical. But other factors should also be considered. For example, the ability to use conventional handsets may, from a marketing perspective, be important to many companies. Rural services are generally provided at a loss or by way of some form of subsidy. Providing rural subscribers with state-of-the-art cellular phones may substantially spur demand, so the net increased loss (or subsidy) may exceed the initial gains made by lesser installation costs.

The ability of many fixed cellular phones to operate in party-line mode could justify the extra cost (particularly if two party-line phones can replace two transportable units). Party-line facilities of eight or more phones are available as a standard option.

Where pay phones are considered, Type II is the only solution. Some administrations have a policy that all fixed phones must have the same handsets; such a policy dictates that Type II phones must be used.

When Type I phones are used, there may be a marketing conflict between rural phones that are provided at a nominal cost and mobile phones (which are the same thing) provided at a premium cost. This may cause conflict even when the rural service is restricted to the local cell.

NETWORKS WITHOUT EXTERNAL TERMINAL EQUIPMENT ANTENNAS (TYPE I)

A network where the terminal equipment does not use an external antenna is one where, in most instances, a subscriber is normally equipped with a conventional transportable or handheld telephone. A few instances of more distant subscribers with external antennas may occur.

The usable range of base stations meant to operate this way is necessarily limited, but they are inherently capable of supporting a conventional "mobile" telephone population. The subscriber densities that economically can be supported depends on the base-station coverage. Because telephone density is normally quite low in rural areas, it is important to seek out base-station sites which are high. Table 31.2 gives the expected range and area covered for different base-station heights.

From Table 31.2, a few important deductions can be drawn about the economics of customer densities and the importance of tower height. Since the density is low, it is important to get

BASE STATION HEIGHT (m)	AMPS		TACS		NMT900	
	RANGE	AREA (km^2)	RANGE	AREA (km^2)	RANGE	AREA (km^2)
30	9.3	271	8.2	211	5.5	95
50	12.6	498	11	380	7.3	167
70	15.6	764	13.6	581	8.9	248
100	20	1256	17	907	12	452
150	26	2123	23	1661	14	615
200	31	3019	27	2290	19	1134
300	35	3848	32	3216	24	1809
450	43	5808	37	4300	30	2827

Table 31.2   Range and area of a base station to a conventional mobile

maximum range out of the system. To determine if high towers are economical, let's examine a particular case. Assume the following:

• The cost of a base station with a 30-meter tower and 50 channels equals $ 970,000.
• The cost of tower extension beyond 30 meter equals $1,000/m.
• The number of subscribers is directly proportional to the area covered.
• An AMPS system is used.
• The base capacity is 1500[1] customers and has 50 channels.
• The cost of power supplies (assuming local power is available) and huts is included. The cost of links is not included.

It is clear from Table 31.3 that this very generalized case study implies that the cost per square kilometer covered decreases fairly rapidly with base-station height. Because a low uniform density is assumed, it follows that at low-density sites the cost per customer is less if greater tower heights are used instead of extra bases. Table 31.3 can be used to directly convert to cost per customer, once the customer density is known.

In practice, of course, this approach must be tempered with the reality of the actual terrain, which may not permit increased range with increased antenna height. But in general, it can be said that tall towers or high sites need to be a feature of rural cellular systems (notice that the opposite applies to urban systems). As the customer density increases, the need for high towers decreases.

Type I installations use conventional transportables and handhelds and so have lower terminal end-costs. Typically, the terminal end-cost is $1,000 each (June 1989)—the mobile unit is assumed to be a transportable. Type I installations are more economical in higher-density environments, where large towers are not required.

---

1.   Note that with AMPS/TACS systems, it is possible to use up to 1/2 of the total spectrum at one site. For AMPS, this is 395 channels (1/2 of 832–42), which would allow about 8000 subscribers from a single base if frequency reuse is not a problem. For NMT systems, it is possible to use all of the available spectrum at one site.

TOWER HEIGHT (m)	ADDITIONAL COST OF TOWER ($ 000)	COST OF COMPLETE BASE ($ 000)	AREA COVERED (km^2)	AREA COVERED ($/km^2)
30	0	970	271	922
50	20	990	498	542
70	40	1010	746	388
100	70	1070	1256	254
150	120	1120	2123	174
200	170	1170	3019	139
300	270	1270	3848	135
450	420	1420	5800	115

Table 31.3   Network cost per area covered by bases with towers higher than 30 m. Note that the cost is for the base station only. Switch and transmission costs have not been included.

NETWORKS WITH EXTERNAL TERMINAL EQUIPMENT ANTENNA (TYPE II)

An alternative to using conventional mobiles is to use specially configured "fixed" mobiles with high-gain (and elevation) antennas. This configuration is more economical in very low-density regions. It also precludes an effective mobile network from operating in the same district as a spin off (except in a hybrid arrangement as discussed later).

For relatively low antennas (< 10 m), the height gain is approximately 6 dB when the height is doubled.

Because the price of an external antenna installation is not strongly dependent on height (a single pole is assumed for the structure), it is reasonable to assume that all external installations are mounted on a 7-m pole. The height gain for this height (over the conventional 1.5 m for vehicle mounting) is about 12 dB. Further, it is reasonable to use a high gain Yagi antenna, (for example, 9 dB gain compared to the 3-dB conventional omni antenna). Provided low-loss feeders are used (less than 3 dB), the net gain now is $12 + 9 - 3 = 18$ dB. Table 31.4 shows the corresponding range and area and dollars per area, assuming the additional 18-dB height gain for an AMPS system.

It is immediately obvious that this configuration is much more economical in coverage costs than the Type I; in fact, a 30-meter tower can produce a smaller cost per kilometer than a 450-meter tower in the Type I example. Table 31.4 considers only network costs and does not reflect net costs. A Type II network has more costs at the subscriber's end, however, and these must be added before the two systems can be compared. The systems must also be studied further to determine how they compare.

CALLING RATES AND CUSTOMER DENSITY

The calling habits of fixed subscribers differ markedly from those of conventional mobile subscribers and do tend to show higher peak-hour traffic rates, even when the average calling rate

TOWER HEIGHT (m)	COST OF BASE ($ 000)	RADIUS COVERED	AREA COVERED (km²)	AREA COVERED ($/km²)
30	970	28	2463	101
50	990	34	3631	74
70	1010	37	4300	67
100	1070	43	5808	55
150	1120	51	8171	42
200	1170	57	10207	41
300	1270	67	14102	37
450	1420	78	19113	35

Table 31.4   Network cost per area covered for different tower heights, assuming a 7-m fixed antenna and an AMPS system with a 9-dB Yagi antenna on the mobile

is about the same. The calling rates of rural subscribers are very hard to predict, but in general they increase as the cost of the installation increases. For example, if a phone costs $10,000 to have installed, the rural users will organize to buy fewer units than if they cost $1,000 each; but the fewer $10,000 units are likely to be used much more heavily.

If it is assumed that the maximum capacity of an AMPS cell is 1500 subscribers, then some limits can be placed on tower heights for a given customer density. Note that in rural areas much bigger bases may be used. (It has already been shown that the maximum tower height yields the cheapest cost/km).

For any tower height, the maximum customer density is 1500/base.

Table 31.5 shows the maximum customer density per kilometer for one cell. Costs shown in the table include those for mobile and base stations only. Net costs would need to include switch and trunk costs and so would be a few hundred dollars higher than these figures. The additional costs would be the same for both Type I and Type II systems.

The cost of a Type II system compared to a Type I system is based on the following assumptions:

Cost of 7-meter pole and Yagi	$1,000
Cost of Type II mobile	$1,500
	$2,500
Cost of Type I mobile	$1,000
Difference	$1,500

From Table 31.5 it is obvious that when mobile costs are included, a Type I installation is clearly cheaper than a Type II installation. This applies even to very low densities, but below a density of 0.25 customers/km², it is worth noting that if a large area is involved, it may be necessary to use many more Type I bases than Type II to cover the same total area. In this instance, Type II systems may be cheaper. It can be shown that because of this effect, the break-even density (where Type I costs equal Type II costs) for a large area is about 0.15 customers/km². This figure should be used as a guideline only, because at these low densities other solutions may be more appropriate. It can thus be seen that, on economic grounds, Type I clearly is the cheapest, but there may be good reasons to use Type II systems despite the cost.

TOWER HEIGHT (m)	TYPE I TOWER MAX DENSITY CUSTOMER/km^2	TYPE II TOWER MAX DENSITY CUSTOMER/km^2	NETWORK $/CUS-TOMER (200 USERS)	TOTAL $/CUSTOMER TYPE I	TOTAL $/CUSTOMER TYPE II
30	5.50	0.61 ↑	646	1646	3146
50	3	0.41	660	1660	3160
70	2	0.35	673	1673	3173
100	1.19 ↑	0.26 ↓	713	1713	3213
150	0.71	0.18	746	1746	3246
200	0.50	0.15	780	1780	2320
300	0.39	0.11	846	1846	3346
450	0.26 ↓	0.08	946	1946	3446

Table 31.5   Maximum customer density/km for one cell

HYBRID SYSTEMS

Because it is generally cheaper to use a Type I system (that is, one in which the subscriber uses a conventional mobile telephone), the possibility of also using the system for mobiles arises. In practice, however, a hybrid system is designed basically as a Type I system, with good mobile

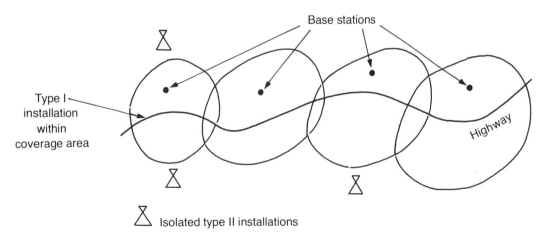

Figure 31.2   A hybrid, fixed/mobile network

coverage planned but with some Type II installation expected for fixed services. A typical application is shown in Figure 31.2.

In US rural areas, defined as 38 km from the nearest MSA, 500-watt ERP transmitter power has been approved. This high ERP is usually used with high-gain grid-pack antennas to give directional coverage along highways. The directivity, to some extent, precludes the use for fringe area coverage, as the beamwidth is quite narrow.

Chapter 32

PREPARING INVITATIONS TO TENDER

Preparing an invitation to tender consists of writing a document that sets forth the commercial and technical specifications required from a supplier of equipment and/or services. An operator extends an invitation to tender in order to elicit several detailed bids so that the most suitable supplier can be selected. The bid selected will not always be the lowest one, a possibility that is usually stated in the tender offer. Price, of course, is for most operators a major consideration, but many other factors, such as the type and quality of the offer, as well as support and service, may affect the final choice. Invitations to tender may be issued on an open basis or only to select suppliers. Government agencies often require open bidding. Open invitations may be published in major newspapers.

TECHNICAL PREPARATION

It is important that the person preparing the invitation to tender has a clear idea of the type and quantity of equipment and services required. The invitation is not the place to "feel out" the market to find out what is available. The person(s) writing the invitation to tender should do the required homework and, if necessary, "feel out" the market before the invitation to tender is written.

Responses from suppliers are usually detailed, extensive, and reflect the original tender offer. Most suppliers prefer a clearly defined invitation to tender that specifies exactly what is to be supplied and leaves no loose ends. An operator cannot expect a supplier to guess what is really needed. An operator who is not confident of the system defined in the invitation to tender can invite suppliers to offer alternative configurations in addition to the one specified. This gives the tenderer a chance to input new or additional ideas without being disadvantaged with respect to other suppliers and without running the risk of having a nonconforming tender. It should be realized, however, that the preparation of a tender response is quite time consuming and generally does not leave time for detailed analysis of alternatives.

It is to the operator's benefit to prepare a precise invitation to tender, as will become obvious when the bids are received. If the invitation to tender is vague and imprecise, then the responses will be likewise. Ultimately it is up to the operator, when assessing the bids, to draw conclusions about the best one. If an invitation to tender is not precise, then the responses are likely to be so different as to make comparing them difficult or impossible.

A clearly written invitation to tender invites a more serious response because it is obvious to the supplier that the operator is serious. Suppliers may decline to bid if they do not feel confident that they understand what the operator wants. Remember that preparing a cellular tender bid is an expensive, time-consuming process that is not undertaken lightly by a supplier.

An invitation to tender is usually divided into four parts, as seen in the sample, later in this chapter. This sample is for an imaginary company called *Supercell*, which intends to provide a system with a capacity of 6000 subscribers with 296 RF voice channels in a town called *Mobile Town*. This invitation to tender assumes that *Supercell* will provide and select sites, and that it will also provide the microwave links, towers, shelters, and power (including rectifiers and batteries).

The four parts of the invitation to tender are as follows:

- Terms of tender

  This part details the terms under which a bid will be accepted, the documentation required, and how the contract will be awarded. It also defines the terms used.
- General conditions

  This part defines in more general terms deliveries, timing, acceptance of deliveries, schedules, payment, default, warranties, and other conditions that apply to the bid once it is accepted.
- Technical specification

  This part gives technical specifications of the goods or services to be supplied. In the sample in this chapter, for example, this section clearly sets forth the number of subscribers (6000), bases (10), and voice channels (296).
- Appendices and specifications

  Any additional information or attachments are included here.

All the features required are detailed so that the supplier has a clear idea of what is required. This specificity helps both supplier and operator.

The following sample tender is included only as an example, and its application to a particular case may be limited. The first two parts, "Terms of Tender" and "General Conditions," should be evaluated by legal experts in the country of application to ensure that local legal requirements and liabilities are covered. The technical part must agree with the actual technical requirements of the operator.

SAMPLE TENDER OFFER

I. TERMS OF TENDER

1. INVITATION TO TENDER

Suppliers are invited to tender for the installation and provision of a Cellular Mobile Telephone System based on the AMPS system for *Supercell* to be installed in *Mobile Town*.

2. ALTERATIONS TO TENDER

No alterations to the tender documentation will be permitted but alternate proposals that offer enhancements may be considered.

3. CLOSING DATE

3.1.

Tenders submitted up to *12 A.M. October 10, 1990* will be considered valid.

Tenders should be sealed and sent by verifiable means. No responsibility or allowance for delays will be made by *Supercell* for any tender documents sent by mail.

The tender should be delivered to:

> *Supercell*
> *100 Hustler Street*
> *Mobile Town*
> *Phone* ....
> *FAX* ....
> *TELEX* ....

Enquiries should be directed attention:

> *Mr. Fast Bucks*

The original plus one copy should be sent.

3.2.

Late and incomplete tenders may not be considered.

3.3.

The tender document must be signed by a duly authorized person acting on behalf of the tenderer.

4. DOCUMENTATION

4.1.

The tenderer will answer each point listed in the tender and provide either the required information, or note each as either "fully complied with," "partially complied with," or "not complied with." In the last two instances, details should be given of the degree of non-compliance.

4.2.

All attachments and technical drawings will have drawing numbers that clearly identify them.

5. PRICING

Separately itemized prices are required for the following. Equipment prices are FOB.

- a. Switch equipment:
    - i.   At initial 6000 line capacity
    - ii.   For 50,000 lines
    - iii. For small (approx. 5000 line max.) switch(es)
    - iv.   For expansion on per subscriber basis and detailing any step functions
    - v.   Recommended peripherals

- b. Cell equipment for the initial sites:
    - i.   10 voice channels, omni

        30 voice channels, omni

        3 x 12 voice channel, 120-degree sectored sites
    - ii.   Expansion on a per channel basis
    - iii. Any step functions in expansion (for example, increases in switch capacity)
    - iv.   Site controller expansion

- c. Installation costs, including cabling
- d. Rectifiers, batteries, and power control panels (optional)
- e. Software packages and support
- f. Billing package(s)
- g. Management information packages
- h. Training courses, to take place on *Supercell Premises*
- i. Prices for all recommended spare parts needed for an 18-month period
- j. Any specialized test equipment
- k. Cost and arrangements for repair
- l. Travel and living allowances

Prices should be presented as an itemized price list for the system tendered and additional information (break points and expansions) presented separately.

6. LOWEST TENDER NOT NECESSARILY ACCEPTED

Supercell will not necessarily award the tender to the lowest bidder.

7. CURRENCY OF TENDER

The prices should all be FOB prices in US dollars.

8. COSTS OF PREPARATION OF TENDER

The costs of preparation of tender will be borne totally by the tenderer and will in no part be the responsibility of *Supercell*.

9. SUBCONTRACTORS

If the tenderer intends to use subcontractors for any part of the work, this should be clearly indicated.

10. VALIDITY OF OFFER

The offer will be valid for a period of four months after the closing date but may be extended by mutual agreement of both *Supercell* and the tenderer. During that period, prices offered are binding.

11. REFEREE

The tenderer should indicate any references who have purchased similar equipment and would be willing to confirm satisfactory operation.

12. INQUIRIES

All inquiries will be directed to the address indicated in clause 3.1.

13. RESPONSIBILITY FOR INFORMATION

It is the tenderer's responsibility to ensure adequate information for submission of the tender not withstanding any information contained herein or afterwards supplied by *Supercell*.

14. AWARD OF CONTRACT

Supercell will evaluate the offers during the period *October 10, 1990 to December 10, 1990*.

A short list of tenders will be produced on or about *December 15, 1990* and short-listed tenderers will be invited to final negotiations.

On or about *February 15, 1991* the successful tenderer (the Contractor) will receive a letter of intent.

Final negotiations will take place on or about *February 20, 1991* and the successful tenderer shall attend a final negotiation meeting. The representatives of the tenderer must have full negotiation powers over any part of the Contract.

Tenderers who do not receive notification after this date should assume they have not been successful.

15. DEFINITIONS

Supercell. Means *Supercell PTY LTD*, the purchaser.

Contract Price. Means agreed price payable to the Contractor by *Supercell* under the Contract for full and proper performance by the Contractor in the execution of the works and provisioning as agreed.

Contractor. Means the Contractor named in the Contract and includes any successors and permitted assignee(s).

Contract Amendment. Means the document that changes prices, date of completion, and supplied items pursuant to clause 13.

Tender. Means the offer submitted by the Contractor to *Supercell*.

Tenderer. Means one of the invited bidders. The successful tenderer becomes the Contractor.

Contract. Means the agreement concluded between *Supercell* and the Contractor named herein incorporating these conditions and includes:

a.   Any Contract amendment
b.   All agreed specifications, plans, drawings, and other documents that are prepared pursuant to the said agreement

II. GENERAL CONDITIONS

1. ACCEPTANCE

1.1.

Acceptance will be deemed to have occurred 90 days from the issuance of a duly signed acceptance certificate.

1.2.

Each base station and the switch will be separately accepted.

2. REJECTIONS

2.1.

Supercell reserves the right to reject goods that are damaged, spoiled, or do not meet specification.

2.2.

Upon written advice of rejection (by TELEX, or FAX), the supplier will undertake to replace the rejected goods. The replacement is to be completed within 45 days. The Contractor will advise by return TELEX or FAX the expected delivery dates.

2.3.

The Contractor will be liable for all shipping costs of rejected goods.

2.4.

In the instance where the replacements are not made good as required by 2.1., 2.2., or 2.3. then *Supercell* reserves the right to:

a.   Terminate the Contract and claim liquidated damages.
b.   Claim liquidated damages but replace the rejected goods with other suitable goods.
c.   Claim liquidated damages and terminate the Contract where the same item has been rejected and so has its subsequent replacement.

3. TIME OF ESSENCE

3.1.

Upon receipt of a written order from *Supercell*, the Contractor shall supply and deliver in accordance with the delivery schedule.

3.2.

Where delivery dates are doubtful, the Contractor shall inform *Supercell* at the earliest time of any foreseen delays.

4. DELIVERY SCHEDULE

4.1.

Switch—160 days from receipt of order.

4.2.

Base Stations—120 days from receipt of order.

4.3.

Other items—120 days from receipt of order.

5. PACKING AND DOCUMENTATION

All items delivered shall be suitably packed for transportation and will be accompanied by a packing list detailing each part number and item delivered.

6. SUBCONTRACTORS

The Contractor will not, without written consent from *Supercell*, transfer any part of the Contract to subcontractors. All subcontractors will be wholly the responsibility of the Contractor.

7. LIABILITY FOR PERSONNEL

7.1.

The Contractor shall indemnify *Supercell* from any liability for injury to its staff or any staff employed or subcontracted by the Contractor while performing work on *Supercell's* sites or on behalf of *Supercell*.

7.2.

Evidence shall be produced that the Contractor has suitable insurance to cover the liability mentioned in 7.1.

8. ACCEPTANCE OF DELIVERIES

8.1.

Supercell will provide a note of acceptance on all deliveries made to it.

8.2.

Goods not rejected within 40 days of unpacking by *Supercell* will be deemed accepted.

9. TERMS OF PAYMENT OF EQUIPMENT

9.1.

Supercell will pay 80 percent of the costs of shipped goods within 10 days of receipt of all of the following:

    a.   The goods
    b.   Delivery documents detailing items delivered, order number, and price
    c.   Shipping documents, including certification of contents by a suitable authority (to be agreed to between the Contractor and *Supercell*)

9.2.

A further 10 percent of the total price will be paid after acceptance of the goods by *Supercell* (as per clause 8).

9.3.

The final 10 percent, being the balance, will be paid within 30 days of the *system* acceptance.

10. PAYMENT FOR SERVICES

10.1.

Upon completion of each service, the Contractor will invoice *Supercell*.

10.2.

Supercell will, upon receipt of the invoice, confirm that the service has been satisfactorily performed and will pay 80 percent of the costs within 30 days.

10.3.

A further 10 percent will be paid within 30 days of the issuance of an acceptance certificate.

10.4.

The final 10 percent will be paid after acceptance as defined in clauses 1.1. and 1.2.

11. INSTALLATION SCHEDULE

11.1.

All base stations will be installed and ready for acceptance within a period of 120 days from delivery.

11.2.

The switch and peripherals will be installed within 120 days from delivery.

12. PERSONNEL QUALIFICATIONS

The tenderer will be required to submit the curriculum vitae of all supervisory personnel to *Supercell* for approval. Only approved supervisors will be employed.

13. DEFAULT

13.1.

If the Contractor fails to deliver any items within the specified periods (clause 4) or fails to obtain an acceptance note (clause 8), then *Supercell* may:

   a. Contact the supplier and require a satisfactory (to *Supercell*) explanation and assurance of delivery within 14 days, or
   b. Terminate the Contract forthwith, without penalty to *Supercell* by notification in writing to the Contractor. *Supercell* reserves all rights to claims of breach of contract and may obtain the outstanding items from alternate suppliers.
   c. Make the Contractor liable to any costs incurred in rectifying the shortcomings.

13.2. Penalty

Supercell may decide to take action on Contract termination due to late deliveries or failure to obtain an acceptance. In this instance, a penalty of 3 percent of the total Contract price per full 30 days of delay may be applied.

14. ALTERATION OF CONTRACT

Supercell reserves the right to alter the Contract at any time. In the event of an alteration, a mutually agreeable amendment to price, and where relevant, delivery times, will be made by *Supercell* and the Contractor. Any such amendment will be agreed to and confirmed in writing by both *Supercell* and the Contractor before being considered valid.

15. BANKRUPTCY

In the event of bankruptcy by the Contractor or such other events as detailed below, *Supercell* may summarily terminate the Contract.

   a. If the Contractor shall at any time be adjudged bankrupt, or shall have received an order for receivership or for administration of Contractor's properties or shall take any proceedings for liquidation or compensation under any Bankruptcy Ordinance for the time being in force or make any conveyance or assignment of Contractor's effects or compensation or arrangement for the benefit of Contractor's creditors or purports to do so, or

b. If the Contractor, being a company shall pass a resolution or the Court shall make an order for the liquidation of its assets or a Receiver or Manager shall be appointed on behalf of each creditor or circumstances shall have arisen which entitle the Court or debenture holders to appoint a Receiver or Manager.

Provided always that such determination shall not prejudice or affect any right or action or remedy which shall have accrued or shall accrue thereafter to *Supercell*.

16. WARRANTY

16.1.

The Contractor will warrant *Supercell* for a period of 12 months, from the issuance of Acceptance as defined by clause 1.1., against all defects in workmanship and materials.

16.2.

The Contractor shall make good without undue delay all defects in workmanship and materials as are brought to Contractor's attention by *Supercell* within the warranty period.

16.3.

The warranty will cover all costs except freight and insurance charges for the shipping of items sent for repair.

16.4.

Clause 16.3. does not apply to any items delivered but not accepted as per clause 8. Such items will be returned at the Contractor's expense.

16.5.

The Contractor will undertake to make good all warranty claims with all possible speed and in any case within 60 days of notice in writing from *Supercell*.

16.6.

In the event of the failure to meet clause 16.5., *Supercell* may undertake to remedy any defects at the risk and cost to the Contractor.

16.7.

The Contractor will be liable to *Supercell* under the warranty terms, regardless of whether or not the goods as supplied were in part or whole manufactured by Contractor.

The supplier shall satisfy *Supercell* that any goods or services supplied by subcontractors or other suppliers carry the same warranty as required by *Supercell*.

16.8.

Software support for a period of 12 months will also be included in the warranty.

17. HARDWARE OR SOFTWARE UPDATES

The tenderer will describe the procedure for the release of hardware and software upgrades and their implementation.

18. COPYRIGHT INDEMNITY

The Contractor shall indemnify *Supercell* against any infringement of copyright, patents, trademarks, or registered design rights related to the equipment, system, or design supplied in relation to this Contract and shall affirm this in writing.

The Contractor will be liable for all costs of such infringement including negotiation and litigation costs.

19. SUPPORT

The Contractor will guarantee support for any hardware and software provided for a period of eight years from acceptance.

20. ARBITRATION

20.1.

In the event of any dispute or controversy in a matter of interpretation, implementation, and enforcement of this Contract, the same shall be resolved through arbitration by the Chamber of Commerce of the City of *Mobile Town* and the decision will be final.

20.2.

In the event that both parties shall, in writing, waive the right to avail themselves of the arbitration clause, then any controversy or dispute shall be referred to the Court of the City of *Mobile Town*.

21. SETTLEMENT OF DISPUTES

The laws of the land applicable to the City of *Mobile Town* shall apply in the interpretation, implementation, and enforcement of this Contract. In the event of any court action that any party may bring against the other under this Contract, the venue of such court action shall be in the Court of the City of *Mobile Town*.

22. CONFIDENTIALITY

During the term of this Contract and for a period of four years after its fulfillment, neither party shall intentionally disclose or permit to be disclosed to any third organization any information of a confidential or propriety nature concerning the other party.

23. PERFORMANCE BOND

The Contractor shall issue, within 60 days of acceptance of the Contract, an irrevocable Performance Bond (Letter of Credit) acceptable in the City of *Mobile Town* to the value of 10 percent of the total Contract value and to expire 12 months after the date of Acceptance.

24. GOVERNMENT REGULATIONS

24.1.

The Contractor will be responsible to ensure that all relevant government and local authority requirements and regulations are complied with in full.

24.2.

The Contractor will indemnify *Supercell* against any penalty or loss arising from the Contractor's non-compliance with any such regulations, laws, or requirements.

25. ORDER OF PRECEDENCE

The order of precedence in the Contract is:

    a. Part I, Terms of Tender
    b. Part II, General Conditions
    c. Part III, Technical Specification
    d. Appendices and Specifications

III. TECHNICAL SPECIFICATION

1. GENERAL

Supercell is seeking an AMPS system to be installed in the City of *Mobile Town* which conforms to the following US specifications:

    a. FCC - TITLE 47 CFR PARTS 2, 22
    b. EIA - IS-3-D, IS-19-B, IS-20-A

The system will operate on the AMPS B band and be capable of operation on the extended B band.

Provision will be made for an initial installation of 6000 subscribers with 10 base stations and one switch.

Expansion to 50,000 lines is anticipated. Future rural expansion will require the provision of additional switches with capacities of about 5000 subscribers. Fully automatic roaming is to be provided between the switch areas.

A 7-cell frequency pattern will be employed (that is, N=7).

Supercell will provide only the following:

    a. Site selection, clearing, base station, switch shelters, towers, mains power, rectifiers, and batteries

b. Microwave 2 Mbits (or 2/8), 32-channel links, with channel equipment between bases and the switch

c. A 34-Mbit link from the switch to the PSTN

2. THE SWITCH

2.1. General

a. A stored program, full-availability (non-blocking) switch will be provided. Automatic roaming and interswitch handoff will be provided.

b. The switch will be designed to interconnect at trunk level to the PSTN.

c. Modularity of construction will be a switch feature.

d. Digital voice announcement capability of at least ten 30-second announcements will be provided.

e. Dual processors with complete redundancy will be provided. Automatic changeover in the event of processor error will occur and a detailed record of the changeover and its causes will be available.

Provision will be made to take one processor off-line at any time for service or other purposes.

f. A voice message facility with DTMF priority control will be provided. At least five categories of message will be available together with five retrieval categories.

g. Both a positive and negative validation file will be provided.

h. A printer will be provided which will:

i. Print out all alarm details.

ii. Print out all information input to the switch.

iii. Print out details of any system reconfigurations whether automatic or manually instigated.

i. All information directed to the printer will also be recorded on magnetic tape. This tape will also record details of all calls and call attempts with full call accounting (that is, A & B subscriber number, call duration, and details of unsuccessful attempts). The call duration and airtime shall be separately recorded.

j. A full billing package is to be provided and samples of billing package outputs will be submitted.

k. A management information system (MIS) will be required. Full details of compatible management information systems available will be submitted.

l. A means of loading programs and patches and adjusting any system parameters will be provided. A detailed description of the associated hardware and software is to be provided.

2.2. Facilities

The following minimum facilities are required:

2.2.1.

Real-time billing will be provided and facility to access a subscriber's billing details either at the switch site or at a remote location will be provided. Full account details of any subscriber will be available on request.

2.2.2.

Roaming subscribers will be able to be marked for particular attention and in particular all roaming subscribers whose calling rate exceeds a predetermined level will be identified. This facility may be part of a real-time billing system.

2.2.3.

Facilities to read and alter subscriber details including validation, block validation, and negative file at both the switch and at least one remote location will be provided.

2.2.4.

Call tracing will be provided, and details of any current call including base station, channel number, circuits used, and called or calling party, will be provided.

The tenderer should state if historical records are available of the above details and the period for which those details are available. It is envisioned that such details could be useful for fault tracing and malicious call traces.

2.2.5.

A priority class subscriber category will be provided.

2.2.6.

Call interception will be available.

2.2.7.

Voice encryption will be available for some subscribers. The tenderer should give details of voice encryption operation and the proposed means of activation/deactivation.

This feature will be available to subscribers at extra cost. The tenderer should state the degradation in signal-to-noise (S/N) that will result from the use of the proposed encryption and the degree of security provided.

2.2.8.

A facility to automatically access the cellular alarm and diagnostic system will be provided. It is envisioned that subscribers would dial an access number which will connect them to a suitable modem and then to a 300/1200 baud modem for land-line operation.

2.2.9.

The following subscriber facilities will be offered as optional features to *Supercell's* customers.

    a. Call Waiting

        Some indication that another call is waiting will be provided.

    b. Call Waiting Cancel

        The subscriber can cancel the call-waiting feature (a.) from the mobile phone.

    c. Follow-Me

        The subscriber can redirect incoming calls to any other number. This will have two modes, one permanent and the other which automatically cancels at midnight on the day of activation.

    d. Three-Way Conference Call

        A subscriber can call a third party while holding an existing call. All parties can then interconnect in the conference mode.

    e. Call Hold

        An existing call can be placed on hold.

    f. Voice Mail Transfer

        The customer can transfer calls to the voice mail system.

    g. Call Charge Information

        A means of providing real-time billing information to the customer will be available. This will preferably be fully automatic.

    h. No Answer Transfer

        Calls are automatically diverted to another number if the mobile phone is not answered after a predetermined period.

    i. Call-Received

        A means of alerting the subscriber that a call was received while the mobile was unattended.

2.2.10.

Both automatic periodic switching-in of redundant equipment and manual changeover will be available.

2.2.11.

Alarms shall be categorized and a facility to output major alarms by audio-visual means will be provided.

2.2.12.

A facility to monitor the activity of any trunk or link will be provided which will include:

    a. Audio trace
    b. Call details
    c. Loop test
    d. Frequency-response test

2.2.13.

Full details of diagnostic programs available will be supplied.

2.2.14.

The system will handle local numbers of a minimum of 15 digits and international numbers of 24 digits

2.3. Details to be provided

2.3.1.

A floor plan of a typical fully equipped switch room with all required peripheral equipment will be submitted. The plan will show the initial 6000 line installation and the subsequent expansion to 50,000 lines.

2.3.2.

Power consumption, voltage, and heat dissipation for both the initial 6000 and ultimate 50,000 line switch will be stated.

2.3.3.

Maximum floor loading and physical dimensions for each type of rack will be stated.

2.3.4.

Operational temperature range for each type of equipment will be stated.

2.3.5.

The number of alarms that can be handled from each base station and from other peripheral equipment is to be stated.

2.3.6.

The degree to which the system can be reconfigured as a result of a fault will be detailed.

2.3.7.

Details of the MTBF of the switch and of each of its component parts (such as boards and racks) will be provided.

2.4. Management Information Systems (MIS)

The MIS should be integral with the switch and will collect data for proper management of the system.

2.4.1.

The minimum requirement for the MIS is that it will provide the following data:

a.   Traffic information on all circuits incoming and outgoing to the switch
b.   Circuit availability
c.   Data error rates on links to base stations
d.   Base station transceiver performance (including availability)
e.   Channel blocking both manual and automatic
f.   Control channels usage and outages
g.   Call attempts and call success rate
h.   Call holding times
i.   Total traffic and total traffic per cell as a function of time of day

Details of other MIS data available should be provided.

2.5. Hardware

2.5.1.

The switch will be of modular design and be capable of upgrades without causing any system loss. In particular, upgrades of the following will be available without interruption unless the maximum capacity is provided at start up:

a.   Processor capacity
b.   Number of input/output ports and peripherals
c.   Memory capacity
d.   Switch capacity

The tenderer shall state the start-up and ultimate capacity of the above items.

2.5.2.

The tenderer will be responsible for the compatibility of any peripheral equipment offered and for the commercial availability of spare parts and consumables for a period of five years from delivery.

2.5.3.

The central processor shall be capable of performing simultaneously all housekeeping and switching functions in both the installed and ultimate capacity (50,000) without overload resulting in lost calls.

2.6. Software

2.6.1.

The software shall be fully documented and training provided for *Supercell* engineers to enable them to make any necessary patches. An agreement between the Contractor and *Supercell* to allow approved patches will be entered into.

2.6.2.

The software should be of modular design and capable of partial updates without system downtime.

2.6.3.

Support and updates of software for 12 months will be provided as part of the warranty.

2.6.4.

All software updates shall be backward-compatible, tested, approved, and well documented.

3. BASE STATIONS

3.1. General

The base-station equipment to be provided consists of a site controller, transceiver racks (each of which contains at least eight channels), rectifiers, batteries, and associated equipment.

Supercell will provide all shelters, which will be transportable buildings with a floor plan of 3.1 x 8 meters and floor to ceiling height of 2.9 meters. The buildings will have a maximum floor loading of 1000 kg/m and be provided with single phase 240, VAC, 50 Hz power. *Supercell* will also provide towers, antennas, and feeders.

3.2. Facilities and Equipment

3.2.1.

The base stations shall be provided with back-up batteries sufficient for three hour's operation when fully equipped (45 channels) including air-conditioning load. Automatic and manual changeover to batteries will be provided. The tenderer will specify the required battery capacity for the equipment.

3.2.2.

Complete equipment monitoring facilities will be available at the site controller. The tenderer will state how manual access is gained to the controller and what additional hardware is needed.

3.2.3.

The ability to quickly replace faulty channels will be required. The tenderer should state the method used to tune a replacement transceiver.

3.2.4.

All critical components will be monitored by alarms that are accessible at the base station and remotely at the switch. At least four auxiliary alarms, such as entry and power, will be required.

3.2.5.

The control channel will use a voice channel as a standby control channel in the event of control channel failure. The changeover will be able to be effected both automatically and manually.

3.2.6.

Continuous field-strength monitoring will be provided. It is preferred that a standard channel unit (that is, transceiver or receiver) be used for field-strength monitoring so that special equipment is not required. It is preferred that a voice-channel receiver can be used as a standby signal strength receiver. The supplier will state details of signal strength measurement.

3.2.7.

Diagnostics of local equipment faults will be carried out by the site controller and automatically sent to the switch.

3.3. Information Required

The tenderer will provide the following information:

3.3.1.

a. Power consumption (for each base-station configuration), voltage, and air-conditioning load of the base station
b. Floor loading, maximum weight per rack
c. Rack dimensions and empty weight
d. Environmental operational constraints
e. Degree of redundancy
f. Typical floor plan of a fully equipped base
g. MTBF of a base station and each of its modular parts and details of the method of calculation

3.3.2.

The degree of remote control of base station parameters will be clearly stated. Also, whether all base station alarms and diagnostics are forwarded to the switch.

4. TRAFFIC AND PLANNING CONSIDERATIONS

4.1. General

4.1.1.

The traffic capacity will be based on an average subscriber having a calling rate of 30 mE and a call holding time of 130 seconds. The grades of service will be 0.05 on the RF path and 0.002 on the switch to PSTN path.

4.1.2.

Based on 6000 subscribers at 30 mE, the total PSTN traffic of 180 E will be based on 212 circuits to the PSTN and 296 RF channels.

4.1.3.

The base stations will be provided with the following circuit capacities:

2 x omni sites with 10 channels each	20
2 x omni sites with 30 channels each	60
6 x sectored sites with 12 channels per sector	216
Total voice channels	296
Total control channels	20

4.1.4.

The links between the base stations and the switch will be via 2-Mbits digital radio links (30 voice and 2 control channels).

5. SPARE PARTS

The tenderer will provide a list of recommended spare parts for a period of 18 months after commissioning. Special attention should be given to consumable parts (fuses and lights), which will be provided at least at 100 percent of provisioning.

6. TEST EQUIPMENT

The tenderer will list in detail any specialized test equipment needed to maintain the bases or switches. A separate list of other recommended test equipment will be provided.

7. REPAIRS

A repair contract will be sought for faulty boards and components. The tenderer will give details of repair facilities, including in particular:

    a. Minimum number of years for which repair by the supplier can be guaranteed
    b. Cost basis of repairs
    c. Expected turn-around time on a board swap basis

8. TRAINING

8.1.

Training will be required for *Supercell* engineers in the following aspects:

    a. System overview (1 week)
    b. Switch operation, maintenance, commissioning, and software (approximately 8 weeks)

   c. Base-station installation, operation, commissioning, and maintenance (approximately 3 weeks)

   d. Billing and MIS (1 week)

8.2.

The tenderer will give details of proposed courses, costs, and maximum number of attendees per course.

8.3.

All course material and presentation will be in English.

9. COMMISSIONING

9.1.

Supercell engineers will participate in the commissioning tests that will form part of the Acceptance test.

9.2.

Full commissioning test sheets will be provided to *Supercell* not less than six weeks before commissioning begins.

9.3.

Supercell may require additional or modified commissioning procedures.

10. INSTALLATION

10.1.

The Contractor will be fully responsible for installation but will undertake to give on-site training to three *Supercell* staff in installation practices. The *Supercell* staff will be assigned to the tenderer for the duration of installation at no cost to the tenderer. All staff so assigned will have previously done the installation and commissioning course offered by the tenderer under clause 9.

10.2.

PERT and bar charts will be provided that detail:

   a. Activities from the award of contract

   b. Time interval for the various activities up to and including acceptance testing

   c. Delivery schedules

10.3.

The tenderer will state the number of staff hours for each activity phase and the installation team size for each activity.

10.4.

The tenderer shall submit monthly progress reports with details of any slip in timetable.

11. ACCEPTANCE

11.1.

Acceptance testing will be done by *Supercell* engineers. Conditional acceptance may be given where minor shortcomings are detected. When conditional acceptance is given, it will be subject to the rectification of the outstanding problems within 30 days (or as agreed).

11.2.

The tenderer will submit a copy of normal acceptance procedures but should be advised that *Supercell's* own test sheets (copies of which are available on request) will be used for Acceptance.

11.3.

Acceptance will be deemed completed either after the issuance of an Acceptance Certificate or after the rectification of any outstanding, specified problems on a conditional test sheet.

IV. APPENDICES AND SPECIFICATIONS

Attachments which *Supercell* should include with the offer to tender are:

1. Maps of the area to be covered with proposed coverage area
2. Signaling format (in detail) to PSTN
3. Coverage proposed from each site*
4. Coordinates of each base-station site*
5. Site plans*
6. Shelter drawings*
7. Power provisioning (in volt amps)*
8. Tower heights and AMSL height*
9. Tower plans and feeder size*
10. Switch room location and plans*

Note: The items marked (*) are not required for a turnkey system.

Chapter 33

DIGITAL CELLULAR

Evidence to date suggests some compelling advantages of digital cellular systems over their analog counterparts. The enormous research cost to develop a digital system does, however, probably preclude the early release of a cheap consumer digital mobile system for some time.

Although a number of digital cellular radio systems are being developed worldwide, the only ones that currently have been well-defined are Pan-European Groupe Special Mobile (GSM) and the US digital. Conceptually, these systems have many similarities, but they are very different in design philosophy. The US system has been designed to complement and integrate with the analog AMPS network; GSM, on the other hand, is meant to overlay and replace the analog networks so no attempt has been made to work with the existing analog structure.

As a result, the introduction of US digital will be more graceful and much less costly than GSM. Countries adopting GSM have ambitious plans to implement large systems quickly throughout the country. The cost of doing so is such that only government or large multicompany conglomerates can afford the infrastructure costs involved. The real driving force behind GSM was the European Economic Community (EEC), which believed that reaching agreement on a compatible pan-European cellular system was politically imperative.

In the analog environment, the diverse number of incompatible systems in Europe meant that achieving even a reasonable degree of inter-country roaming was not practical. A potential roamer was compelled to purchase a number of different phones. In practice, however, it usually meant that the cellular subscriber just had to come to terms with the fact that the mobile phone was usable only in the subscriber's home country.

The Scandinavian countries proved the value of roaming with their successful analog NMT networks, the first analog networks to incorporate wide-scale roaming; this feature was one of the first specified for GSM. Indeed, GSM is a system specified in such fine detail that many manufacturers feel that they are limited in the extent to which they can provide special and additional features. This limitation is not obvious in the US digital specification, so it is likely that US digital will be more "interesting" than GSM to system engineers.

The diversity of incompatible systems seen in the world of analog cellular radio is not likely to be lessened with the introduction of digital; the same commercial pressures that led to analog diversity are present in the digital arena, possibly even at a higher level because the potential commercial rewards are now more evident than they were in the early analog days. Digital cellular systems are also under development in Japan and South Korea. The Koreans hope to market their system in competition with GSM/US digital worldwide.

The recently announced "personal networks" under development in the UK are really just second-generation digital systems. These networks, however, are not strictly cellular, and some proposals are not even digital. In the UK, it is planned to operate GSM and the new digital mobile systems in a competitive environment. The lack of roaming capability due to an absence of uniform standards may be the biggest limitation of these second-generation systems.

In complexity (which can be crudely measured by the necessary memory capacity of the mobiles), the first-generation digital systems are about four times more complex than the current analog mobiles and require about 250 K of memory. As memory becomes cheaper, however, it can be used more freely, so the cost factor probably weighs more heavily on the memory capacity installed than does complexity.

Naturally, new digital systems will require additional spectrum, of which none has been allocated thus far. Much work remains to be done in this area if there is to be any hope of standard worldwide frequency allocations. This task has been traditionally undertaken by the ITU, but it appears that this group cannot move fast enough to satisfy the needs of the cellular radio industry. In fact, even the current analog systems are operating in bands nominated by the CCIR for other purposes.

Initially, the first-generation digital systems will operate in the same bands as the current analog systems and so will be competing for spectrum. This means that in many cities the economic life of the analog mobiles will be somewhat limited. The problems of how to recover the needed spectrum and who will pay the costs remains unresolved.

Digital cellular mobiles are widely forecast to wholesale at around $700, which is about the same price for which analog mobiles wholesaled in 1988. Thus, they will probably arrive on the market at a price around double that of their analog counterparts.

The network costs on a per-mobile basis are predicted to be less than the current price of approximately $1500. This prediction may well be realistic as many administrations are planning to install systems with initial capacities of hundreds of thousands or, in some cases, even millions, of lines. These quantities are large enough to ensure good prices.

In order to achieve implementation within a reasonable amount of time, and to contain development costs, second-generation digital systems will necessarily be adaptations of their first-generation counterparts. As recently as 1980, the problems of operating a mobile digital radio were beyond the scope of affordable technology. The main problem was synchronizing the data stream as the mobile moved through it in a multipath environment. Digital systems read the incoming data in blocks called timeslots. The decoder must recognize the synchronization bits and then extract the information contained within the timeslots. Figure 33.1 shows that in the multipath environment, bursts of data from different paths travel different distances and therefore have different propagation delays. Accommodating this requirement was easily accomplished on

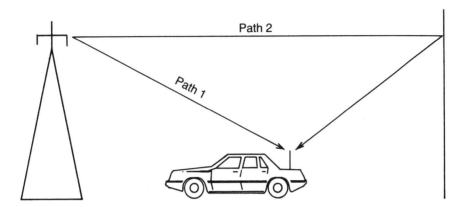

Figure 33.1   The two radio paths are different in length and so will have different time delays.

fixed services by introducing the appropriate delays. The multipath environment, however, causes big differences in the propagation delays experienced by signals received over the different paths. These time delays must be equalized in real time for each data frame. How to do so was not immediately obvious in the mobile environment.

The first thing that must be done in the mobile environment is to make the timeslots short because large differences in propagation delays can occur over very small distances (and so can occur quickly). Next, the propagation delay must be accounted for in real time. This problem has now been solved with the introduction of adaptive equalizers that effectively continuously adjust the equalization as required.

The solution to this problem brought with it an unexpected bonus. Not only was the adaptive equalizer able to set the equalization for each burst of code, it also could do so in a multipath environment. In fact, it performed better in a multipath environment than in a single-path environment because for each burst of code, it synchronized with the best of the multipath alternatives. This produced bit error rates of an order of magnitude better than could be achieved without use of the multipath as a diversity mode.

The first-generation digital systems have been designed to perform at least as well in RF as the analog counterparts, and most systems are expected to perform even better. The US digital system, which has been specified to co-exist with analog, requires a similar RF path performance. Second-generation systems will be deliberately designed to enhance small-cell operation and so will not be so constrained by the need to comply with analog compatibility.

DIGITAL CAVEAT EMPTOR

Digital is a "buzz word" today, and those who use it as such seem to have little or no sense of history. Digital is not new (it's been around now for half a century). Despite this fact, there are always legions who cite digital as the only way to the future. They are probably correct in this, but it is the rate of progress that needs to be viewed with some caution.

Integrated Switched Digital Network (ISDN) is much touted as the ultimate solution for the telecommunications industry (if only someone can find the problem).

I remember seeing my first ISDN switch in Sydney in 1975. It was a most impressive sight, with rows of flashing lights making it look like the control deck of the Starship *Enterprise*. The people running this machine spoke in hushed reverence about the capabilities of this "ultimate universal switch" and how it would revolutionize data communications. This miracle switch had only two problems: first, it didn't really work; second, no one seemed to be able to identify any real customers who could use it even if it did work!

Cellular digital, almost 20 years later, features ISDN compatibility. But ISDN is also still largely a buzz word, and many years will pass before there is wide consumer acceptance of switched data networks.

The moral here is that just because a system is digital does not mean it is necessarily better; and it is worthwhile to note that the two first-generation digital systems have been specified to be "at least as good as analog." But because these digital systems are due to reach the market almost a decade after the analog systems, this hardly represents a great leap forward. Initially, digital mobiles can be expected to cost more, will be much more difficult to maintain, and will be bigger than their analog counterparts. So don't expect too much from digital cellular, especially from the first-generation equipment.

NO DUPLEX FILTER

Digital systems promise eventually to remove much of the bulk from cellular mobiles because an RF duplex filter will no longer be required. In analog systems, to have duplex speech paths, it is necessary to insert a duplex filter to enable the simultaneous operation of the transmitter and receiver. Digital systems can operate sequentially (that is, transmission and reception in different timeslots), so the filter is no longer needed. Recent advances in duplexer design (and size), however, have minimized this advantage, particularly as smaller, simpler, and cheaper duplexers are becoming available. But even small, low-cost filters take a heavy toll on battery talk time because of their high power loss.

BATTERY TALK TIME

One of the greatest shortcomings of analog systems is that, despite improvements in battery technology, the handheld talk-times are limited; most, even moderately heavy users, find it is always necessary to carry a spare battery. Digital systems can significantly improve the battery talk and standby times.

The removal of the duplexer is a major factor, as the insertion of a duplex filter typically results in a 3-dB loss. The removal of this component alone cuts the power requirements in half during transmission without reducing the ERP.

Because digital paging transmissions can be made synchronous, it is not necessary for the mobile to "listen" continuously; the receiver can "sleep" during periods when data transmission is not relevant. Using this technique, it is possible for a mobile on standby to "awaken" for only a small amount of the standby time, thereby reducing the standby current drain to a small fraction of the consumption of its analog counterpart, which must "listen" continuously.

Digital systems are designed for small cell operations, so low mobile ERPs are not only possible but are necessary to reduce interference problems. Taken together, removal of the duplexer and use of low mobile ERPs can be expected to improve the effective battery discharge cycle by a factor of at least five.

GSM PAN-EUROPEAN CELLULAR

The GSM specifications are comprehensive and voluminous. Manufacturer compatibility has been assured by precisely specifying the system at a number of interface levels. Perhaps this preciseness has something to do with GSM's having originated with a 1981 French-German study and later evolving into a Pan-European standard. The specification has been almost 10 years in the drafting and is still not complete.

THE RADIO FREQUENCY INTERFACE

The RF interface, also known as the Um interface, has been designed to use the same spectrum currently allocated in Europe for the existing analog systems. The RF power levels are similar to the existing systems but spread spectrum techniques are used so that each carrier is 200-KHz wide. The RF interface has been specified because of the need for high capacities and fast, accurate handoffs. For this reason, both the mobile and the base station are involved in handoffs.

Some RF parameters are as follows:

- Frequency range: 935.2–959.8-MHz base transmit; 890–915-MHz base receiver.
- Channel spacing: 200 KHz for 8 channels or 16 half channels.
- RF power: 32-watts base station/carrier; 13-watts mobile transmit.
- Sensitivity for 1 percent BER: 103 dB (approximately).
- Power steps: 15 of 2 dB each.
- Carrier to interference target: 10–13 dB.
- A total of 124,200 KHz radio channels are available.
- Operates at C/N of 10–12 dB.
- 1000 full-rate (16 Kbit/second) or 2000 half-rate channels.
- Traffic capacity: 40 Erlangs/km^2.
- Because of high-speed handoff and good interference immunity, microcells are possible using very low power, and low-elevation transmitters which could achieve densities of 200 Erlangs/km^2.
- A 3/9 cell or 4/12 (125 channels) cell structure is planned.
- The GSM system adds 10 MHz to the existing 900-MHz systems (TACS and NMT900) to allow for the gradual implementation of GSM on top of the existing systems.

The RF interface uses slow frequency hopping (217 hops/second). The resulting frequency diversity results in a reduced bit-error rate at a given incident S/N and improved immunity to interference. The interface also provides for full-rate channels, designated Bm, and half-rate channels, designated Lm. The half-rate channels are expected to use codecs (digital speech encoders), which would enable them to be used with almost no degradation in speech quality.

The propagation delay equalization can allow for up to a 233 microsecond absolute time delay. This is necessary to ensure data synchronization. The system can also cope with a 16 microsecond dispersion.

It is emerging that a C/I (carrier-to-interference ratio) of 12 to 14 dB is more realistic than the 10 to 13 dB originally specified. The degradation of C/I immunity is very marked as the threshold is approached, and GSM is quite sensitive to C/I and certainly less robust than analog systems. The high densities possible with GSM, however, will make it practical eventually to replace analog systems completely. This transition period will be a most difficult one.

Co-channel interference will still be a problem with GSM, and it will be necessary to ensure that channels at any one site are separated in frequency by one channel width (200 KHz).

SUPPORT FOR GSM

Europe is the main platform for GSM, and intergovernmental agreements are now in place to ensure its success. The countries that to date have made a commitment to GSM are Austria, Belgium, Denmark, France, Finland, Germany (West and East), Greece, Iceland, Ireland, Italy, Luxembourg, Netherlands, Norway, Portugal, Spain, Sweden, Switzerland, and the UK. The commitment includes meeting the schedule for the early introduction of coverage of all major cities and major highways. As of late 1989, five countries (Finland, France, Germany, Sweden, and the UK) have confirmed their intention to have a second operator.

Outside of Europe, some markets may be interested in GSM. Of these, the most likely would be countries physically adjacent to Europe, such as the Eastern-Bloc nations, and those that see Europe as a natural roaming base. Some non-European countries, such as Australia, are known to be actively evaluating GSM, but most are wisely taking a wait-and-see approach to this rather daunting new technology. The reasoning behind this caution is largely to let Europe "pay" for debugging this system and to allow time to properly evaluate the alternatives once they are working commercial devices. GSM is scheduled to be in operation in 1991, with nation-wide roaming and coverage in 1993, and handheld portables by 1993.

MODULATION

Modulation rates of 16 Kbit/second have been successfully employed to produce a subjective voice quality equal to that of a 64-Kbit PCM system. To obtain low bit-error rates (and hence, good-quality signals) in the hostile mobile radio environment, a good deal of encoding redundancy is required. Typically 40 percent of the bits are redundant code (error correcting) so that modulation rates (including redundancy) of about 27 Kbit/second will be required. Acceptable voice quality (similar to analog cellular) can be obtained at 9 Kbit/second data rates.

The speech codec mode is called Regular Pulse Excitation with Long Term Prediction (RPE-LTP). The speech is interfaced to the encoder via a 13 bit A/D converter at a data rate of 8000 bits/second. The encoder then produces the output at 13 Kbits/second in 20 millisecond bursts containing 260 data bits each.

A battery-saving feature of GSM is that it uses Discontinuous Transmission (DT), in which a voice-detection device is used so that the transmission can be turned off when speech is not present.

The periods of quiet that occur during the time that the transmitter is turned off were found to be disturbing to the mobile user, and so it was necessary to inject "comfort noise" in the quiet periods. Figure 33.2 illustrates this injection, as well as the main speech processing functions.

FREQUENCY-HOPPING

Frequency-hopping is a technique that improves the signal quality. Each code burst can be sent on a different frequency so that the benefits of frequency diversity can be used. It is possible, using redundant coding and repeated transmissions, to correct an occasional "bad burst." Because the error-correction algorithm can allow for some corrupted data, frequency-hopping provides immunity from local interference and propagation anomalies.

GSM TERMINOLOGY

An essential part of the GSM specification is the standardization of the terminology used to describe the component parts of the system. The multinational nature of GSM has made the use of an extensive uniform vocabulary a practical necessity. But mastering this terminology then

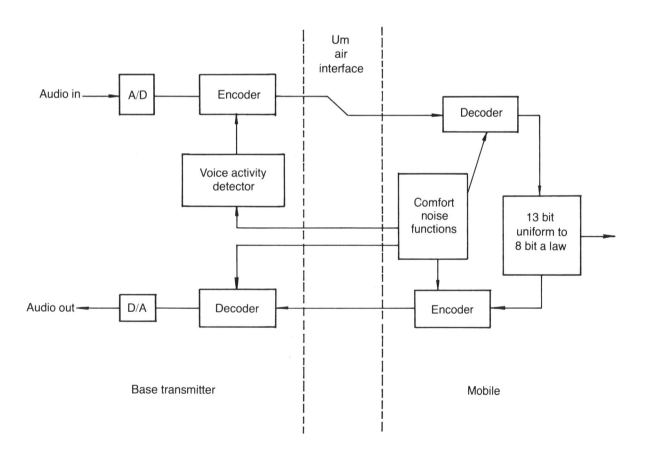

Figure 33.2 Speech processing includes the injection of "comfort noise." The speech path is first passed through an A/D converter and then encoded. At the encoder, "comfort noise" is added to give an impression of path continuity during the "blanking" that occurs when no voice activity is present.

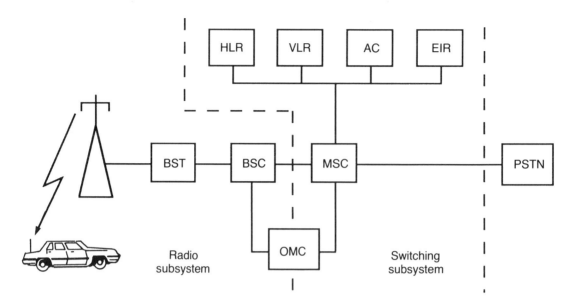

Figure 33.3   GSM system architecture

becomes essential in order to follow discussions about the system. Figure 33.3 illustrates the logical parts of the GSM system with the appropriate terms.

These are the functional parts of the GSM system:

- MSC (Mobile Switching Center). The mobile switching center controls switching and much of the system processing. It may jointly be a local or trunk switch. The other switching subsystem components are designed to be operated either as part of the switch or as physically separate entities. The MSC will probably have been designed originally as a PSTN switch; the basic functions are the same as those required in a trunk switch. The MSC controls a number of base-station controllers.
- HLR (Home Location Register). The home location register contains the permanent records of the subscriber and is the main user database. Full access is available via the operations-and-maintenance (OMC) center. Typically, this register has a capacity of 100,000 to 500,000 subscribers.
- VLR (Visitor Location Register). This is a temporary register accessed by all callers. It contains a temporary file on current users and accesses information from the respective HLR as required.
- AUC (Authentication Center). Here the "authentication" parameters are generated for subscribers access, and a cipher key is obtained for encryption of data communications between the system and the mobile.
- EIR (Equipment Identity Register). This register is a database of mobile stations and performs identification functions. It will identify unauthorized attempts to access the

network and trace and control the use of unauthorized equipment. It can be fully accessed by the operation and maintenance center (OMC).

- BSC (Base-Station Controller). The base-station controller is the interface between the MSC and the base stations. It may serve one or more cells and allocates radio link resources. It also controls handoffs, and it may internally control handoffs to its own sectors.
- BTS (Base Transceiver Station). This station contains all the RF hardware, including transmitters, receivers, and the channel-encoding functions.
- OMC (Operations and Maintenance Center). This is the overall control center and the operator's link to the system.

GSM will use CCITT no. 7 signaling between network components (that is, the Mobile Application Parts, MAP). Because large-capacity systems are planned, it is expected that GSM switches will have subscriber capacities well in excess of 100,000. Refer back to Figure 33.3 for details of the GSM interface specification.

BASE STATIONS

GSM base stations are state-of-the-art devices with military-like specifications on a commercial budget. One of the earliest problems encountered was that of how to provide frequency-hopping, which must be done simultaneously at the base station by each BTS transceiver for 8 channels. The net rate of change of frequency is 1700 times a second.

At RF frequencies, this type of performance can be achieved but not cheaply. The most favored solution today is to fix the BTS channels while switching the data streams. Although there are no commercial BSCs available today, those in development vary enormously in size and it is likely that some of these will be commercially unattractive. The size has a lot to do with the degree of chip customization. It is the specialized chip and software production that makes GSM implementation an extremely expensive exercise.

THE LOCATION REGISTRATION

GSM networks will consist of a number of geographical locations zones. Within these zones, the mobiles will listen out on the Broadcast Common Control Channel (BCCH) for the identity of its location. When changes are noted, the mobile will issue a Location Update Request (LUR) that will register the mobile in the new location. Consequently, the customer base in each location, although needing continuous updates, can be held at a controllable size. If the mobile roams into an area supported by a different VLR, then the VLR will issue a new Mobile Subscriber Roaming Number (MSRN). The HLR is then given the information about the mobile's new MSRN. Alternatively, the VLR can inform the HLR of the current MSC location and the MSRN can then be transacted between the MSC and the HLR. At certain time intervals, the mobile will be requested to provide Periodic Registration (PR) to update the file on its location and status.

A feature of GSM, and probably of all future digital systems, is that it is the mobile and not the base stations that reports on the field strength. Digital systems can use the idle time between active timeslots for other purposes. The mobile scans the adjacent channels during the idle time.

DIGITAL FLEXIBILITY

Voice channels can be derived from normal 64-Kbit channels so that each one can carry four separate voice channels. This transcoding means that digital links can be much more efficiently used. The channels have been specified to have an ISDN-like flexibility and allow for future data add-on services. The services that can be added are an alpha-numeric message service, electronic mail, and Group 3 FAX (with some possibility of group 4).

When used in the various data modes, the bit rate is automatically reconfigured to optimize the spectrum use. This flexibility extends to the polling rate and type of services that are available to the PSTN:

- Circuit switched, asynchronous channels from 300 to 9600 bits/second
- Circuit switched, synchronous channels from 300 to 9600 bits/second

This flexibility also extends to the Packet Switched Public Data Network (PSPDN):

- Packet synchronous channels with 2400 to 9600 channels
- Packet asynchronous channels with 300 to 9600 bits/second

It is envisioned that there will be a demand for high reliability data services, so both transparent and non-transparent modes have been specified. The non-transparent modes have a built-in sophisticated error-correcting code that gives enhanced protection. When nontransparent modes are specified, the maximum data rate becomes 4800 bits/second.

The GSM system is fully specified to BSC–MSC level so that the signaling between these subsystems will be standardized, thereby enabling connection between equipment from different suppliers. GSM has an overall propagation delay of about 90 msec. Because an echo is noticeable at 20–25 msec, echo-suppression techniques must be applied.

GSM MOBILE UNITS

Mobile units are under development by a number of companies and will probably be fairly large by the standards of state-of-the-art analog sets. Initially, only in-vehicle units will be produced; handhelds will be commercially available about 1993.

A typical mobile unit will likely have the following specification:

- GSM class 2 (8-watts peak power)
- Volume, 1 liter
- Weight, 900 grams

A typical handheld unit will likely have the following specification:

- GSM class 4 (2-watts peak power)
- Volume, 2.300cc
- Weight, 300 grams

Concern has been expressed that the schedule for the introduction of commercial type-approved GSM mobiles is very tight, and some manufacturers have suggested adopting an interim type approval. A problem arises due to the requirement of mutual recognition of type approvals across Europe. The procedures to ensure mutual recognition are not yet fully in place, and so an interim type approval would assume that mutual approval follows. This is just one example of the complex web of engineering and politics that are very much an integral part of GSM.

INTELLECTUAL PROPERTY RIGHTS

The Intellectual Property Rights (IPR) issue has become a major stumbling block that will not readily be solved. The GSM specification calls for the rights to intellectual property (essentially, but not limited to patents) to be readily available on a reasonable and non-discriminatory basis. Although the intent of this specification is clear, the path to implementation is not. The bulk of the property rights are held by Motorola and Philips.

US DIGITAL CELLULAR OPTIONS

The US digital cellular system is notable for its spectral efficiency, much of which is a consequence of its modern efficient digital voice encoders. The initial installation will enable three channels to be used in the same bandwidth as a current single channel. In the near future, it is expected that codecs able to encode with a spectral efficiency of 3 bits/second/Hertz will be used. Such encoders take intelligent advantage of the idle time and redundancy in normal speech, and are thus able to compress the spectrum needed to transmit voice intelligence.

It is proposed that the digital system will use the same band as the AMPS network, with no additional spectrum initially allocated. In some areas, where only 666 channels of the expanded 832-channel band have been used, there is a relatively easy path to digital by first using the spare channels. Where there are no spare frequencies, the initial change to digital means that some frequencies will need to be recovered. In small cities, this is not much of a problem, but in larger places (for example, Los Angeles), the necessary recovery will lead to increased interference and dropped calls on the analog network. Additional frequency allocations are under discussion by the FCC but a decision is probably a few years off.

DIGITAL TIME DIVISION MULTIPLE ACCESS

In the US, the debate on the future digital format has settled in favor of the Time Division Multiple Access (TDMA) system, albeit with the following provisos:

- The digital system will use the same channel allocations as AMPS.
- The system must be in service no later than 1991.
- Because of the huge infrastructure of AMPS equipment by that date, it is preferable to have a system compatible with the analog bases.
- Additional capacity is the most important objective.
- Additional features and enhancements are required.
- No degradation in quality relative to analog (speech, quality, range, and so on) will be permitted.

Figure 33.4 illustrates the concept of coexistence of digital and analog systems. Mobiles with dual digital/analog capability will be a feature of the future digital scene.

The US digital system also envisages ongoing compatibility with AMPS, so initially, only cities with spectrum shortages, such as Los Angeles and New York, will need to convert to digital with some urgency. Smaller cities and rural areas can thus opt to remain analog until it is economically desirable to change to digital.

Current predictions are that the economic break-even date (when a digital mobile costs about the same as an analog mobile) will be around 1995. This lag in time will lead to an interesting marketing problem in the interim (1991–1995), namely, how to convince consumers to purchase a digital phone when analog units are both smaller and cheaper. Achieving a smooth transition from analog to digital is a very important feature of US digital. Compatibility at the base station is planned so that it will be possible to convert a single analog channel to three digital channels without changing racks or RF coupling equipment.

Adding digital channels means that more space will be needed at the bases. Because base stations are usually just big enough for the existing equipment, the need for more space could present problems. AMPS operators should now be allowing for spare rack space for digital expansion. The concept is to build three digital channels in the space of one analog channel. Naturally, it will be necessary to recable the audio path for the additional channels. The digital channel will simply replace the analog one, so there will not even be the need to retune the cavities.

The digital base will have potentially three times the voice channel capacity of the old analog equipment, so there will be a need for many operators to upgrade their links. This is particularly true where small-capacity microwave or copper cables have been used to the cell sites.

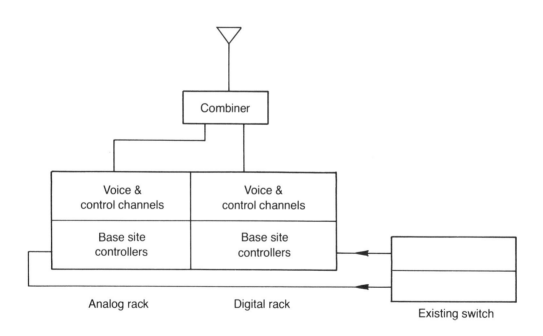

Figure 33.4   An AMPS digital system could use a common switch.

Base-station floor-space could also be a problem, because at least the early digital versions will have bulky controllers and may require additional power.

TDMA VERSUS FDMA

TDMA has long been used in the point-to multipoint environment, but there have been doubts until recently that it would be economically cost effective in the mobile environment. Military mobile applications had been in use for some time, but only as late as 1989 was mobile TDMA demonstrated as a cost effective and competitive alternative, with doubts only about cross channel interference remaining.

TDMA was chosen over FDMA (Frequency Division Multiple Access) because once the necessary LSIs (Large Scale Integrated circuits) have been developed, TDMA will have better voice quality and greater simplicity. Development of LSIs will be an essential part of the process, as the construction of the technology out of existing integrated circuits involves a great deal of space weight and complex construction. A great deal of development is needed to make the chip count reasonable. Figure 33.5 shows the TDMA and FDMA systems.

TDMA has the following advantages and disadvantages:

Advantages

- Fewer radio channels
- Simpler mobile IF (variable bandwidth is not required)
- No duplexer required
- Lower power consumption
- Historical continuity (other systems such as coaxial cable, microwave, and satellite systems all started as FDMA but evolved to TDMA)

Disadvantages

- The need for additional logic
- The need to key the transmitter quickly on and off (currently expensive)
- Higher development costs and longer lead times

A very interesting battle ensued before US cellular finally chose between FDMA and TDMA. The strongest argument for FDMA was that the technology was simpler and that, if chosen, it would require less development time than TDMA. Both systems were developed for evaluation, and both were shown to meet the objectives. The compelling advantage of TDMA for US Cellular was that it would be more compatible with future digital technology. TDMA's effective demonstration in late 1989 by Ericsson tipped the balance.

TDMA transmissions are divided into frames. Each frame represents the transmission of the three timeslots. The timeslots themselves have between them a guard band of 10 bits. As illustrated in Figure 33.6, the timeslots are composed of synchronization bits, followed by information bits, and finally by parity (or error-correction information).

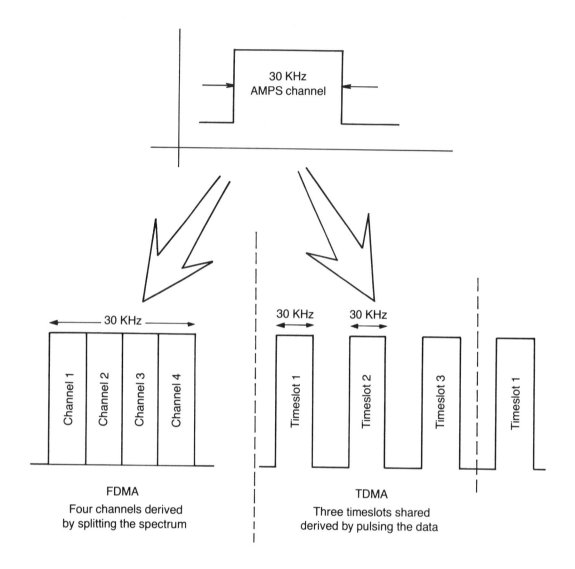

Figure 33.5 FDMA and TDMA. TDMA uses three timeslots to derive three channels from the existing 30-KHz channels; the FDMA proposal divides the spectrum into four channels.

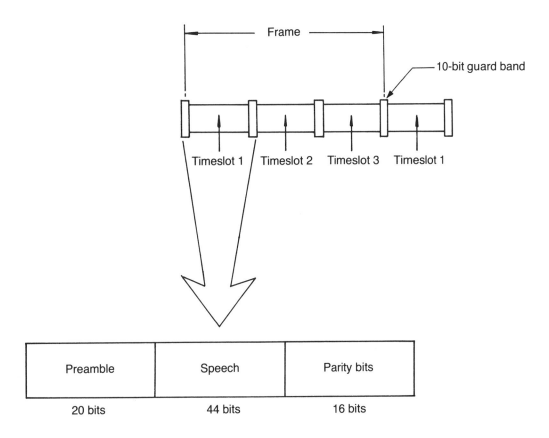

Figure 33.6   TDMA timeslot structure

THE RF ENVIRONMENT

Although the main vehicle for additional traffic density was the splitting of the channel bandwidth, another objective was to improve the C/I ratio (the ratio between the level of interference that can be tolerated to the desired signal level). If the cells can be made more interference tolerant, then they can be operated closer together; and more efficient frequency reuse follows.

It does appear that a C/I level of 17 dB is achievable. This level represents about a 5-dB improvement over the analog counterpart. This additional interference immunity will make the 4-cell pattern much more attractive. One consequence for operators of analog cellular systems using the 7-cell pattern is that 4-cell base sites have more channels per site, and analog systems that have been shoehorned into a small site may not be able to fully exploit this advantage.

TDMA has been designed to be compatible with the existing 4- and 7-cell patterns. This is a vital point because the digital overlay will only be viable if the digital channels can be added without adversely effecting the service area or interference immunity.

DUAL-MODE MOBILES

The compatibility of the US digital system with analog extends to the mobile. The Dual-mode Digital Cellular Standard specification EIA-IS54 was released in December 1989 by the Telecommunications Industry Association (TIA). This specification calls for the initial introduction of dual-mode (analog and digital) mobiles to allow subscribers to the digital network to preferentially access a digital channel but to use the extensive existing analog network where digital is not available.

US DIGITAL COMPARED TO GSM

When the US digital and GSM systems are compared, it seems, at least on paper, that the less complex US system wins in nearly all areas. Much of this advantage has to do with the more advanced codecs used in the US system; these were not available when the codec specification for GSM was agreed upon.

Table 33.1 compares the GSM and US digital systems. When the systems are compared, the US digital system, with its greater spectral efficiency (more than two times higher) but simpler technology and backward compatibility, has a greater conceptual attractiveness.

Although it is widely expected that GSM will use the 3-cell pattern, Motorola has announced that it is working on a 2-cell pattern. The intent of this project is, of course, to increase the spectral efficiency, but further development of this scheme is being hampered by the huge demand for resources in the development arena. It is claimed that the extra spectral efficiency can be obtained without a degradation in C/I.

The dominant position of the US in cellular radio is unlikely to be lost to GSM. The backward compatibility of US digital will make implementation easier and less costly. The US now has about 50 percent of the world's cellular units and should retain that lead for some time to come. This dominance is illustrated in Figure 33.7.

SYSTEMS COMPARED		
	GSM	**US DIGITAL**
Reuse plan	3/9	4/12
Channel bandwidth	200 KHz	30 KHz
Erlangs/sector	43	93
Codec frequency	16 Kbits (8 Kbit half rate)	8.7 Kbits
Modulation	G MSK	Q PSK

Table 33.1   Comparison of GSM and US digital

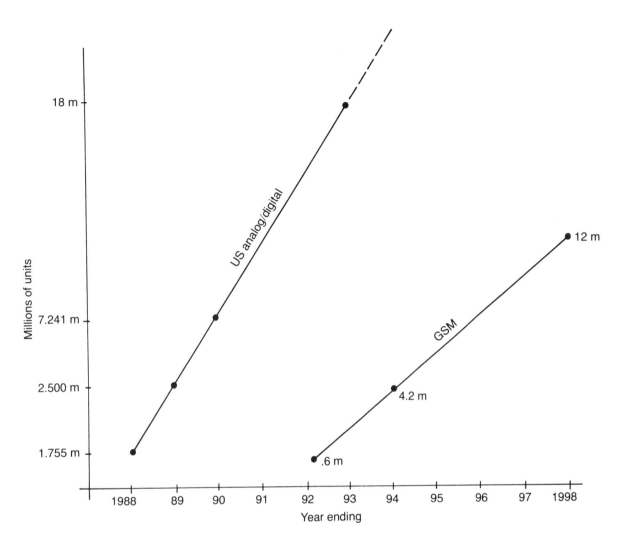

Figure 33.7   Forecasting cellular markets in the US versus Europe

JAPANESE DIGITAL

The Japanese have not revealed their specifications for digital cellular, but since the formation of the special committee of the Ministry of Posts and Telecommunications Office in June 1989, a few parameters have emerged. The Japanese are studying the applications of the 800-, 1500-, and 3000-MHz bands and are considering using 25-KHz channel spacing, employing both FDMA and TDMA. The encoding will be at 8 kbits/second. The Japanese will also use two-branch diversity, which has already been successfully employed in the latest analog versions. The committee's target dates are May 1990 for standardization of the specification and 1992 for implementation of the system.

THE FUTURE OF ANALOG CELLULAR
IN A DIGITAL WORLD

The future for cellular is exciting and dynamic. Even as some existing urban analog systems are today rapidly reaching maximum capacity, it is evident, by extrapolation, that the first-generation digital systems, soon after their introduction, will face the same fate. But even analog has a future for some time to come.

The ever-decreasing cost of mobile units is set to continue as more and more customized chips are manufactured for cellular purposes. The earliest handheld units were made from semi-customized chips that were really developed for other industries, notably the computer and control industries. As speed was more important than volume for most of these applications, the chips generally were parallel-bus devices. As a result, interconnection required the use of high-level multilayer printed circuit boards, and 10-layer boards were not uncommon.

Because a significant demand for cellular phones is now evident, a number of chip manufacturers have begun to produce specialized cellular customized chips that generally use serial data buses to reduce the total number of connections. One such set recently released by Philips claims to cut the total chip count in half and to enable the use of only 4-layer boards. These chips come with TACS and AMPS protocols built in, which represents major savings in the chip count. Lower chip counts mean lower fault rates and lower costs.

Developments in passive duplexer technology are also leading to cheaper and smaller RF hardware. The volume reduction in handhelds is clearly illustrated by the Motorola product line. Figure 33.8 shows that the volume has halved in two years. At the time of writing, the Motorola line included the world's smallest handheld, the Micro-TAC, which is shown in Figure 33.9. The Micro-TAC weighs just under 0.2 liters. It is available in AMPS and TACS and has a slimline as well as a larger, long-life-battery option. The mouthpiece folds up to protect the keyboard.

There is no reason to believe that there are technological limitations to slow this trend, so it can reasonably be expected that by the time of introduction of digital cellular, analog mobiles with a volume of about 0.6 liters, and handhelds somewhat smaller than 0.2 liters, will be the norm.

This situation will present an interesting market challenge for the purveyors of digital, not perhaps unlike the dilemma now being faced by manufacturers of digital audio tapes. The acceptance of the newer technology may be difficult when confronted with a well-developed and sophisticated analog market that adequately serves the needs of its users. Except in cities where the analog capacity has reached saturation or where government legislation supports the introduction of digital, the first years of digital cellular will probably make only a small impact on the demand for analog.

Like handhelds, analog base stations are also getting smaller, but unfortunately, unlike handhelds, they are not getting cheaper at the same rapid rate. Based on Ericsson's figures, at current rates of development base-station volume per Erlang is decreasing by 50 percent about every two years. This means more compact base stations, lower gross floor loads (a major factor in bases installed in high-rise buildings), and smaller air-conditioning loads.

The expansion of analog networks to rural areas is influenced to a large extent by the cost of a minimal (say five channel) base station. This cost is now high enough to discourage extensive

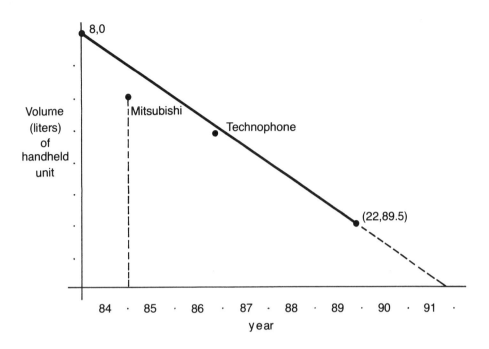

Figure 33.8   The volume of handheld cellular phones has decreased since the introduction of the AMPS system.

investment in rural areas. And there is little prospect, however, that, at least in the early years, digital systems will do anything to improve this situation.

FUTURE TRENDS

As high-density cellular systems become commonplace, there will be a trend away from the concept of a home cellular switch. Today, roaming is achieved by using the subscriber's number to identify the home switch and then exchanging information between the switches for billing and call facilities to enable connection. As the number of switches increases, even in a single city, this method will become inefficient. A centralized database network, accessible to all switches, will replace the current localized database.

Figure 33.9   The world's smallest handheld. Photo courtesy Motorola Communications and Electronics, Inc.

HANDHELD TELEPHONES

Handhelds will continue to get smaller and have longer battery lives. In 1988, the average size of a handheld was approximately 0.6 liters and the talk time of a battery was around one hour. One year later, a unit below 0.2 liters with a one-hour talk time was available. It is expected that a digital handheld could be as small as 0.2 liters and have a battery life of days on standby. However, early in the introduction phase, digital handhelds are expected to be about 0.3–0.6 liters (the size of today's analog units). Although battery technology will improve, the main savings will be through lower power outputs, higher efficiencies, and lower duty cycles of both the transmitter and receiver.

DIGITAL ADVANTAGES

The driving force behind digital cellular is spectral efficiency, but there are some other advantages foreseen by developers. The advantages include greatly enhanced privacy; greater data-rate capability (up to 48 Kbits on a 30 KHz channel); enhanced voice quality; and enhanced services, such as voice recognition and voice messaging.

DIGITAL CELLULAR OPERATIONS

The digital cellular operator will have to contend with a system that was developed in an environment where time was of the essence. Digital cellular will require state-of-the-art equipment, in contrast to analog cellular, which was much longer in planning and development, and which used largely well-proven techniques and often even well-proven equipment. By contrast, every part of the digital network will be specifically purpose-developed. The operators will need to have higher technical skills in the digital environment.

Analog cellular was generally introduced at a civilized rate, which gave the operator a chance to grow and learn with the system. Such is not the case with digital, where in many instances the initial capacities being installed mean that unforeseen system "bugs" and propagation problems will be far more troublesome than in the past. The GSM system, which was designed from scratch, will be particularly vulnerable to bugs and propagation problems.

The current state-of-the-art test equipment is just not good enough or flexible enough for digital. The new-generation test equipment is bound to be expensive and initially at least quite complex to operate. Again, it is the hasty introduction of the technology that is at the heart of this problem.

The complexity of digital, and the consequent complexity of the potential faults in that equipment, is a magnitude greater than analog for GSM. Even for US digital, the problems are not insubstantial. Manufacturers who are anxious to be in the marketplace on time will surely be looking for shortcuts, and diagnostics are sure to suffer. You only have to look at the crude diagnostics available in current state-of-the-art analog to see that this area receives low levels of attention from virtually all manufacturers. Typical diagnostic output reads like it was written by a space alien. The cryptic abbreviations written by the first system developers are still used. Diagnostic outputs are not designed to be user-friendly, and manufacturers do not seem to be able to find the time to change them. How much more of a problem will this be with digital?

PERSONAL COMMUNICATIONS NETWORK

The Personal Communications Network (PCN) is a concept that promises to incorporate digital (or perhaps analog) cordless phone technology with the digital cellular network. The most advanced implementation of this concept is in the UK, but there are ongoing trials in many countries.

The CT2 system (basically a cordless phone with outgoing access only) was introduced with some haste into the UK in the 1980s, to mixed reception. The name CT2 signifies "second-generation cordless phone." Obviously, the lack of an incoming call facility led to the general conception that the CT2 system was second rate. Coupled with the relatively low cost of cellular terminal equipment, this concept led to low market penetration for CT2. The CT2 system is based on small (about 200 meters) service areas with relative low terminal equipment costs (around $350 each), as well as a low call charge rate of around 18 cents per minute.

The concept of a cellular alternative did, however, draw a strong following and so CT3 and many other advanced developments of the cordless phone concept have appeared across Europe. The very diversity of these systems is itself a major weakness. It appears that the only future for this type of equipment is as a supplement to conventional cellular systems.

The Department of Trade and Industry (DTI) of the UK has called for "expressions of interest" from companies interested in developing and operating a new PCN based on the GSM and Digital European Cordless Telephone (DECT) systems. This system is due to be operational by 1993 and will be capable of working with the conventional GSM network or with the cordless phone system, depending on the availability of the services in the usage area.

It is envisaged that the wide-area coverage available with GSM will fill the gaps that are inevitable in the cordless phone networks. The radio interface is to be in the public domain so mobiles can be purchased from a number of manufacturers.

A number of European countries are studying the possibly of a PCN based on GSM combined with the CEPT cordless phone, which comes in two versions. These two versions—known as CEPT1 and CEPT2—conform to the Conference Europeene des Posts et Telecommunications (CEPT) standards. Both of these CEPT standards are under consideration for PCN. There are, however, a number of ways to implement the PCN concept and the situation is far from clear at this time.

SECOND-GENERATION DIGITAL AND BEYOND

The traditional mobile band (below 1 GHz) is not only very heavily occupied, but it also has services such as TV operating within the band. Work has recently been done on mobile applications up to 5 GHz, which is currently regarded as a useful upper limit for mobile communications. That this limit has been moving progressively upward is indicative of the advances in technology that have made operation at these frequencies economical.

Wide band services, like TV, consume vast amounts of spectrum. With the current worldwide growth in mobile services, it is obvious that there will be a good deal of competition for spectrum over the next few decades. Fortunately, the ever-increasing density required by mobile systems makes the use of higher frequencies attractive, because they offer even greater prospects for frequency reuse.

The International Telecommunication Union (ITU), through CCIR, has established a committee to set an international cellular standard for the year 2000 and beyond. In late 1985, Study Group 8 of CCIR, which is responsible for all mobile services, formed a special international group to identify the requirements of standardized Future Public Land Mobile Telecommunications Systems (FPLMTS). The group, called the Interim Working Party 8/13 (IWP 8/13), was formed largely because of the alarming diversity of analog systems that have sprouted in the last few years. IWP

8/13 is considering both fixed and mobile applications and is studying the evolution of mobile and cordless telephones in both the long and the short terms.

IWP 8/13 has passed a special resolution to organize a WARC on services in the 1–3-GHz band before 1992. Some further considerations being looked at by IWP 8/13 include how to minimize transmit power and implement distributed control throughout the network, frequency reuse models in three dimensions, safety aspects including RF radiation, and quality and security of the service provided.

Frequency standardization is a major problem, and most Western countries have already consumed most of the spectrum allocated to cellular services. Without frequency standardization, roaming on an international scale will not be practical. Because of existing commitments, it is unlikely that any standardization (worldwide) below 1 GHz is practical. If the frequencies rise, then the application of wider band TDMA will increase, because the higher operating frequencies offer more bandwidth.

Forecasts of at least 100 million cellular users by the year 2000 are considered conservative. Clearly, the need for international standards will be acute by that time. (Note that all of the current generation analog cellular standards use spectrum outside the CCIR recommendations.)

Both GSM and US digital systems are meant to operate in channel space that is adjacent to that of existing analog systems. As a result, some problems can be expected with interference from the digital into the analog systems. Providing guard bands to prevent this interference will probably be necessary.

Second-generation systems are likely to be intelligent cellular systems with real-time adaptive equalization. They will be able to reconfigure in order to make maximum use of the available spectrum. Artificial intelligence is likely to be used as the generation of suitable algorithms to take care of all the potentially different base-station environments becomes too complex. This way, the base will gradually learn how to survive in its environment, and it will adapt to the changes in the network around it.

Cellular switches perform basically the same tasks as their trunk fixed-network counterparts, so special switches won't be necessary. This situation is unlikely to change in the near future.

THE POTENTIAL OF GHz CELLULAR

Although small cells at frequencies around 1 GHz have been proposed, there are more problems than just high-speed handoff and large-cell capacity. As has been previously mentioned, at current cellular-radio frequencies, the propagation is not "ray-like" and the refraction, diffraction, and reflection modes are quite significant. Given this situation, it will be difficult to select sites that can reasonably guarantee high-density reuse in central business districts because the extraneous propagation that can cause interference will be difficult to foresee (and with the new high-rise building rate in some CBDs, perhaps impossible).

By going to very high frequencies, where true ray-like propagation occurs, small cells can be contained. Of course, the opposite problem (unpredictable "dead spots") could well mean that a mix of the two techniques may be necessary. High-density coverage would be provided by very high-frequency cells with second choice lower-frequency umbrella cells provided to overcome the dead spots.

It has been suggested that the optimal frequency range for high-density systems is the 59–66-GHz band, where atmospheric attention peaks at around 15 dB/km and refraction and diffraction effects are minimal. Such frequencies would be ideal for large office buildings and other high-density areas.

At least 2 GHz of spectrum could be made available. With existing channel spacing (25 KHz), this amount of spectrum would accommodate 80,000 channels. Of course, digital systems can achieve satisfactory voice performance over channels much narrower than this. These channels could be reused 0.5 km away without any real prospect of interference.

Additional interference protection is provided by the inability of these frequencies to pass through obstacles, which at lower frequencies are virtually transparent. As little as 3 cm of wood, concrete, or brick would form an impenetrable barrier. One of the most difficult problems at this frequency is that the human body is similarly opaque. Therefore, diversity transmission is needed to avoid the situation where the mobile transceiver is carried on a person in such a way that the person's body is between the base and the transceiver. Because of these difficulties, the use of umbrella cells at lower frequencies would be essential.

Transceivers operating at 60-GHz frequencies present technological problems which have yet to be solved. The antenna could be a phased array and the array would only be a few centimeters across (1 wavelength = 0.5 cms). Diversity could be provided by placing a second array, 10 wavelengths away (5 cms). Steerable arrays are used today but the cost is prohibitively high. A specialized monolithic integrated circuit would need to be developed to perform this function.

GHz BASE-STATION ANTENNAS

Omnidirectional antennas are difficult to make and not really practical at these frequencies for reasons previously discussed.At frequencies of 60 GHz, normal leaky feeders are very lossy; and some new types of transmission medium will be required. A type of transmission known as "image line," with characteristics similar to an optical fiber, appears most promising.

Figure 33.10 shows a typical GHz+ installation. Naturally, the usable range of a system operating at 60 GHz would be very limited, so any practical application would have to work in conjunction with some type of overlaying digital cellular system. The natural extension of this concept is to have an additional satellite overlay to provide almost universal coverage.

The network would be always programmed to select the highest-density option, so that the basic function of high-penetration components are mainly to fill in areas that otherwise would be difficult to cover. Handoffs between systems would greatly enhance the applications and traffic capacity available. Handoffs would require the integration of three technologies into a single handset and this integration is probably not fully possible for the next decade.

Building windows may be able to "radiate" in the future, using a technique that is opposite to the technique for preventing radiation from microwave-oven windows. Future buildings will need to be "wired" for the new technologies, and leaky cable will become a standard part of construction.

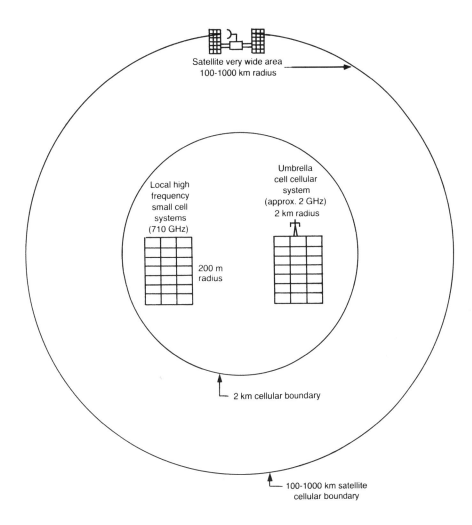

Figure 33.10 The ultimate PCN. Local coverage using GHz+ frequencies can be complimented by digital cellular radio that, in turn, is complimented by satellite coverage. This blend will be the ultimate PCN (once handoffs between systems become possible).

MOBILE UNITS

The mobiles would operate on a low band for cellular and satellite access (about 2–15 GHz) and on a high band (60 GHz) for the immediate vicinity and with automatic switching as in a conventional handoff.

SATELLITE SYSTEMS

Satellite based mobile systems have in the last few years been extensively tested with encouraging results. A number of companies have announced plans to release a commercial product by the first few years of the 1990s. The initial commercial targets for these services are aircraft operators and the long distance trucking industry. Both of these operators will have little difficulty accommodating the somewhat bulky antenna systems that are necessary for reception. These antennas would require gains of around 12 dB and would need to be steerable.

Satellites offer virtually universal coverage, but the price is paid in low spectral efficiency and high equipment costs. Imarsat has announced that a briefcase-sized "personal" satellite communicator will be available by late 1990. The unit will, however, only provide data facilities because these can be sent in a much noisier environment than speech. As a further enhancement of the umbrella concept, frequencies for the umbrella cells could be chosen to be compatible with the up and down links of a satellite system. Other satellite operators are known to be conducting their own satellite evaluations.

A certain block of spectrum could be reserved for satellite operation so that the mobile has three choices:

- 60 GHz: local (1st choice)
- 1.5–10 GHz: umbrella (2nd choice)
- 2–10 GHz: satellite (3rd choice)

This segregation of the spectrum could ensure worldwide coverage and is probably indicative of cellular radio beyond the year 2000. More development is needed to develop an effective mobile-satellite system that is comparable in convenience to a cellular phone.

Brute power in space is only a partial answer. Although high-power satellites with enormous gain antennas are practical, the high power will limit the ability to reuse frequencies. Conversely, high-gain spot beams will enhance frequency reuse and dynamic assignment of spot beams is promising.

Although mobile transceivers are tending towards lower powers, a power booster, with consequent high battery drain, will probably be required for satellite operation.

UNIVERSAL MOBILE RADIO

There is much talk in the industry about the convergence of the mobile technologies (such as paging, packet radio, data, and cellular) into a single personalized communicator. This vision acknowledges that, given the current state-of-the-art, universal communications are not economically practical for the general population, but it fails to accept that the demand for improvements in the existing technologies is so high that standardization is being neglected. With so many variations and basic incompatibilities within the same technology subgroups, it is possibly a little optimistic to expect that other mobile technologies will halt development long enough to achieve integration. Figure 33.11 illustrates this concept.

The mobile "pie" is big and lucrative and there are many takers for a piece of it. The market is also very immature and, until the technology is capable of supplying the underlying long-term demand, most of the research dollars will be directed toward enhancing capacity.

Allocating large slices of spectrum to mobile services may in the long term prove to be imperative. It would save much costly development because the "new" systems could just be enhanced versions of the first-generation digital systems that have been moved in frequency. The

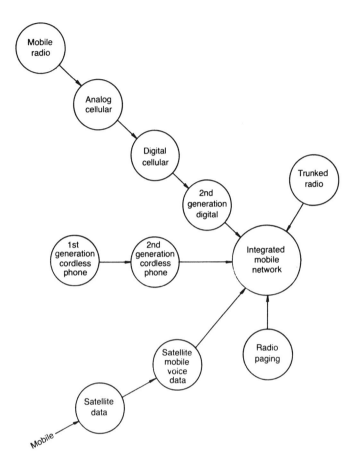

Figure 33.11 Ideally, future mobile technology will be integrated. The convergence of the mobile technologies is a worthwhile goal—but is it just a dream?

shortage of spectrum in general, and the lack of uniformity of spectrum that can be made available, will cause continuous pressure in those bands where roaming is possible.

Already in major Western countries, cellular connection rates account for about one in five of all new telephone installations. Forecasts that most new connections will be mobile within the next decade are very believable.

The concept of everyone being assigned a number at birth for the universal PCN network seems a bit like something out of Orwell's 1984 and indeed it is. Even though it may be possible for everyone to have their own PCN number, a lot of people just don't want it. We should leave some room for individuality.

The cellular dreamers are having a field-day speculating on the options and potentials of equipment that doesn't even exist in commercial forms today. It will help, perhaps, to establish a more realistic perception—to recognize that many of the AMPS mobiles on the market even today do not meet the full AMPS specification and that, as a result, they will fail to function properly (and sometimes even at all) in a system that uses functions not properly provided for in those mobiles.

Chapter 34

OTHER MOBILE PRODUCTS

Acellular operator will eventually begin to look at the potential of other mobile products. The similarity between Public Mobile Radio (PMR), paging, and cellular radio cannot easily be overlooked, and cellular operators would do well to evaluate the opportunities presented by each.

Voice-message/mail, systems which are telephone network value-added services, are often added to cellular operations. In 1988, *Cellular Marketing* found that 35 percent of cellular users in the US also were paging customers. A recent decision by the FCC has allowed paging over cellular as a standard feature. As a result, paging can be expected to be an integrated feature in the near future.

Although many operators are enthusiastic about the "one-stop shop" concept for mobile products, others are not. From the customer's perspective, the "one-stop shop" has advantages. So why are some operators so cool on the idea? The answer has much to do with the degree of similarity. Admittedly, these other services have a lot in common with cellular operations, but they also have a lot that is not similar. Operators tend to regard cellular as the "senior service" and the other activities as add-ons. As a result, operators tend to overlook the opportunities presented by these other activities. Without properly evaluating the resources to operate these new services, it is easy to underestimate and consequently underfund them.

A cellular operator usually has a fairly substantial infrastructure network of shelters, communications links, emergency power supplies, and so on, all of which are necessary for the successful operation of other mobile services. By using these already established features, the cost of adding "other services" can be about half what it would be if the other service was provided completely independently.

As the staff skills and mix are also closely related, the ongoing operating costs of "other services" can also be integrated in a very cost-effective way. This chapter explores these options and examines their implications for the cellular operator. The advantages of expanding into new services include better use of resources (both staff power and tangible resources), and lower start-up and maintenance costs than for non-cellular operators.

PUBLIC MOBILE RADIO

Because many cellular operators might consider PMR to be a competitive system, they may be reluctant to enter the field. In practice, many large cellular operators not only also operate PMR but are themselves extensive users of the system. To understand why, it is necessary to look at the differences between cellular radio and PMR from the customer's viewpoint.

PMR includes simple two-way, as well as trunked and wide-area, radio systems. Among the advantages of PMR are the following:

- The annual operating cost is substantially lower.
- Group calls can be made.

   All vehicles, or some subgroup thereof, can be called simultaneously. This capability is essential for many operations, such as taxi and parcel-pick-up-and-delivery services. Note that this feature was specifically excluded from the recent FCC ruling on extended cellular features.
- Although telephone access is usually not as convenient as with cellular, it is much cheaper on PMR.
- PMR is more suitable for fleet operation.
- An incoming local call is identified and charged as a local call only.

PMR can consist of a simple, conventional two-way radio repeater, or it can be a much more sophisticated (and expensive) trunked network. In all cases, automatic telephone access and selective calling can be provided.

Selective calling is the ability to call either a particular vehicle (by alerting it with a ring tone) or a predetermined group of vehicles. Normally, the system is configured so that only the called group can hear (or take part in) the conversation. Cellular radio systems can be thought of as having selective call capability to only one mobile unit (car telephone) at a time.

Group calling can be used to call a predetermined group. For example, a cellular-telephone company can call, as a group, all technicians or all salespeople or all staff. A large number of groups and subgroups is possible. For example, a group of all technicians and all salespeople could also be called. Of course, individual calls are also possible within this scheme.

Automatic telephone access is also available as a feature of PMR systems. There are three modes of operation: full-duplex, half-duplex, and simplex. In all cases, fully automatic in-dialing (to the mobile) and out-dialing (from the mobile) are possible.

FULL-DUPLEX MODE

Full-duplex mode is the same as that used in a car telephone. It enables simultaneous talk-and-receive signals in the same way as a conventional telephone. Although this mode is the one preferred by most users, it is not the usual mode for PMR because full-duplex requires the simultaneous operation of the transmit-and-receive functions of the mobile radio. This capability results in a higher-cost mobile (because continuous transmit duty requires more elaborate heat sinks) and requires the addition of a duplexer to enable transmission and reception simultaneously from one antenna, which is not only costly and bulky but significantly reduces the useful range of the mobile.

HALF-DUPLEX (TWO-CHANNEL SIMPLEX) MODE

Half-duplex is the usual compromise adopted by PMR operators. A press-to-talk (PTT) switch is needed in this mode because simultaneous transmit-and-receive on the radio path has been eliminated. In half-duplex, separate frequencies are used for transmission and reception (this allows the repeater function to be used, as described earlier).

When telephone conversations are conducted on a half-duplex system, only one party can talk at a time. This is because the mobile can only receive or transmit, but it cannot do both at one time. In some systems, control is given to the mobile, so that if PTT switched is pressed, land subscribers cannot transmit. In most systems, however, land subscribers can still transmit on top of the mobile, though the transmission cannot be heard. Half-duplex telephone interconnections, although useful, cannot provide the same quality of service as full-duplex or as a cellular system.

SIMPLEX MODE

In simplex mode, only one radio frequency is used for transmission and reception and so only one path of conversation is possible at any one time. The direction for telephone interconnect is usually chosen by a voice voting system so that once a signal is heard (for example, voice), then that direction has priority until the conversation ceases. In less sophisticated systems, this can be somewhat embarrassing when, for example, a recorded announcement with very few pauses is dialed, and the system locks-up on that announcement. The system, however, can have land party or mobile party priority.

Such systems are often referred to as VOX (Voice Operated Switch, X = transmit as in Tx). Although they can work well if they are carefully set up, they can be the source of some frustration if the VOX operates erratically. VOX operation can clip the first syllables or even short words of a user's sentence if it is incorrectly adjusted (too long attack time prevails).

Simplex mode is the mode least-preferred by both users and most operators, but it is most-preferred by frequency-management authorities because it uses only one frequency.

TRUNKING

Trunking offers more sophisticated features than a simple mobile repeater system and is used for circuit efficiency and wide-area coverage. As discussed earlier, an advanced trunking system has most of the features of cellular radio but can be quite expensive to implement.

The mobiles used for trunking are usually more sophisticated and more expensive than those used on a simple repeater. Trunked systems become economical only when used on a fairly large-scale basis and usually only in high-density areas. Because a trunked network is very similar to a cellular network, it is the type of system most suited to be included as part of a network package. Trunked systems may become economical when five or more mobile channels are justified on the basis of traffic.

OTHER WIDE-AREA SYSTEMS

When traffic density is low, other methods of obtaining wide-area coverage (more than the area covered by one base) can prove just as effective as trunking and are somewhat cheaper.

Simulcast (also known as quasi-synchronous) can be used to parallel a number of transmitters/receivers on one frequency. This technique fell out of favor some years ago when it was found to be difficult to synchronize the transmitters in a way that would avoid "dead spots" caused by interference, particularly due to audio phase drifts. Current technology can handle these problems so that simulcast is now the cheapest wide-area system. It is effective, but of the techniques it still requires the most maintenance. Figure 34.1 illustrates a simulcast transmission.

Race voting is also popular and offers the wide-area coverage of simulcast without the attendant difficulties in synchronization. Race voting, however, consumes more frequencies and requires more complex multichannel, microprocessor-controlled mobiles. This technique is illustrated in Figure 34.2.

Race voting uses base stations with different transmit but identical receive frequencies. In this way, the need to phase synchronize the base transmitters is removed. The term "race voting" is used to signify the way in which the base receivers are selected. When a mobile transmits, it may be received by one or more base receivers. Each receiver measures the field strength, and then attempts to seize control at a speed that increases with increased received field strength. In this way, the receiver that has the highest receive level seizes the line and locks out all others. This "race" occurs each time the mobile presses the PTT switch, or it can be forced to occur at fixed time intervals for duplex systems.

The receiver that wins the "race" inhibits all other receivers on the seizure line and then takes control of the audio both to land lines and to the other transmitters. Ordinarily only three base transmit frequencies are required.

A more sophisticated voting system using a centralized comparator to compare the S/N levels at each base station can be used. This system requires individual links back to the voting system from each base station. Although these extra links make the centralized voting more costly, in practice centralized voting has proven more effective and reliable than the race-voting technique.

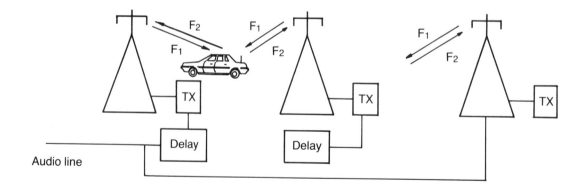

Figure 34.1   In simulcast transmission, all bases use the same frequencies. Destructive interference presents a challenge when coverage overlap at similar levels occurs. The delays ensure that signals that overlap signals of similar levels are in phase.

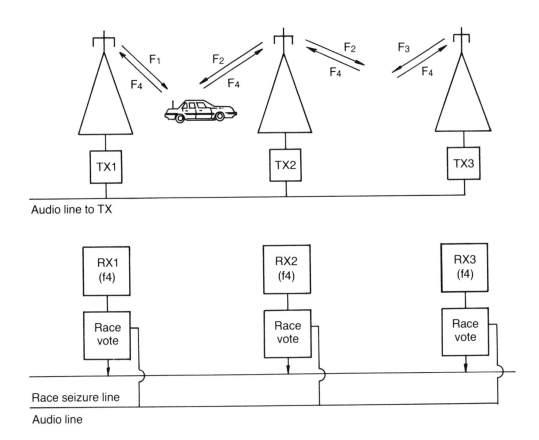

Figure 34.2   Race voting typically uses three base-station transmit frequencies and one Tx frequency. The race-vote device seizes the line at a rate that increases with receive level. All receivers are on the mobile Tx frequency, so the one with the best receive level seizes the line and controls the audio for the whole system.

MARKETING

As has already been noted, PMR appeals to a different market segment than cellular customers, and so no strong conflicts are expected. A plus for marketing strategy is that target groups identified for cellular radio, if not in the market for cellular will probably be in the market for PMR, and vice versa. The potential customer is usually (today) the owner of a small- to medium-sized business who undoubtedly will benefit from some form of mobile communications. Thus, whether the potential customer is targeted by advertising, direct marketing, displays, or any other medium, the same potential customer has a good chance of wanting one of the two systems. Hence, marketing costs can be contained.

Indeed, a good sales force should be able to adapt very readily to the additional technology and, although a specialized salesperson would probably still be the norm, each salesperson should be able to sell the other product when the opportunity arises. The main difference in marketing emphasis is that the PMR salesperson typically seeks the fleet customer, and the cellular salesperson usually targets the single-unit buyer. This distinction alone justifies a separate sales structure, even though cross-sales can be anticipated. PMR also has a greater diversity of options that tends to require specialist sales attention.

Any serious PMR operator should have an experienced, technically qualified sales engineer who can discuss in detail the system requirements of the customers. Most customers will be relatively unclear about what they really want and will have only the vaguest idea of the available options. The sales engineer should therefore back up sales personnel and directly handle only the bigger, more technical customer accounts.

FINANCING

The financial structure of a PMR system is very similar to that of a cellular operation. The operation is rather capital-intensive, with the PMR mobiles costing about the same as, or even more than, a cellular mobile. The marginal cost of expansion is similar (on a per channel basis). One advantage of PMR is that large customers often contract for extensive systems in a single order so that the uncertainty of demand is much lower than cellular.

Billing is sometimes on a straight monthly fee per mobile unit, but it is often on an airtime basis. Telephone calls are usually charged separately. The cellular billing system can usually be adapted to PMR operators.

The income per mobile is significantly less than that expected from a cellular system, but this shortfall is balanced by lower call airtimes, which means that more customers can be accommodated per channel. As a general rule, the income per customer will probably be between 20 and 35 percent that per cellular customer, but the network cost per customer is also about 30 percent that of a cellular customer.

ENGINEERING

Not all cell sites make ideal PMR sites because the frequency reuse requirement of cellular radio leads to less prominent sites. On the other hand, PMR usually uses large cells, and since it usually uses half-duplex mode, very large cells can be achieved.

With some care, a good number of cellular sites can be used for PMR, but some additional, special sites will probably be required. To some extent this requirement reduces the attractiveness of operating both systems. Notice, though, that there is no conflict in rural areas, where cellular radio also uses high sites.

The technology used to operate and maintain a PMR network is very closely related to the cellular radio requirements and, with a minimum of additional training, an engineering staff can handle both systems. The test equipment employed for cellular radio installation and maintenance should be readily applicable to PMR. Radio survey and installation techniques are similar in both systems and could use the same staff and facilities. PMR repair, however, probably requires a specialist technician, but this work can be contracted.

PAGING

Paging is a relatively mature industry compared to cellular radio. Because many of the resources needed to run a successful paging business are closely related to those needed for cellular radio, most operators of either business at some time will probably consider the possibility of entering the other domain. Figure 34.3 shows a typical paging configuration.

Paging can come in alert (tone) only, numeric, or alphanumeric forms. Recent advances in

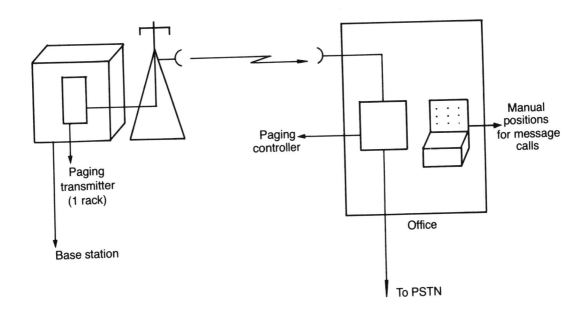

Figure 34.3   A basic paging configuration

paging have concentrated on reducing the size and capacity of the paging receiver. The price has decreased only slowly over time and ranges from about $100 for a simple alert pager to about $500 for a sophisticated message pager (at retail prices).

Recent developments include NEC's PGR9000, which can scan multiple frequencies, and a new pager from Telefind (of Florida), which has an 80,000-character memory and scanning ability. Motorola and Timex have recently released numeric pagers that incorporate a digital watch. Messages are recorded and time-stamped. Uniden and AT&T are working on wristwatch pagers.

When considering paging, the cellular operator must look closely at the economies of scale without losing sight of the differences. This analysis is best done by looking at marketing, financing, and engineering functions separately.

MARKETING

Most marketing studies have shown that, although each paging customer may be a potential cellular customer, the two customers have very different needs and expectations, and that migration from paging to cellular, if it occurs, is very slow. It is best to consider the markets as independent, with paging being a much larger market than cellular. Indeed, while a third of cellular customers are also paging users, it is obvious that the applications are to some extent complementary.

The cellular market, is still new and growing rapidly, and therefore the marketing thrust in cellular must be to gain an ever-increasing share of that market. On the other hand, paging grows slowly, although the growth is consistent. The aim of paging operators is to maintain their existing customers while adding more. The type of market and the target customer is hence somewhat different between the two systems. For this reason, advertising is one area where there is little opportunity to combine resources.

Organization of sales needs to be carefully considered; the present sales-force structure may or may not be adequate to handle both products. Because the markets can be regarded as complementary, the potential for extra sales with very little extra sales effort should be considered. (This possibility applies more to "cold calling," where the needs of the potential customer are not necessarily well known in advance.) The margins on pager sales are relatively low and sales costs may need to be subsidized by future revenue. The need for high sales volume is therefore very pronounced.

Some users have a pager and a handheld cellular phone. The phone can be kept switched off until it is needed. By eliminating idle battery consumption waiting for incoming calls, longer talk time per battery is available. The pager alerts the user who to call.

FINANCING

The complementary financial nature of cellular and paging operations is definitely worth investigating. While cellular radio has a marginal cost of $1,200–$2,000 per subscriber, the marginal cost for paging is $20–$50. Paging is therefore very much a cash flow business, and because very few cellular operators can report a positive cash flow, paging may be very attractive indeed.

Financial managers must understand the fundamental differences between the two systems, but they should have little difficulty managing both systems.

Integrated billing systems are not commonly available, but because of the very simple nature of paging billing, it could easily be added to an existing cellular billing system. Pagers are typically billed either a straight monthly fee or a monthly fee plus airtime.

ENGINEERING

The technology needed to successfully operate paging is closely related to that needed for cellular and, if it is planned in advance that both systems are to be operated, test equipment suitable for both can be purchased. As with a cellular base, the cost of paging base-station equipment usually represents only a small proportion of the establishment costs. The cost of sites, accommodation, power, batteries, towers, and links are usually dominant.

When a cellular system exists, provision can be made to add on paging at a very low marginal cost. By careful selection, it is usually possible to use some of the cell sites as paging sites, thus minimizing infrastructure costs. Although cell sites will not be optimally located for paging, it is usually possible to find a workable combination from the sites. When infrastructure costs are considered, this combination is usually significantly cheaper than an optimal site solution.

The test equipment used to maintain cellular bases can generally be used with paging equipment. Radio survey and installation requirements are similar in each instance and can use the same resources.

If possible, analog tone signaling paging should be avoided, because to successfully operate it, dedicated phase-synchronized links are necessary, and this adds a significant cost to the network. For new operations, there is no need to consider analog-tone paging. If an existing system is brought out, however, it may come with a group of analog-tone customers who must either be supported or transferred to digital. With digital systems, time and group-delay equalization are necessary, but these capabilities can be provided using the same links as those for the cellular system. In the near future, integrated cellular/paging switches will be available.

OPERATING

A paging system ordinarily requires a manual operator, which can be a major expense. If the cellular system already has an operator, these functions can be combined. Approximately one operator per 150 manual subscribers is required.

PAGING FORMATS

Four paging types may need to be supported:

- Alert
- Alert and Voice
- Numeric
- Alphanumeric

The Alert paging type usually gives an audible alarm that indicates a page. There may be more than one tone so that different conditions can be indicated. Most Alert pagers have either one, two, or four separate tones. Some also have a vibrating mode, so that calls can be received in meetings or other situations without disturbing others. Many users agree that the most valuable function of the Alert pager is the test switch that simulates a paging call and can be used as an excuse to escape from especially trying meetings to attend to an "urgent call." Early versions of these pagers were analog (tones such as two of five, or two of seven), but more recent systems are digital.

The Alert-and-Voice paging type gives the audible alert followed by a voice message. This type usually, but not necessarily, involves a manual operator who takes a message and relays it. At times, a most inappropriate message can be broadcast over these systems. For example, a quote may be relayed while the customer is within hearing range. Alert-and-Voice pagers must be used with care. Some systems have a message-storage facility that works much like a telephone answering machine.

A Numeric pager operates much as the Alert pager, but it can also register a number. This number can be sent either by a manual operator or from a conventional touchtone telephone. The most common use is to transmit the telephone number to which the return call is expected.

Alphanumeric (*alpha* refers to letters and *numeric* to numbers) systems initially call in the same way as an Alert pager, but the caller can leave an alphanumeric message. These pagers usually have memories, and some can store more than 20 messages or 400 characters. The message can be entered by a manual operator or, in some cases, from a personal computer via an appropriate modem. The personal computer usually needs specialized software to access the paging controller.

VOICE MESSAGING/VOICE MAIL

As of 1987, voice messaging in the US was growing at the rate of 60 percent per year, and the market is predicted to continue growing at that rate for the foreseeable future.

VOICE MAIL

Voice mail enables messages to be stored and retrieved at the user's convenience. Because more than half of all business calls are known to be unsuccessful, it makes sense for cellular users to have access to a voice mail service. This service can be likened to a sophisticated answering machine.

As an additional service, it is normal to charge additional fees for voice mail. To enhance the image of the voice-mail service, it may be necessary to charge additional fees; as a free service, it may be seen to be of little or no value.

Additional revenue is earned from the charges for additional airtime, so it may not be necessary to also charge for storage or access. This decision depends on airtime charges, but usually they are adequate to cover voice-mail costs with only a nominal rental charge of, for example, $5–$10 per month. In the US, it is mainly wireline operators who have ventured into voice mail, but non-wireline companies are showing increasing interest.

Outside the US and other developed countries, a major consideration must be the availability of DTMF (tone phones). Virtually all cellular phones offer DTMF overdial facilities, but the system is considerably limited if the user cannot access the message system from any phone. Although handheld, acoustically coupled, DTMF dialers are available, they are not widely used and users are reluctant to carry yet another gadget.

The concept and value of voice mail (even like the concept of cellular itself) can be difficult to explain to the general public. Advertising may be of limited value unless a "free trial" is offered so that the users can become familiar with the service. In the US a trial period of 90 days seems to be the norm.

The voice mail system should be co-located with the switch, because, like the switch, it requires 24-hour attention. Cellular switches can work directly with voice-messaging systems; operators who have added this feature report cellular customer demand at around 30 percent.

The voice message system is basically a voice mail box; it differs from an answering machine only in its flexibility. Incoming messages can be categorized as urgent, recent, or even filed according to a particular caller. Once a message has been received, it can simply be stored to await an access request or to initiate a pager or telephone call.

Automatic purging of the files after a specified period means the system holds only up-to-date information. Recorded announcements have also been exploited; systems can carry information about the weather, news, sports, the stock market, traffic, restaurants, theaters, and other items of interest, all available on request from a menu. Selections are made using touchtone dialing to indicate the menu item sought.

PACKET RADIO

Packet radio systems are used to send "packets" of data over mobile radio equipment. From a control console, an operator can send information specifically addressed to one or more receivers. As dedicated land links are quite expensive, shared alternative data links can be attractive. In its present form, packet radio systems are suitable for point-to-point-to-multipoint thin line (low-

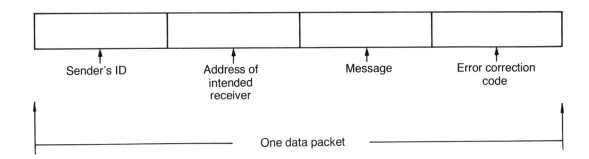

Figure 34.4   Typical data packet. Notice that the actual message may be broken up into a number of packets.

capacity, usually single-channel) systems. For this reason, they are especially attractive to smaller companies and to operations in rural areas, where high capacity is not a requirement.

Reliability in these systems is built into the error-correction code and is quite high. Often, however, the system is used to replace dedicated land links, and conventional mobile two-way equipment is used for the RF link. This equipment has an MTBF (Mean Time Between Failure) that is low compared to leased lines. As a result, for high security applications, it is necessary to build in a good deal of redundancy (which usually means 100-percent duplication and hence doubling of costs).

The data sent is synchronous and requires a handshake reply (that is, a confirmation that the data has been correctly received and decoded). This precaution ensures very low data-error rates together with positive assurance that the message was received.

Packet switching can be accomplished in a single hardware unit that forms the interface between the computer terminal (usually an RS 232 bus) and the radio transceiver. A packet of information contains the sender's identification, the receiver's address, the message, and a block of redundant error-correction code. The process of assembling and disassembling data for transmission is called "padding." Data bursts typically last 250 seconds; this short period offers some measure of security. Figure 34.4 shows a typical data packet.

The cost of adding this technology to an existing mobile system is quite low (less than $1,000 without the personal computer). The cost of a complete dedicated network, however, is such that payback periods of five years (when compared to leased lines) should be anticipated.

Cellular mobile operators often have their own mobile radio networks (to provide an independent operational link between the base stations and switch) and may find incidental internal uses for data-packet technology.

Chapter 35

GLOSSARY OF TERMS

AIRTIME

The total time that a channel is occupied, including call time, call-set-up, and cleardown time.

AMPS

Advanced Mobile Phone System.

ANTENNA GAIN

The gain of an antenna compared to a dipole or quarter-wave antenna.

AREA CODE

Usually a two- or three-digit number that identifies the area of a telephone outside the home area of a caller.

BASE

A site that contains the cellular radio equipment. It can have one or more cells.

BOUNDARY (OF COVERAGE)

The usable limits of a particular cell. Usually defined to be 39 dBμV/m for AMPS systems.

CAVITY

A resonant device, usually drum (or cylinder) shaped, that acts as a filter in cellular systems.

CBD

Central Business District (city center).

CELL

A group of co-located channels that cover the same area. A base can have one or more cells by using directional antennas.

CELL EXTENDER

A cellular repeater that repeats base-station channels. It is essentially a linear amplifier.

CELL SITE

The location of the cell.

CELLULAR OPERATOR

The owner and/or operator of a cellular network.

CHANNEL

A pair of frequencies used by the mobile (that is, one send and one receive frequency).

COAXIAL CABLE

A pair of conductors consisting of a central conductor surrounded by an outer conductor. These cables are used because of their immunity to interference and relatively low power losses at high frequencies.

COLLINEAR ANTENNA

A gain antenna with dipoles stacked vertically.

COMBINER

A device for combining several (usually four or five) transmit channels.

COUPLER

A device for connecting two or more sources of RF energy to a single cable or port.

COVERAGE

The area over which the service is of an acceptable standard.

dB (DECIBELS)

A unit for expressing the relative intensity of sound. This is equal to:

$$10 \times \log \frac{Power\ referred\ to}{Power\ of\ a\ reference\ level}$$

or

$$20 \times \log \frac{Voltage\ referred\ to}{Voltage\ of\ a\ reference\ level}$$

dBd

Gain relative to a dipole antenna.

dBi

Gain relative to a hypothetical isotropic antenna.

DEVIATION

The amount of frequency change in a fully modulated FM system (expressed in KHz or 1000 Hz).

DIFFRACTION

Propagation around an obstructing object.

DTMF

Dual Tone Multi Frequency. The signaling used on modern push-button telephones. Combinations of two tones represent various numbers.

ERLANG

A unit of telephone traffic such that 1 Erlang is one occupied circuit per hour.

ERP

Effective Radiated Power. The power, expressed in watts, that is radiated in the direction of maximum antenna gain calculated by multiplying the power at the antenna terminals by that gain.

ESN

Electronic Serial Number that is contained in the NAM (Number Assignment Module).

EXTENDED CELL

A cellular repeater that uses frequency translation.

FCC

Federal Communications Commission. The regulatory body in the US.

FDMA

Frequency Division Multiple Access. Cellular systems are FDMA systems. In these systems, the users are assigned a particular pair of frequencies (channels) on request for the duration of a particular call.

FEEDER

A coaxial cable or waveguide connecting a transmitter/receiver to an antenna.

FEEDLINE

Same as FEEDER.

FM

Frequency Modulation. A very common analog modulation technique noted for its excellent signal-to-noise (S/N) characteristics. The frequency of the carrier varies in proportion to the amplitude of the modulating signal.

FREQUENCY

The rate at which the electric and magnetic fields of a radio wave vibrate per second. Frequency is usually expressed in MHz (1,000,000 Hz); 1 Hz (Hertz) = 1 cycle per second.

FREQUENCY TRANSLATION

The process of converting a signal to a different frequency. Often used in repeaters.

FSK

Frequency Shift Keying. A modulation method using frequency changes (in steps) to transmit data. Usually only two frequencies are used.

GAIN

The factor, usually expressed in decibels (dB), by which the signal received is amplified or improved (in the case of an antenna).

GROUND PLANE

The area directly below a quarter-wave or other unbalanced antenna. It should be of low resistance and at least a quarter wavelength in radius from the antenna.

GSM

Groupe Speciale Mobile. The Pan-European digital system.

HANDOFF

The ability of a cellular mobile to be able to move through the coverage area, handing off from cell to cell in order to maintain a good signal quality. The handoff is ideally not perceptible to the user.

HANDS FREE

A voice-operated system that allows hands-free (no handset) telephone operation (most loud-speaker telephones are of this type).

HEXADECIMAL

A number system based on 16.

IC

Integrated Circuits. The building blocks of modern electronic devices.

INTERFERENCE

The reception of unwanted signals that are impressed on the desired signal.

ISOLATOR

A unidirectional RF device.

ISOTROPIC

The same in all directions; in antennas, equal radiation in all directions.

KHz

Kilohertz. One-thousand (kilo) cycles per second; 1000 KHz = 1 MHz.

LEAKY CABLE

A cable that is designed to deliberately leak RF energy. This is often used to provide coverage in tunnels and basements.

MAINTENANCE

Restoring a unit to working order by replacing it or replacing an integral module (for example, a panel).

MAST

A guyed structure meant to support antenna(s).

MEMORY

Integrated circuits that store information such as telephone numbers.

MILLIWATT

One-thousandth of a watt.

MIS

Management Information System. A software package containing billing and management information. Usually sold as a package to the cellular operator but sometimes sold as a service by third parties.

MODEM

Modulator/demodulator that converts binary to analog signals and analog to binary signals. Used to connect digital devices like computers over analog telephone lines.

MODULATION

The method by which the transmitted signal is impressed on the carrier.

MSA

Metropolitan Statistical Area (US). Basically, the major city cellular service areas.

MTBF

Mean Time Between Failure.

MULTICOUPLER

A tuned device that couples 2 or more (usually 16 in cellular) channels into or out of one feeder or antenna.

MULTIPATH

The interference patterns created by the addition of signals from more than one path.

NAM

Number Assignment Module. A PROM 32 x 8 bits long that contains subscriber, system, and options details about a cellular telephone.

NMT

Nordic Mobile Telephone.

OMNIDIRECTIONAL ANTENNA

An antenna radiating energy equally in all directions (horizontally) around it.

PCM

Pulsed Code Modulation. A digital transmission that uses a number of channels over the same bearer in different timeslots.

PM

Phase Modulation. An analog modulation form related to FM in which the phase of the carrier is varied with the amplitude of the modulating signal.

PMR

Public Mobile Radio (two-way radio).

PORTABLE

A handheld cellular telephone.

POT

Plain Old Telephone.

PTT

Push To Talk. A radio switch (usually part of the microphone) that must be pushed before the user can transmit. Usual in two-way radio (PMR).

REFRACTION

Propagation other than in a straight line due to bending of the path by some material medium (for example, air or water). Refraction occurs when a change in density of the medium exists.

REPAIR

Restoring a unit to working order by replacing or reconfiguring some internal component, usually at the board or component level. This usually involves bench work with test equipment.

RF

Radio Frequencies. Varies from 10 KHz to 300,000 MHz.

ROAMING

Using a cellular phone through a system other than the usual "home" switch.

RSA

Rural Service Area (US). Small cellular service areas.

SECTOR ANTENNA

A directional antenna that produces coverage of one or more sectors of the total base-station coverage.

SIGNAL-TO-NOISE RATIO

S/N. The power ratio between the received signal source and the noise source.

SINAD

Similar to signal-to-noise, but it adds the distortion products (SIgnal-to-Noise And Distortion) to the noise power.

SWITCH

In cellular radio, the connecting switch between the telephone network and the radio base station. Also called the Mobile Exchange, MTSC, MTSO, MTX, and other names.

TACS

Total Access Cellular System.

TDMA

Time Division Multiple Access. A digital (usually radio) system that allows a number of users to use the same system by being dynamically assigned a particular timeslot on request. Often used to describe rural radio telephone systems that use this mode.

TOWER

A self supporting structure intended to hold an antenna(s).

TRAFFIC

Calls in progress.

TRANSCEIVER

A transmitter and receiver in one unit such as a mobile telephone.

TRANSPORTABLE

A cellular mobile telephone, complete with battery pack that enables portable operation. This unit can also be used in a vehicle, running from the vehicle battery. It is usually significantly larger and has higher output power than a true portable.

UHF

Ultra High Frequency. The radio frequency band from 300 to 3000 MHz.

WAVELENGTH

The distance from a point on a radio wave to the same point on the next wave. Cellular wavelengths are about 0.3 meters long.

WIRELINE

Refers to a cellular carrier who also provides fixed telephone services.

Index

The author, Neil Boucher, is currently Chief Technical Adviser, Cellular Radio, for the International Telecommunication Union (I.T.U.) and as such, Mr. Boucher has presented dozens of technical papers and workshops. Additionally, he has held top engineering and management positions with Telecom Australia, and has been instrumental in the creation and implementation of cellular systems throughout many Pacific Rim countries.